T0140123

Advances in Neurobiology

Volume 20

Series Editor
Arne Schousboe

More information about this series at http://www.springer.com/series/8787

Rita Sattler • Christopher J. Donnelly
Editors

RNA Metabolism in Neurodegenerative Diseases

 Springer

Editors

Rita Sattler
Department of Neurobiology and Neurology
Dignityhealth St. Joseph's Hospital
Barrow Neurological Institute
Phoenix, AZ, USA

Christopher J. Donnelly
Department of Neurobiology
University of Pittsburgh
Pittsburgh, PA, USA

Live Like Lou Center for ALS Research
University of Pittsburgh Brain Institute
Pittsburgh, PA, USA

ISSN 2190-5215   ISSN 2190-5223   (electronic)
Advances in Neurobiology
ISBN 978-3-030-07832-4   ISBN 978-3-319-89689-2   (eBook)
https://doi.org/10.1007/978-3-319-89689-2

Printed on acid-free paper

This Springer imprint is published by the registered company Springer International Publishing AG part of Springer Nature.
The registered company address is: Gewerbestrasse 11, 6330 Cham, Switzerland

Preface

Neurodegenerative diseases are rapidly increasing in prevalence in the growing population of the elderly worldwide. With this comes an increased financial and social burden to care for the affected individuals. There are no effective treatments for these diseases and the mechanistic underpinnings of these progressive and often fatal disorders remain unclear. This emphasizes the urge to better understand the underlying disease mechanisms, which will then allow for the development of new and more effective treatments. RNA metabolism denotes a variety of distinct processes RNA undergoes to promote cellular health and function. It has become apparent in recent years that altered RNA metabolism significantly contributes to disease pathogenesis in numerous neurodegenerative disorders.

Dysfunction in RNA processing is a prevailing theme in neurodegeneration. This was first highlighted as a key pathobiology following the discovery of TDP-43 neuropathology in nearly all amyotrophic lateral sclerosis (ALS) patients and almost half of patients diagnosed with frontotemporal dementia (FTD). In healthy cells this protein is predominantly nuclear; however, in ALS/FTD patients TDP-43 is aggregated in the cytoplasm within the affected neurons. This neuropathology has since been identified in 30% of patients diagnosed with Alzheimer's disease (AD) and nearly 80% of patients diagnosed with chronic traumatic encephalopathy (CTE). Subsequent studies uncovered a number of autosomal dominant mutations in ALS patient populations within the *TARDBP* gene, which encodes the TDP-43 protein. TDP-43 is an RNA-binding protein that functions in a variety of aspects of RNA processing, including splicing, mRNA transport, RNA stability, and micro-RNA processing to name a few. Since this discovery, nearly two dozen mutations in genes that encode RNA-binding proteins have been identified in ALS patients. FUS protein, an RNA-binding protein that exhibits remarkable functional similarity to TDP-43, demonstrates a similar neuropathology in a subset of FTD patients and autosomal dominant mutations in this gene have been identified in ALS patients. In addition to mutations in genes encoding RNA-binding proteins, microsatellite repeat expansions have recently been identified as a causative mutation in subsets of ALS/FTD patients. One mechanism underlying their pathobiology is the generation of noncoding RNA species that are considered "toxic" since they disrupt a variety of RNA/

RBP interactions. This was initially described in patients with myotonic dystrophy but has since been observed in other genetic neurodegenerative disorders including spinocerebellar ataxia type 8 (SCA8) and C9orf72 ALS/FTD. The production of noncoding RNA species is, in part, due to pathogenic epigenetic modifications that allow for the transcription of these RNA species. A-to-I RNA editing defects have been identified in ALS neurons and this can disrupt the function of calcium channels and promote hyperexcitability, a known pathophysiology of ALS motor neurons. Beyond ALS/FTD, RNA mis-splicing and improper ribonucleoprotein (RNP) assembly are common features of spinal muscular atrophy (SMA) due to the genetic mutation responsible for this developmental neurodegenerative disease. Abnormal RNP assembly/disassembly and RNA stability are similarly implicated in a wide range of neurodegenerative disorders, including ALS/FTD, AD, and Huntington's disease (HD) to name a few. Together, these highlight an intrinsic deficit in RNA metabolism across a wide range of genetic and nongenetic neurodegenerative conditions. This common etiology supports the need for a better understanding of these dysfunctions to uncover novel therapeutic targets for drug development for a large patient population with unmet needs.

With this book series on *RNA Metabolism in Neurodegenerative Diseases* we aim to provide review articles addressing the different aspects of RNA pathobiology that contribute to neurodegeneration. These articles summarize the most recent and novel studies highlighting our current understanding of pathologic RNA metabolism. Specifically, we focus on epigenetic changes that alter coding and noncoding RNA expression; abnormal splicing and editing events; impaired RNA transport, RNA stability, and RNP assembly that promotes protein misfolding; the genetic mutations that generate toxic RNA species or dysfunctional RBPs; and deficits in the translation of mRNAs.

The first three chapters of this series provide examples of how direct RNA modifications contribute to disease pathogenesis in varying neurodegenerative diseases. In the first chapter, Veronique Belzil and colleagues review the critical aspects of epigenetic regulations in the ALS and FTD disease spectrum (Chap. 1). Epigenetics generally defines modifications of gene expression without the alteration of the underlying primary nucleotide sequences. Increasing evidence suggests that cell-specific and tightly regulated aberrant RNA processing events of noncoding RNAs are contributing to the epigenetic changes observed during neurodegenerative disease development and progression. This can result in the production of coding and noncoding RNA species that perturb neural function. The following review by Ravindra Singh and Natalia Singh provide a comprehensive overview on spinal muscular atrophy (SMA), the most common genetic cause of infant death. Specifically, they focus on how mutation-induced RNA splicing deficits of the *survival motor neuron 1 (SMN1)* gene lead to disease pathogenesis (Chap. 2). This chapter focuses on the role of cis-elements and transacting factors regulating *SMN* splicing and highlights how these exemplary studies led to the first FDA-approved therapeutic treatment for SMA patients. Notably, this recent therapeutic breakthrough is the only effective treatments to significantly halt disease progression for any neurodegenerative disorder and functions by targeting SMN2 RNA. The next chapter to address RNA modifications, authored by Rita Sattler and colleagues (Chap. 3), summarizes the role of RNA editing in neurodegenerative diseases including ALS, AD, HD, and others. RNA editing increases the diversity of translated protein variants and can thereby significantly alter the function of target genes, which is critically relevant for many aspects of central nervous system (CNS) function, as discussed in this review.

The next four chapters summarize mechanisms of RNA transport and stability in neurodegenerative diseases. The transport of RNA from the nucleus to the cytoplasm is required for the translation of mRNAs. Recent evidence suggests that nucleocytoplasmic transport is impaired in a variety of genetic and nongenetic neurodegenerative disorders. The chapter by Boehringer and Bowser reviews common cellular mechanism of nucleocytoplasmic RNA transport and further discusses how these pathways are perturbed in neurodegeneration (Chap. 4). RNA stability is of greatest necessity to ensure cellular function and is therefore tightly regulated via numerous complementary mechanisms. Kaitlin Weskamp and Sami Barmada provide a comprehensive overview of these mechanisms and summarize exciting studies presenting evidence to support that aberrant RNA degradation and turnover contribute to neurodegenerative disease pathogenesis (Chap. 5). Ribonucleoprotein complex assembly is a critical step in the formation of mRNP transport particles and RNA containing membraneless organelles, including stress granules. In the next chapter Wilfried Rossoll and colleagues discuss the mechanisms underlying abnormal RNP transport granule assembly in SMA. They also discuss the consequence of dysfunction of this pathway on physiological cellular growth, neuronal maturation, and synaptic plasticity (Chap. 6). In the final chapter of this section, Ross Buchan and colleagues describe the role of ribonucleoprotein granules in ALS (Chap. 7). During cellular stress, stress granules (SGs) form and function to halt translation until the stressor is removed. Many ALS-causing mutations are found in genes whose protein products comprise these RNP granules and, when incorporated, alter SG dynamics. Persistent SG formation is hypothesized to induce protein misfolding and a key event in the seeding of TDP-43 neuropathology.

The final four chapters provide in-depth reviews of disease-causing mutations that disrupt RNA metabolism. Auinash Kalsotra and colleagues first describe how noncoding microsatellite repeat expansions alter RNA processing due to sequestration of RBPs as observed in myotonic dystrophy (Chap. 8). Interestingly, similar intronic repeat expansions were recently identified as the most common genetic cause of ALS/FTD and are thought to act through similar mechanisms. The following chapter by Janice Robertson and colleagues discusses a variety of proposed mechanisms on how the RNA-binding protein TDP-43 might contribute to disease pathogenesis in ALS and FTD (Chap. 9). Mutations in TDP-43 lead to ALS, but the interesting fact is that up to 97% of ALS patients and up to 45% of FTD patients exhibit cytoplasmic aggregation of wild-type TDP-43, independent of disease etiology. This discovery has made TDP-43 one of the most studied RNA-binding proteins in neurodegenerative diseases and different mechanisms leading to these aggregations have been proposed since. This chapter provides a comprehensive overview of the nuclear and cytoplasmic function of TDP-43 and how defects in these functions might lead to neurodegeneration. Craig Bennett and Albert LaSpada next provide an in-depth discussion on the consequence of mutations in the Senataxin protein on neural health (Chap. 10). Senataxin is a DNA-RNA helicase and mutations in the *SETX* gene are associated with a variety of neurodegenerative diseases, including juvenile-onset ALS and cerebellar ataxia with oculomotor apraxia type 2. This exciting chapter describes the role of Senataxin on RNA processing. We conclude this book with a chapter by Erik Lehmkuhl and Daniela Zarnescu with a focus on the final step in mRNA processing—translation (Chap. 11). Here, the authors provide an in-depth review on the mechanisms of translational inhibition in the context of a variety of neurodegenerative disorders and the implication of dysfunction of this process on cellular proteostasis.

In conclusion, these chapters provide a current view into the rapidly growing areas of investigation of the pathobiology of RNA metabolism and RBP dysfunction in neurodegenerative diseases. This book includes in-depth reviews on a variety of RNA-binding proteins across multiple disorders. We hope the contents of this book stimulate provocative discussions and inspire novel avenues of investigation to further broaden our knowledge of pathologic RNA metabolism with the ultimate goal of translating these discoveries into therapeutic programs for patients.

Phoenix, AZ, USA   Rita Sattler
Pittsburgh, PA, USA   Christopher J. Donnelly

Contents

Contributors

Sami J. Barmada Neuroscience Graduate Program and Department of Neurology, University of Michigan School of Medicine, Ann Arbor, MI, USA

Veronique V. Belzil Department of Neuroscience, Mayo Clinic, Jacksonville, FL, USA

Department of Neurology and Neurosurgery, McGill University, Montreal, QC, Canada

Craig L. Bennett Department of Neurology, Duke University School of Medicine, Durham, NC, USA

Department of Neurobiology, Duke University School of Medicine, Durham, NC, USA

Department of Cell Biology, Duke University School of Medicine, Durham, NC, USA

Ashley Boehringer Department of Neurobiology, Barrow Neurological Institute, Phoenix, AZ, USA

School of Life Sciences, Arizona State University, Phoenix, AZ, USA

Robert Bowser Department of Neurobiology, Barrow Neurological Institute, Phoenix, AZ, USA

J. Ross Buchan Department of Molecular and Cellular Biology, University of Arizona, Tucson, AZ, USA

Mark T. W. Ebbert Department of Neuroscience, Mayo Clinic, Jacksonville, FL, USA

Nichole Eshleman Department of Molecular and Cellular Biology, University of Arizona, Tucson, AZ, USA

Nikita Fernandes Department of Molecular and Cellular Biology, University of Arizona, Tucson, AZ, USA

Auinash Kalsotra Department of Biochemistry, University of Illinois, Urbana-Champaign, IL, USA

Carl R. Woese Institute of Genomic Biology, University of Illinois, Urbana-Champaign, IL, USA

Rebecca J. Lank Department of Neurology, University of Michigan, Ann Arbor, MI, USA

Erik M. Lehmkuhl Department of Molecular and Cellular Biology, University of Arizona, Tucson, AZ, USA

Feikai Lin Department of Biochemistry, University of Illinois, Urbana-Champaign, IL, USA

Ileana Lorenzini Department of Neurobiology, Barrow Neurological Institute, Dignity Health, St. Joseph's Hospital and Medical Center, Phoenix, AZ, USA

Chaitali Misra Department of Biochemistry, University of Illinois, Urbana-Champaign, IL, USA

Stephen Moore Department of Neurobiology, Barrow Neurological Institute, Dignity Health, St. Joseph's Hospital and Medical Center, Phoenix, AZ, USA

Interdisciplinary Graduate Program in Neuroscience, Arizona State University, Tempe, AZ, USA

Dmytro Morderer Department of Neuroscience, Mayo Clinic, Jacksonville, FL, USA

Phillip L. Price Department of Neuroscience, Mayo Clinic, Jacksonville, FL, USA

Department of Cell Biology, Emory University, Atlanta, GA, USA

Janice Robertson Tanz Centre for Research in Neurodegenerative Diseases and Department of Laboratory Medicine and Pathobiology, University of Toronto, Toronto, ON, Canada

Wilfried Rossoll Department of Neuroscience, Mayo Clinic, Jacksonville, FL, USA

Rita Sattler Department of Neurobiology and Neurology, Dignityhealth St. Joseph's Hospital, Barrow Neurological Institute, Phoenix, AZ, USA

Marc Shenouda Tanz Centre for Research in Neurodegenerative Diseases and Department of Laboratory Medicine and Pathobiology, University of Toronto, Toronto, ON, Canada

Natalia N. Singh Department of Biomedical Sciences, Iowa State University, Ames, IA, USA

Ravindra N. Singh Department of Biomedical Sciences, Iowa State University, Ames, IA, USA

Albert R. La Spada Department of Neurology, Duke University School of Medicine, Durham, NC, USA

Department of Neurobiology, Duke University School of Medicine, Durham, NC, USA

Department of Cell Biology, Duke University School of Medicine, Durham, NC, USA

Duke Center for Neurodegeneration & Neurotherapeutics, Duke University School of Medicine, Durham, NC, USA

Anna Weichert Tanz Centre for Research in Neurodegenerative Diseases and Department of Laboratory Medicine and Pathobiology, University of Toronto, Toronto, ON, Canada

Kaitlin Weskamp Neuroscience Graduate Program and Department of Neurology, University of Michigan School of Medicine, Ann Arbor, MI, USA

Daniela C. Zarnescu Department of Molecular and Cellular Biology, University of Arizona, Tucson, AZ, USA

Department of Neuroscience, University of Arizona, Tucson, AZ, USA

Department of Neurology, University of Arizona, Tucson, AZ, USA

Ashley B. Zhang Tanz Centre for Research in Neurodegenerative Diseases and Department of Laboratory Medicine and Pathobiology, University of Toronto, Toronto, ON, Canada

About the Editors

Rita Sattler is an associate professor of neurobiology at the Barrow Neurological Institute in Phoenix, AZ. She received her master's and doctorate degree in neurophysiology from the University of Toronto in Toronto, Canada, where she studied mechanisms of neurodegeneration in stroke. As a postdoctoral fellow at Johns Hopkins University, Dr. Sattler focused her research on studies of synaptic biology and glutamate receptor function. Her current research combines her expertise in neurodegeneration and synaptic function and is aimed at the elucidation of synaptic dysfunction in neurodegenerative disorders, including amyotrophic lateral sclerosis (ALS) and frontotemporal dementia (FTD). The Sattler laboratory primarily uses human patient-derived induced pluripotent stem cells differentiated into neurons and glial cells as a disease model and employs state-of-the-art molecular, biochemical, physiological, and imaging technologies to identify novel disease pathways and therapeutic targets.

Christopher J. Donnelly Christopher Donnelly is an assistant professor of neurobiology and a member of the Live Like Lou Center for ALS Research at the University of Pittsburgh School of Medicine in Pittsburgh, PA. He received his Ph.D. in molecular biology and genetics at the University of Delaware where he studied RNA trafficking and local translation during axon regeneration. As a postdoctoral fellow at Johns Hopkins University School of Medicine, Dr. Donnelly employed patient-derived induced pluripotent stem cell (iPSCs) neurons to study the pathogenic mechanisms underlying mutations that cause amyotrophic lateral sclerosis (ALS) and frontotemporal dementia (FTD). These studies revealed RNA-based mechanisms of neurotoxicity and defects in the nucleocytoplasmic transport pathway as drivers of disease. Dr. Donnelly's lab at the University of Pittsburgh currently focuses on employing human iPSC-derived cultures and Drosophila models to elucidate the pathogenic mechanism that contribute to neurodegeneration. Specifically, his lab studies how genetic mutations alter nucleocytoplasmic transport of RNA and proteins and developed a photokinetic approach to understand the triggers and consequences of intracellular protein aggregation that are pathological hallmarks of neurodegenerative diseases.

Abbreviations

3′UTR	3′ Untranslated region
5′UTR	5′ Untranslated region
5caC	5-Carboxylcytosine
5fC	5-Formylcytosine
5hmC	5-Hydroxymethylcytosine
$5HT_{2c}$	Serotonin receptor subunit 2C
5mC	Methylation of carbon 5 on cytosine
A site	Acceptor site
A/I	Adenosine to inosine
AD	Alzheimer's disease
ADARs	Adenosine deaminase acting on double-stranded RNA
ALS	Amyotrophic lateral sclerosis
AMPA	α-Amino-3-hydroxy-5-methyl-4-isoxazolepropionic acid
AMPAR	AMPA receptor, α-amino-3-hydroxy-5-methyl-4-isoxazole propionic acid receptor
ARE	Adenylate-uridylate-rich element
ASD	Autism spectrum disorder
AUF1	ARE RNA-binding protein 1
c9ALS	Amyotrophic lateral sclerosis patients carrying a pathogenic *C9orf72* repeat expansion
c9FTD/ALS	Frontotemporal dementia and amyotrophic lateral sclerosis patients carrying a pathogenic *C9orf72* repeat expansion
C9orf72	Chromosome 9 open reading frame 72
carboxy-DCFDA	5-(and-6)-Carboxy-2′,7′-dichlorofluorescin diacetate
CBC	Cap-binding complex
CDE	Constitutive decay element
CH	Cerebellar hypoplasia
CHMP2B	Chromatin modifying protein 2B
CNS	Central nervous system
CpG	Cytosine followed by guanine in cis
CPSF	Cleavage/polyadenylation specificity factor

DENN	Differentially expressed in normal and neoplastic cells
DM	Myotonic dystrophy
DM1	Myotonic dystrophy type 1
DM2	Myotonic dystrophy type 2
DMPK	Myotonic dystrophy protein kinase
DNA	Deoxyribonucleic acid
DNMTs	DNA methyltransferases
DPR	Dipeptide repeat
dsRBDs	Double-stranded RNA-binding domains
E Site	Exit site
EEJ	Exon-exon junction
Eif2α	Eukaryotic initiation factor 2 alpha
EJC	Exon junction complex
ELISA	Enzyme-linked immunosorbent assay
ER	Endoplasmic reticulum
eRNAs	Enhancer RNAs
FMR1	Fragile X mental retardation gene 1
FMRP	Fragile X mental retardation protein
FTD	Frontotemporal dementia
FTO	Fat mass and obesity-associated protein
FUS	Fused in sarcoma
FXTAS	Fragile X-associated tremor ataxia syndrome
G_4C_2	GGGGCC nucleotides
GABA*	Gamma-aminobutyric acid
GBM	Glioblastoma
GFP	Green fluorescent protein
GluA2	Glutamate ionotropic receptor AMPA type subunit 2
GRN	Gene encoding progranulin
H2A	Histone 2A
H2B	Histone 2B
H3	Histone 3
H4	Histone 4
HATs	Histone acetyltransferases
HD	Huntington's disease
HDACs	Histone deacetylases
HERV	Human endogenous retroviruses
HMTs	Histone methyltransferases
HpC	Hippocampus
hPGRN	Human progranulin
HRE	Hexanucleotide repeat expansion
hSOD1	Human superoxide dismutase 1
HTT	Huntingtin gene
IP	Inferior parietal lobe
iPSCs	Induced pluripotent stem cells
IRES	Internal ribosome entry site

Juvenile ALS	ALS4
KDMs	Lysine demethylases
KO	Knockout
$K_v1.1$	Potassium voltage-gated channel subfamily A member 1
lincRNAs	Large intergenic noncoding RNAs
lncRNAs	Long noncoding RNAs
m^5C	RNA methylation at cytosine
m^6A	RNA methylation at adenine
m^7G	RNA methylation at guanine
MAPT	Microtubule-associated protein tau
MBNL	Muscleblind
mCA	Methylation of cytosine followed by adenine
mCG	Methylation of cytosine followed by guanine
miRNA	Micro-RNA
miRNAs	Micro-RNAs
MRE	miRNA recognition element
mRNA	Messenger RNA
mtDNA	Mitochondrial DNA
ncRNA	Noncoding RNA
nDNA	Nuclear DNA
NES	Nuclear export sequence
NGD	No-go decay
NLS	Nuclear localization sequence
NMD	Nonsense-mediated decay
NMJ	Neuromuscular junction
NPC	Nuclear pore complex
NSD	Nonstop-mediated decay
NXF1	Nuclear export factor 1
Orf/c9orf72	Open reading frame
P site	Peptidyl site
PAB2	Polyadenylate binding protein 2
PARs	Promoter-associated RNAs
P-bodies	Processing bodies
PCH1	Pontocerebellar hypoplasia type 1
PCR	Polymerase chain reaction
PD	Parkinson's disease
PERK	PRKR-like ER kinase
piRNAs	Piwi-interacting RNAs
poly(A)	Long chain of adenine nucleotides
pre-miRNAS	Precursor miRNAs
pri-miRNAs	Primary miRNAs
PRMTs	Protein arginine methyltransferases
PSP	Progressive supranuclear palsy
PTC	Premature stop codons
RAN	Repeat associated Non-AUG

RAN translation	Repeat-associated non-AUG translation
RISC	RNA-induced silencing complex
RNA	Ribonucleic acid
RNAi	RNA interference
RNAseq	RNA sequencing
RNP	Ribonucleoprotein particle
ROS	Reactive oxygen species
RRBS	Reduced representation bisulfite sequencing
RRM	RNA recognition motif
rRNA	Ribosomal RNA
RUST	Regulated unproductive splicing and translation
S6K1	S6 kinase 1
sALS	Sporadic amyotrophic lateral sclerosis
SBS	Staufen-binding site
SCA8	Spinocerebellar ataxia type 8
SCI	Spinal cord injury
SG	Stress granule
siRNA	Small interfering RNA
siRNAs	Small interfering RNAs
SKAR	S6 kinase 1 Aly/REF-like target
SLBP	Stem loop-binding protein
SLC1A2	Solute carrier family 1 member 2
SMD	Staufen-mediated decay
SMTG	Superior middle temporal gyri
SOD1	Superoxide dismutase 1
sRNAs	Short RNAs
Stau1	Staufen-1
SUnSET	Surface sensing of translation
SURF	Surveillance complex
TARDBP	TAR DNA-binding protein 43
TDP-43	TAR DNA binding protein—43 kilodaltons
TE	Transposable element
TRAP	Tagged ribosome affinity purification
TREX complex	Translation export complex
tRNA	Transfer RNA
TTP	Tristetraprolin
uORF	Upstream open reading frame
UPR	Unfolded protein response
UTR	Untranslated region
VCP	Valosin-containing protein
VEGF	Vascular endothelial growth factor A

Chapter 1
An Epigenetic Spin to ALS and FTD

Mark T. W. Ebbert, Rebecca J. Lank, and Veronique V. Belzil

Abstract Amyotrophic lateral sclerosis (ALS) and frontotemporal dementia (FTD) are two devastating and lethal neurodegenerative diseases seen comorbidly in up to 15% of patients. Despite several decades of research, no effective treatment or disease-modifying strategies have been developed. We now understand more than before about the genetics and biology behind ALS and FTD, but the genetic etiology for the majority of patients is still unknown and the phenotypic variability observed across patients, even those carrying the same mutation, is enigmatic. Additionally, susceptibility factors leading to neuronal vulnerability in specific central nervous system regions involved in disease are yet to be identified. As the inherited but dynamic epigenome acts as a cell-specific interface between the inherited fixed genome and both cell-intrinsic mechanisms and environmental input, adaptive epigenetic changes might contribute to the ALS/FTD aspects we still struggle to comprehend. This chapter summarizes our current understanding of basic epigenetic mechanisms, how they relate to ALS and FTD, and their potential as therapeutic targets. A clear understanding of the biological mechanisms driving these two currently incurable diseases is urgent—well-needed therapeutic strategies need to be developed soon. Disease-specific epigenetic changes have already been observed in patients and these might be central to this endeavor.

Keywords Amyotrophic lateral sclerosis · Epigenetic modifications · Frontotemporal dementia · Methylation · RNA-mediated regulation

Mark T. W. Ebbert and Rebecca J. Lank contributed equally to this work.

M. T. W. Ebbert
Department of Neuroscience, Mayo Clinic, Jacksonville, FL, USA

R. J. Lank
Department of Neurology, University of Michigan, Ann Arbor, MI, USA

V. V. Belzil (✉)
Department of Neuroscience, Mayo Clinic, Jacksonville, FL, USA

Department of Neurology and Neurosurgery, McGill University, Montreal, QC, Canada
e-mail: Belzil.Veronique@mayo.edu

© Springer International Publishing AG, part of Springer Nature 2018
R. Sattler, C. J. Donnelly (eds.), *RNA Metabolism in Neurodegenerative Diseases*, Advances in Neurobiology 20,
https://doi.org/10.1007/978-3-319-89689-2_1

1.1 Introduction

Amyotrophic lateral sclerosis (ALS) is the most prevalent motor neuron disease, causing progressive degeneration of upper and lower motor neurons in 2–3 individuals per 100,000 worldwide [1, 2]. Clinically, ALS is characterized by rapidly progressing muscle weakness and spasticity leading to paralysis and eventually respiratory failure [3]. Patients present with symptoms at a mean age of 55 years and die within 2–4 years [4].

Patients with ALS are also frequently affected by the second most common form of early-onset dementia known as frontotemporal dementia (FTD). While patient symptoms will start with either ALS or FTD, 13–15% of patients will eventually develop symptoms for the other disease [5–9]. This strong link between ALS and FTD was observed before researchers were able to link them genetically. FTD has a prevalence of 15–22 individuals per 100,000 worldwide [10] and is characterized by disturbances in behavior, personality, and language as a result of neurodegeneration in the frontal and temporal lobes. Cognitive impairments are usually detected around age 58, and it is fatal within 6–10 years after symptoms onset [11].

Historically, researchers have extensively searched for the underlying genetic etiologies for these conditions, yet, despite strong efforts to uncover genetic causes, the vast majority of ALS/FTD patients have no known genetic cause and no family history. The discovery of the *C9orf72* DNA GGGGCC (G_4C_2) repeat expansion in ALS and FTD patients brought to light the possibility of RNA-mediated toxicity in disease which could be controlled by epigenetic mechanisms. In this chapter, we will summarize the major epigenetic regulation mechanisms. Then, we will review the current knowledge of epigenetic regulation as it relates to ALS and FTD (Fig. 1.1).

1.2 Epigenetic Mechanisms

The epigenome is generally defined as the collection of heritable but dynamic mechanisms that largely dictate access to the DNA for templated functions such as gene transcription, DNA synthesis, and repair without altering the primary nucleotide sequence; there is, however, a valid argument that chimeric DNA, which does modify the primary nucleotide sequence, is facilitated by the epigenome. These intricacies regarding how far-reaching the epigenome's involvement is in nucleotide sequence will be continually clarified as the field matures. Here, we will focus on general regulation processes.

Post-transcriptional regulation of RNA function and metabolism—by noncoding RNAs (ncRNAs), such as microRNA, or chemical modifications to ribonucleotides—is now also considered part of epigenetic control. Whereas an individual's genome is mostly established at conception, the individual's inherited epigenome continues to change throughout embryonic, fetal, and postnatal life in response to

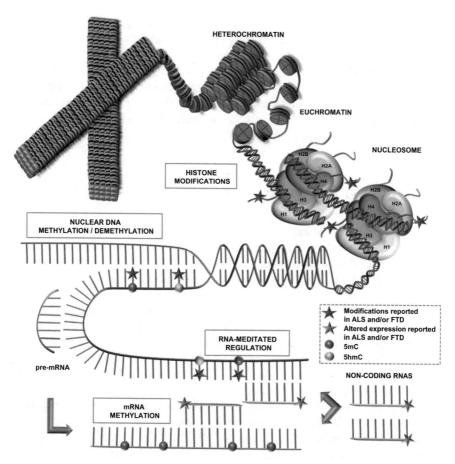

Fig. 1.1 Schematic representation of the major epigenetic regulatory mechanisms and their interactions. The major epigenetic regulatory mechanisms are framed. Red stars mark epigenetic changes known to be involved in ALS/FTD. Orange stars indicate that altered expression of these RNAs has been reported in ALS/FTD. Purple spheres represent 5mC marks while green spheres represent 5hmC marks

cell-intrinsic mechanisms and environmental input. Epigenetic changes are cell-specific and are presumed to be adaptive responses to internal and external stimuli.

As next-generation sequencing is becoming more commonplace, the epigenome is garnering greater attention in many fields, from developmental biology to environmental epidemiology, because it offers a potential mechanistic link between the environment and phenotype. For ALS/FTD researchers, epigenetics is a particularly interesting field that may explain the clinical variability within families sharing ALS/FTD-causing genetic variants, and the cause of disease in patients affected sporadically.

Epigenetic control of genomic functions occurs by regulating nucleosome position and density via histone chaperones and chromatin remodeling complexes [12].

Nucleosomes are the fundamental repeating unit of chromatin, and consist of nucleoprotein complexes containing two copies each of four core histone proteins (H2A, H2B, H3, and H3) and 147 base pairs of DNA wrapped 1.67 turns counterclockwise around this globular structure [13]. Key signals directing nucleosome positioning and chromatin structure include chemical modifications to the DNA and histones, as well as ncRNA species. Epigenetic regulation is also exerted via posttranscriptional regulation of RNA function and metabolism. This section will briefly review our current understanding of these epigenetic mechanisms.

1.2.1   Nuclear DNA Methylation

DNA methylation is the biochemical addition of a methyl group to a nucleotide, often resulting in altered transcription factor binding, chromatin remodeling, and ultimately altered gene expression. Methylation of the carbon 5 on cytosine (5mC) is the most studied and best understood DNA modification.

5mC most frequently occurs at cytosine-guanine nucleotide sequences (mCG), often referred to as CpG sites. Genomic regions enriched in CpG sites, referred to as CpG islands, are often associated with gene promoters. Whether located near transcription initiation sites or distant from annotated promoters, CpG islands play a central role in destabilizing nucleosomes and recruiting proteins that initiate chromatin remodeling [14–16]. Generally, methylation in a CpG island is associated with the reduced expression of a nearby gene.

Recent research has found 5mC marks at cytosine-adenine sequences (mCA). It is believed that mCA may be particularly important in the central nervous system, as the number of mCA and mCG marks detected in the adult brain are approximately equal [17]. Although both mCA and mCG are associated with transcriptional repression, early evidence suggests mCA may be particularly important for more precise transcription regulation—1–2-fold expression changes compared to 100–1000-fold changes observed with mCG [18].

DNA methylation is catalyzed by DNA methyltransferases (DNMTs) 1, 3A, and 3B. While DNMT3A and DNMT3B are associated with de novo methylation, DNMT1 is mainly responsible for maintaining methylation across cell generations by methylating the daughter strand of DNA during cell division, thus preventing hemimethylation of double-stranded DNA [19]. Because of DNMT1's maintenance function, DNA methylation is considered a fairly stable epigenetic mark. Further research into the differentiating roles of DNMT3A and DNMT3B are warranted, as early evidence suggests DNMT3A may be primarily responsible for mCA methylation. Its proper regulation may be especially critical in the brain [20–22].

Bisulfite conversion is currently the preferred method to detect cytosine methylation. Briefly, bisulfite ion deaminates unmethylated cytosines, which result in its chemical conversion to uracil upon alkaline desulfonation [23]. As bisulfite converts 5mCs much more slowly, there is a selective conversion favoring unmethylated cytosines. After DNA amplification, all unconverted cytosines are considered methylated. Importantly, standard bisulfite conversion does not differentiate methylated cytosines from cytosines that have been demethylated, which is the topic of the next section.

1.2.2   Nuclear DNA Demethylation

Although DNA methylation is considered a stable epigenetic change, it is now well understood that it is closely and dynamically regulated not only by epigenetic "writers," such as DNMTs, but also by "eraser" mechanisms, which keep CpGs unmethylated. Importantly, demethylation is of particular interest in neurodegeneration research because, among all organs, its footprints are most abundant in the brain and further increase with age [24].

DNA demethylation can occur as a passive or active process. Passive demethylation occurs when the process of copying methylation marks onto the daughter strand of DNA is impaired. As a result, future cell generations will lose methylation marks by dilution. Active demethylation is mediated by TET family of dioxygenases, which catalyze the oxidation of 5mC in a stepwise manner, resulting in 5-hydroxymethylcytosine (5hmC), 5-formylcytosine (5fC), and 5-carboxylcytosine (5caC). Oxidation is followed by replication-dependent dilution or thymine DNA glycosylase-mediated base excision and repair, resulting in full demethylation and restoration to unmodified cytosine [25]. 5fC- and 5caC-containing regulatory elements show very limited overlap and are therefore believed to play distinct regulatory roles. 5caC sites are more active than 5fC sites, and both are more active than regions decorated by 5hmC [26].

Our current understanding is that DNA methylation and demethylation processes must be highly orchestrated to appropriately regulate the transcriptome. Further research is necessary to elucidate the exact functions of 5hmC, 5fC and 5caC as it is still difficult to estimate the extent to which these events occur and appreciate their full potential as therapeutic targets.

1.2.3   Mitochondrial DNA Methylation

While nuclear DNA (nDNA) methylation is well recognized, mitochondrial DNA (mtDNA) methylation is still controversial. Using restriction enzyme cleavage, early studies estimated that about 2–5% of mtDNA CCGG sequences are fully methylated whereas the remainder of mtDNA is unmethylated [27]. Mass spectrometry confirmed the presence of methylated bases in human mtDNA [28]. Later, using affinity-based methods, both 5mC and 5hmC modifications were found in the D-loop region of human mtDNA at both CpG and non-CpG dinucleotides [29, 30]. Other studies used bisulfite conversion coupled with sequencing or pyrosequencing and claimed that mtDNA methylation may be as high as 2–18% in the D-loop region

[31, 32]. Recent reports however suggest that mtDNA methylation levels may have been greatly overestimated due to the circular structure of mtDNA, which affects bisulfite conversion efficiency [33–35]. Nonetheless, the reported presence of methyltransferases DNMT1 and DNMT3A inside mitochondria suggests that methylation may take place in mtDNA [30, 36]. Since nDNA and mtDNA interact to modulate the transcriptome [37, 38], future studies need to better characterize mtDNA modifications and their potential role in the regulation of gene transcription.

1.2.4 RNA Methylation

Epigenetic regulation goes beyond transcription. RNA is also under epigenetic regulation and can be methylated or otherwise modified. In fact, methylation occurs in most RNA classes, but the exact function of RNA modifications remains unclear due to limitations in detection and quantification.

Whereas DNA methylation has mainly been observed at cytosines, RNA methylation can occur at cytosine (m^5C), adenine (m^6A), and guanine (m^7G). m^7G (N^7-methylguanine) at the $5'$ cap structure is the most widely studied RNA methylation and is necessary for the translation of most messenger RNAs (mRNAs). m^7G cap also mediates nuclear transport of some mRNAs, preserves mRNA stability by protecting it from degradation, and facilitates other processing such as the addition of poly(A) tails to mRNAs [39, 40]. m^7G cap methylation is reversible and can either go through the process of "decapping," where the entire m^7G cap is removed [41], or potentially through m^7G demethylation.

In contrast to the extensively studied 5mC in DNA, the m^5C (N^5-methylcytosine) RNA modification has not been thoroughly explored. While it is still unclear how m^5C is regulated by DNMTs, a study suggested that mouse Dnmt2 RNA methyltransferase may be required for epigenetic heredity [42]. Early studies also demonstrated that m^5C affects interactions of long noncoding RNAs (lncRNAs) with chromatin-associated protein complexes [43].

The most abundant mRNA modification is m^6A (N^6-methyladenine). Dominissini et al. found that silencing m^6A methyltransferase—mostly acting at highly conserved long internal exons, stop codons, and 3'UTRs—alters gene expression and alternative splicing patterns [44]. The same group conducted affinity-based m^6A profiling and identified proteins that bind specifically to m^6A. Their findings suggest that RNA methylation, through the binding of these m^6A-specific binding proteins, may disturb RNA binding proteins' affinity to associate with partner unmethylated RNAs. Importantly, these m^6A-binding proteins recruit additional factors that facilitate functions such as alternative splicing, nuclear export, and mRNA stability [40, 44]. One example is the m^6A-binding protein YTHDC1, which facilitates the export of methylated mRNA from the nucleus to the cytoplasm in vitro. Knockdown of YTHDC1 results in retention of nuclear m^6A-containing mRNA, where transcripts accumulate in the nucleus and are depleted from the cytoplasm [45]. Of interest,

m^6A has been the first reversible modification identified in coding and noncoding RNAs after the discovery that the fat mass and obesity-associated protein (FTO) act as a m(6)A demethylase [46–48], suggesting that, analogous to reversible DNA, reversible RNA methylation may affect gene expression and downstream regulation of RNA-related cellular pathways.

1.2.5 RNA-Mediated Regulation

While 70–90% of the genome is transcribed, only 1–3% of these transcripts encode proteins [49, 50]. As such, the transcriptome is mainly composed of ncRNAs having infrastructural and regulatory functions. These transcripts are gaining increased attention, not only for their recognized roles in transcriptional and post-transcriptional regulation, but also for their targeted effects on gene expression, making them attractive therapeutic targets.

ncRNAs are divided into short (<30 nucleotides) and long ncRNAs (>200 nucleotides), and each subtype of ncRNA fulfills very specific regulatory roles. Specifically, short RNAs (sRNAs) are sub-categorized into micro-RNAs (miRNAs), piwi-interacting RNAs (piRNAs), and small interfering RNAs (siRNAs). lncRNAs are generally sub-classified according to their proximity to protein coding genes— sense, anti-sense, bidirectional, intronic, and intergenic. Apart from their size, sRNAs and lncRNAs have largely different functions in transcription regulation. Two new classes of ncRNAs have also been recently recognized: enhancer RNAs (eRNAs) and promoter-associated RNAs (PARs).

Mature miRNAs are single-stranded 20–24 nucleotide sequences derived from precursor miRNA (pre-miRNAS). These are ~70 nucleotide transcripts with distinct hairpin structures initially derived from nuclear primary miRNA (pri-miRNAs). Mature miRNAs pair with complementary sequences on target mRNA transcripts through the 3′ untranslated region (3′UTR) and repress the targeted mRNA translation after recruiting the RNA-induced silencing complex (RISC) [51, 52]. Mature miRNAs may have several target mRNAs, but their effect on their targets is a modest 1–2-fold change in expression [53]. Although miRNAs seem to initiate modest changes in their targets, changes in expression of miRNAs themselves associate with widespread mRNA expression changes, indicating that miRNAs are central to global expression regulation [53]. In fact, a large screening of the mammalian genome found hundreds of mRNAs with conserved pairing to specific miRNAs, with an enrichment of genes involved in transcriptional regulation [54]. Specifically, several miRNAs target DNMTs that can have far-reaching impact on global methylation levels, which has been implicated in several cancers and autoimmune diseases [55]. The story becomes more complicated when one considers that miRNA expression can in turn be modulated by DNA methylation and histone modifications [55, 56], suggesting that transcriptional regulation needs to be well orchestrated by an epigenetic–ncRNA feedback loop. This highlights the complexity of epigenetics and the interdependence of these mechanisms to regulate gene expression.

piRNAs are 24–31-nucleotide sequences able to form complexes with Piwi proteins of the Argonaute family. They are part of a complex population of small RNAs highly enriched in male gonads [57]. Their main function is to silence transposable elements. These mobile elements are autonomous sequences of DNA that replicate and insert themselves into the genome, potentially introducing detrimental DNA damage. Some results also suggest a role for piRNA in transcriptional regulation and deadenylation-mediated mRNA degradation [58].

Mature siRNAs are 20–24 nucleotide sequences that modulate a given gene's expression after binding to its complementary nucleotide sequence through a process called RNA interference (RNAi). siRNAs mediate post-transcriptional silencing in a way similar to miRNAs, and may also suppress transposon activity in a way similar to piRNAs [59, 60].

lncRNAs are sequences longer than 200 nucleotides characterized by low nuclear expression and low conservation across species. One exception is highly conserved large intergenic noncoding RNAs (lincRNAs), which recruit histone-modifying complexes and transcription factors to transcriptionally modulate targeted chromosomal regions [61, 62]. The majority of lncRNAs are transcribed as complex networks of sense and anti-sense transcripts overlapping protein-coding genes.

eRNAs are sequences of about 800 nucleotides on average that are derived from regions enriched in RNA Polymerase II and transcriptional co-regulators. What makes the genomic sites encoding eRNAs different from those of other lncRNAs is their specific histone methylation signature at histone 3 lysine 4 (H3K4), which is typical of enhancer sites. Similar to lincRNAs, eRNAs are evolutionarily conserved but have a short half-life. eRNA expression levels correlate with the expression of nearby genes, and is thus dynamically regulated upon signaling [63, 64]. eRNAs are believed to act as transcriptional activators.

PARs are short half-life transcripts that can be classified as short or long ncRNAs—their size range from 16 to over 200 nucleotides. PARS are either expressed near transcription start sites or from elements of the promoter [65], from both strands and in divergent orientation with respect to the transcription start site. Most PARS associate with highly expressed genes while being weakly expressed themselves. Increasing evidence demonstrates that PARs may associate with both transcriptional activation and repression [66–69].

1.2.6   Histone Modifications

As discussed above, the density of nucleosomes determines the accessibility of DNA to protein complexes performing templated functions. Heterochromatin contains tightly wound DNA, where gene transcription is repressed. Euchromatin is a looser conformation of the DNA, which is conducive to gene transcription. Chromatin conformation is dynamic and is regulated in part by covalent posttranslational histone modifications, which mainly occur at amino acid residues within unstructured histone "tails." These modifications include acetylation and methylation of lysines, methylation of arginines, and phosphorylation of serines and threonines, among others. Specific functions have been assigned to several modifications, which led to the formulation of the histone code hypothesis [70]. It is our current understanding of the roles of histone modifications that their combinations and association with transcription factors and other chromatin regulators define epigenetic states, which can be discovered and assigned to genomic regions by machine learning [71].

Enzymes such as histone acetyltransferases (HATs), histone methyltransferases (HMTs), protein arginine methyltransferases (PRMTs), and kinases facilitate the addition of chemicals on amino acid residues (writers). For instance, HATs brings in a negative charge that neutralizes the positive charge on the histones. This chemical change decreases the interaction of the N termini of histones with the negatively charged phosphate groups of DNA, resulting in a more relaxed structure. This structure allows transcription factors to access DNA and initiate gene transcription. Other enzymes such as histone deacetylases (HDACs), lysine demethylases (KDMs), phosphatases, and deubiquitylases catalyze the removal of these epigenetic marks (erasers). As such, the process of histone acetylation is reversed by HDACs. Acetyl group removal leads to a more condensed chromatin state and transcriptional gene silencing. Proteins containing DNA methyl-binding domains, chromodomains, bromodomains, and Tudor domains recognize histone modifications and recruit other chromatin modifiers and remodeling proteins to ultimately regulate DNA-dependent processes (readers).

Functional groups affixed to histones can also initiate chromatin remodeling through cis or trans effects. This way, histone marks can modulate the chromatin by directly affecting histone-histone and histone-DNA interactions, or by recruiting non-histone proteins via specific binding domains that recognize particular modifications [72].

1.3   Epigenetic Changes in Amyotrophic Lateral Sclerosis (ALS) and Frontotemporal Dementia (FTD)

The 2011 discovery that an expanded hexanucleotide (G_4C_2) repeat within a non-coding region of the *C9orf72* gene causes both ALS and FTD highlighted the genetic link between these two diseases and recognized this mutation as the most common genetic cause of ALS and FTD identified to date [73–75], yet more than 80% of all ALS and FTD cases remain genetically unexplained [76]. While other unknown genetic mutations are certainly at play, more researchers are recognizing that epigenetic changes may contribute to these two diseases based on their potential to (1) explain the phenotypic variability observed across family members carrying the same ALS/FTD-associated mutation, and (2) explain why the majority of patients develop disease without any family history of ALS or FTD. Very few epigenetic studies have been conducted on ALS and FTD before the finding of the pathogenic *C9orf72* G_4C_2 repeat expansion in 2011. This section summarizes the current knowledge on the epigenetics of ALS and FTD. A full summary can also be found in Table 1.1.

1.3.1 ALS/FTD Epigenetic Studies

1.3.1.1 Nuclear DNA methylation/demethylation

Approximately a decade ago, researchers interested in better understanding the etiology of ALS and FTD started interrogating epigenetic mechanisms using cell models, animal models, and biospecimen obtained from patients.

Early studies on ALS evaluated blood and brain promoter DNA methylation status of a few genes known to be implicated in disease including *SOD1*, *VEGF*, and *SLC1A2* [136, 137], but no changes in methylation were detected.

The methylation status of *SOD1*, *FUS*, *TARDBP*, and *C9orf72* promoters has also been evaluated from the blood of ALS patients carrying not fully penetrant *SOD1* mutations, but again, no methylation variations have been detected at these specific regulatory regions [77]. Of interest, Coppede et al. used an enzyme-linked immunosorbent assay (ELISA) to also evaluate global methylation levels in ALS patients carrying not fully penetrant *SOD1* mutations and observed a significant overall DNA methylation increase [77].

Similarly, others compared brain methylation levels of sporadic ALS (sALS) patients to control cases using Affymetrix GeneChip Human Tiling 2.0R Arrays and identified 38 differentially methylated genomic regions in patients. Further analysis of these 38 regions shed light on specifically altered biological pathways involved in calcium homeostasis, neurotransmission, and oxidative stress [78].

Using ELISA assays, Figueroa-Romero et al. identified global 5mC and 5hmC changes in sALS patients' spinal cords—these changes were not observed in blood [79]. Then, using high-throughput microarrays, the same group conducted genome-wide 5mC and expression profiling and identified loci-specific differentially methylated and expressed genes. The 112 genes identified were highly associated with immune and inflammation responses [79].

In the hope of finding differently methylated regions that might act as modifiers of age of onset in ALS, Tremolizzo et al. evaluated DNA methylation levels in both early onset (<55 years) and late onset (>74 years) ALS patients. They found a global 25–30% increase in DNA methylation levels in whole blood that was independent of age of onset [80]. While no significant difference was found between early onset and late onset ALS, the methylation increase detected in blood is consistent with previous observation in the central nervous system (CNS) of ALS patients [36, 99].

Early epigenetic studies on FTD analyzed the progranulin-encoding gene (*GRN*) known to be mutated in 5–20% of familial FTD and 1–5% of sporadic FTD patients [138–140]. Two groups independently reported that increased *GRN* promoter methylation negatively correlates with *GRN* mRNA levels in FTD subjects [81, 82], a

Table 1.1 Overview of ALS and FTD epigenetic studies

Epigenetic Modification	Overall Findings	Implications for disease	References
Nuclear DNA methylation/demethylation	Overall DNA methylation is increased in ALS patients carrying not fully penetrant SOD1 mutations.	Patients with SOD1 mutations may have unique methylation profiles.	[32]
	Differentially methylated regions are found in sporadic ALS subjects.	Affected genes are associated with calcium homeostasis, neurotransmission and oxidative stress, suggesting a role for these mechanisms in ALS pathogenesis.	[125]
	Global changes in 5mC and 5hmC levels are found in ALS postmortem spinal cords.	Expression changes in 112 genes associate with immune and inflammatory responses.	[51]
	Increased DNA methylation levels are found in ALS blood, independently from age of onset.	While not a modifier of age of onset, DNA methylation levels may act as an ALS marker of epigenetic dysfunction.	[165]
	Increased GRN promoter methylation negatively correlates with GRN mRNA levels in FTD subjects.	Altered methylation may contribute to GRN haploinsufficiency.	[56,11]
	Specific DNA methylation signatures in FTD peripheral blood associate with tauopathy.	Specific DNA methylation signature may be a risk factor for neurodegeneration.	[103]
	DNMT1 and DNMT3A protein levels are increased in motor neurons of ALS patients.	Increased DNMT1 and DNMT3A protein levels may explain the global increase in methylation observed.	[24]
	DNMT3A is present in human cerebral cortex pure mitochondria. Mitochondrial Dnmt3a protein levels are significantly reduced in mice skeletal muscles and spinal cords.	Reduced mitochondria Dnmt3a protein levels may associate with a loss of mtDNA methylation.	[172]
RNA-mediated regulation through miRNAs	Expressing mutant SOD1 (p.G39A) in mouse muscles alters expression of both miRNAs and genes associated with myelin homeostasis in spinal cords.	Results suggest an epigenetic regulation interplay between muscle cells and neighboring neurons.	[41]
	miRNAs are aberrantly regulated in ALS spinal cords.	Mature miRNAs are globally reduced and miRNA processing is altered.	[52]
	Altered microRNAs have RNA targets part of pathways previously associated with ALS.	Altered miRNA expression may result in altered regulation of key pathological pathways involved in ALS.	[52]
	ALS-associated miRNA expression changes in response to nuclear clearance and cytoplasmic aggregation of TDP-43.	Re-localization of TDP-43 alters RNA-mediated regulation.	[52]
	Altered expression of specific miRNAs in SOD1 p.G93A mice and patients carrying TARDBP p.A90V or p.M337V mutations.	miRNAs and/or their targets may potentially serve as biomarkers or therapeutic targets.	[164,113,183]
	miR-29b regulates human progranulin.	miR-29b may be therapeutically targeted to rescue FTD-associated GRN-haploinsufficiency.	[88]
	Expression of FTD-associated CHMP2B in mouse brain decreases miR-124 expression.	Altered regulation of miR-124 may lead to altered regulation of AMPAR receptor subunits.	[57]

(continued)

Table 1.1 (continued)

	Description	Finding	References
C9orf72 Epigenetic Studies	DNA methylation of the C9orf72 promoter and expanded repeat region in c9FTD/ALS.	Hypermethylation of the C9orf72 genomic region may contribute to C9ORF72 loss of function. Evidence for neuroprotective effects and reduced C9orf72-associated pathology.	[174,107,12,177,176]
	5mC and 5hmC are both present at the C9orf72 promoter in C9orf72-associated brain tissues.	Estimates of C9orf72 promoter methylation includes both 5mC and 5hmC modifications.	[49]
	Repressive histone marks at the C9orf72 locus in c9FTD/ALS.	Reduced expression of C9orf72.	[14]
	Differentially methylated regions are abundant in c9ALS and sALS patients and affect many genes and biological pathways involved in ALS.	Altered methylation may result in altered regulation of key pathological pathways involved in ALS.	[45]
	c9ALS and sALS patients have distinct but overlapping brain DNA methylation profiles affecting genes with similar biological functions.	Suggests a conserved pathobiology between c9ALS and sALS.	[45]
Potential drivers of epigenetic changes	Increased incidence of ALS in Chamorro indigenous people of Guam.	Dietary consumption of neurotoxins may induce epigenetic changes through ROS and increase the risk for ALS.	[10]
	Associations between ALS and diet, cycad neurotoxins and ROS.	Neurotoxins may initiate epigenetic changes and increase the risk for ALS.	[16,37,27,134,142]
	Monozygotic twins discordant for ALS.	Supports evidence for environmental contribution to disease onset.	[175,120,180]
	C9orf72 positive identical twins discordant for ALS.		[175,120]
	Stress-induced histone modifications repress transposable elements and other coding and non-coding RNAs.	Acute stress impairs genomic stability and may initiate cognitive impairments.	[82,80,141,48,90,117]
	Exposure to heavy metals such as aluminum sulfate, mercury, lead, and selenium associates with ALS. Heavy metals cause cellular stress and toxicity through ROS.	May drive epigenetic changes, protein denaturation and aggregation and prevent proteasomes from eliminating dysfunctional proteins.	[122,9,30,89,28,50,133]
	Increased risk for ALS in workers exposed to low frequency magnetic fields.	Electromagnetic fields may cause epigenetic changes leading to neurodegeneration.	[70]
	Increased risk for ALS in people exposed to pesticides and herbicides containing organophosphates.	Exposure to organophosphates may initiate epigenetic changes and increase toxicity through ROS.	[115,33,154,124,167,40,89,121,157]
	Increased ALS incidence in athletes and war veterans.	Drug use and ischemia from head injuries may increase ROS and drive ALS.	[25,78,123,137,163]

very interesting finding since *GRN* haploinsufficiency has been recognized as a major cause of FTD [141].

Global DNA methylation levels in blood were also assessed in tau-associated progressive supranuclear palsy (PSP) and FTD patients by Li et al. The major known risk locus for PSP and other neurodegenerative diseases is the H1 haplotype at 17q21.31, a genomic region in linkage disequilibrium with an inverted chromosomal sequence of about 970 kb [142–144]. It was shown that this disequilibrium resulted from an inversion at the H2 haplotype relative to the H1 human reference allele and from a lack of recombination between inverted and non-inverted chromosomes [145]. Li et al. found that the H1/H2 locus may affect the risk for tauopathies through methylation alterations not only at the *MAPT* locus, a region known to be mutated in FTD, but also in at least three neighboring genes. The 17q21.31-associated DNA methylation signature Li et al. identified was unique to tau-associated PSP patients and, to a lesser extent, to tau-associated FTD [83].

As DNA methyltransferases catalyze DNA methylation, researchers have been interested in assessing their expression in disease. Chestnut et al. found that protein levels of DNMT1 and DNMT3A were increased in the motor cortex of ALS patients [36]. Wong et al. went further and confirmed the presence of both Dnmt3a isoform in pure mitochondria of human cerebral cortex and mouse CNS, and 5mcs in mouse mitochondria of neurons and skeletal muscle myofibers, supporting an epigenetic regulation of brain mtDNA [84]. Wong et al. then evaluated mitochondrial Dnmt3a isoform levels in the CNS of different *SOD1* transgenic mouse models including a line hemizygous for a low copy number of hSOD1 -p.G37R mutant allele, a line that expressed high levels of normal wild-type human *SOD1* gene, and a line with skeletal muscle-restricted expression of hSOD1 -p.G37R, -p.G93A, and -wild-type variants. After studying all the lines at presymptomatic or early to middle stages of disease, they found that Dnmt3a protein levels in mouse skeletal muscle and spinal cord mitochondria were significantly reduced early in disease, this even before symptoms' onset. They also found Dnmt1 bound to the outer mitochondrial membrane of the same mice. They observed that 5mC immunoreactivity became aggregated and sequestered into autophagosomes of transgenic mice motor neurons [84].

1.3.1.2   RNA-mediated regulation through miRNAs

It is well established that ncRNAs such as miRNAs are key regulators of RNA functions and metabolism. As such, many have been interested in assessing their potential contribution to ALS/FTD pathogenesis.

Dobrowolny et al. found that selective expression of the human *SOD1* mutation p.G39A after injecting mouse muscles not only leads to hypomyelination in the sciatic nerve, it also alters spinal cord expression of miRNAs and mRNAs known to be involved in myelin homeostasis. This finding suggests that RNA and epigenetic alterations observed in motor neurons may result from changes initiated in neighboring non-neuronal cells [85].

Figueroa-Romero et al. then interrogated spinal cord tissues obtained from post-mortem ALS patients and found that mature miRNA levels are globally reduced. They also identified altered microRNAs having RNA targets part of pathways previously associated with ALS. Knowing that TDP-43 is central to ALS and FTD pathogenesis—TDP-43 pathological signature is observed in about 97% of ALS and 50% of FTD patients [146, 147]—and that TDP-43 plays a central role in miRNA biogenesis [89, 148–152], the same group used transfected cells to determine whether TDP-43 mediates miRNA-induced regulation. They observed ALS-associated miRNA expression changes in response to nuclear clearance and cytoplasmic aggregation of TDP-43, suggesting that TDP-43 pathology may alter the expression or function of endogenous miRNAs and their downstream targets [86].

Several studies have identified alternatively expressed miRNAs in ALS and FTD. Marcuzzo et al. studied the brain of pre-symptomatic and late stage *SOD1* p.G93A transgenic mice and found that expression levels of miR-9, miR-124a, miR-19a, and miR-19b were all altered in late stages of disease. Moreover, the expression analysis they conducted identified miRNA/target gene pairs that were differentially expressed in this mouse model [87]. Toivonen et al. found miR-206 altered in the blood of both *SOD1* p.G93A transgenic mice and ALS patients [88], whereas Zhang et al. found levels of miR-9 decreased in induced pluripotent stem cells (iPSCs)-derived neurons of patients carrying either p.A90V or p.M337V *TARDBP* mutations [89].

Of interest, Jiao et al. found that miR-29b regulates human progranulin (hPGRN, *GRN*) through 3′UTR binding. They demonstrated in vitro that ectopic expression of miR-29b decreased hPGRN expression and knockdown of endogenous miR-29b increased it. Their findings suggest that miR-29b may possibly be therapeutically targeted to rescue the haploinsufficiency observed in FTD patients carrying a *GRN* mutation [90]. Moreover, expression of FTD-associated mutant *CHMP2B* in cerebral cortices of mice has initiated a decrease in miR-124 expression—brain-enriched miR-124 is especially important for the proper regulation of AMPA receptor (AMPAR) subunits. Ectopic expression of miR-124 in the prefrontal cortex of these mice restored AMPAR levels and rescued the behavioral deficits previously observed in the animals [91].

1.3.2   C9orf72 *Epigenetic Studies*

Decreased expression of one or multiple *C9orf72* transcript variants has been observed in various human biospecimen carrying the pathological *C9orf72* G_4C_2 repeat expansion. These biospecimen include frontal cortex, motor cortex, cerebellum, cervical spinal cord, lymphoblastoid cell lines, iPSCs and neurons differentiated from iPSCs, all obtained or derived from ALS and FTD patients [98, 153–159]. In an attempt to better understand the biological mechanism underlying the reduced *C9orf72* expression in *C9orf72*-associated ALS and FTD (c9FTD/ALS) patients, methylation status of the regions flanking or encompassing the repeat expansion has been evaluated.

Using bisulfite sequencing and restriction enzyme assays, Rogaeva's group found that the CpG island upstream of the G_4C_2 repeat expansion is hypermethylated in the brain and blood of about 36% of ALS and 17% of FTD cases [94, 96]. Lee's group not only confirmed these results, but also uncovered that hypermethylation associates with reduced accumulation of intronic *C9orf72* RNA and reduced burden of *C9orf72*-associated pathological signature (RNA foci and dipeptide repeat accumulation). They also found that demethylation increases cell vulnerability to oxidative and autophagic stress, suggesting that *C9orf72* promoter hypermethylation may mitigate downstream molecular aberrations associated with the pathological G_4C_2 repeat expansion [93]. As the methods initially used to estimate DNA methylation in c9FTD/ALS cases could not differentiate 5mC from 5hmC, Esanov et al. were able to confirm the presence of 5hmC within the *C9orf72* promoter in post-mortem brain tissues of hypermethylated patients [97]. This finding suggests that the previous estimates by Rogaeva's and Lee's groups included both 5mC and 5hmC modifications.

However, considering that all c9FTD/ALS patients show a 50% reduction in total *C9orf72* RNA expression [73] and only approximately one third of patients are found with a hypermethylated CpG island, many details were missing. As such, Rogaeva's group attempted to assess whether it is the repeat expansion that is hypermethylated in patients and consequently drives the reduced expression. For this purpose, they developed a new qualitative assay that was independently validated by a methylation-sensitive restriction enzyme assay, and found that the *C9orf72* repeat expansion was indeed hypermethylated in all ALS and FTD cases carrying more than 50 G_4C_2 copies [95]. This finding was later confirmed by Bauer [92]. In addition, Belzil et al. investigated the brain of c9FTD/ALS patients and found that all patients carried repressive histone marks at the *C9orf72* locus [98]. As such, methylation of the repeat expansion together with repressive histone marks at the *C9orf72* locus likely explains the 50% *C9orf72* reduced expression observed in c9FTD/ALS patients.

A recent multi-omic study aimed to better understand the molecular mechanisms initiating RNA misregulation in *C9orf72*-associated c9ALS and sALS combined RNA and DNA methylation data obtained from brain next-generation RNA sequencing (RNAseq) and reduced representation bisulfite sequencing (RRBS). They found an abundance of differentially methylated cytosines in c9ALS and sALS patients, including changes in many genes and biological pathways known to be involved in ALS. They also observed that c9ALS and sALS patients have generally distinct but overlapping brain DNA methylation profiles that differ from control individuals. Of importance, they found that the c9ALS- and sALS-affected genes and biological pathways have very similar biological functions, suggesting a conserved pathobiology between c9ALS and sALS [99].

Several studies aimed to assess whether DNA methylation is a clinical modifier of ALS and FTD but so far, few correlations have been identified. Among these, hypermethylation of the CpG island upstream of the *C9orf72* repeat expansion has been found to correlate with shorter disease duration in ALS [96], but was found associated with longer disease duration and later age of death in FTD [160].

1.3.3   Potential Drivers of Epigenetic Changes

The epigenome is a dynamic, complex machinery that plays a critical role in coordinating cellular functions. It is constantly changing to address cellular needs or to react to environmental threats, such as infections. A prime example is heat-shock proteins. These proteins were discovered in the early 1960s by Ferrucio Ritossa when he noticed a "puffing" pattern—now known to be a sudden increase in RNA transcription—in *Drosophila* cells when one of his lab mates increased his incubator's temperature [161]. Many discoveries have resulted from this unintended finding, but one of the most intriguing discoveries was that the "puff" Ritossa described was observable within 2–3 min of heat exposure, demonstrating how agile the epigenomic machinery is. Understanding not only which epigenetic modifications affect disease, but what drives these epigenetic changes is critical to better understanding human health and disease.

As demonstrated by Ritossa's landmark discovery that "heat shock" can induce an immediate response from the epigenetic machinery, environmental influences are a clear driver. "Environment" has broad implications, however, and can include a cell's internal or external influences, such as neighboring cells, as research has shown that epigenetic changes can be transmitted from cell to cell [162]. Many factors affect the dynamic interaction between environmental influences and the epigenome, including exercise, age, diet, and toxic exposures.

Researchers observed a clear example of environmental factors driving ALS in the indigenous Chamorro people of Guam, who experience high ALS incidence because their diet is enriched in cycad neurotoxins. The Chamorro diet includes the flying fox, which has high levels of cycad neurotoxins because it feeds on cycad seeds [100–102]. Additional studies have found associations between ALS and cycad neurotoxins or reactive oxygen species (ROS) [102–104, 163–165]. Exactly how these neurotoxins are driving disease is unknown, but epigenetic modifications are a primary suspect. Importantly, diet has also been shown to induce epigenetic changes across other diseases [105].

The most striking support for epigenetic contribution is perhaps the reports of ALS-discordant monozygotic twins (monozygotic twins where one has disease and the other is unaffected), implicating environmental and epigenetic factors in disease [106–108]. One study identified monozygotic twins that both carry the *C9orf72* repeat expansion, but only one has developed disease. The other study could not find a clear genetic factor that caused disease. A third study by Young et al. identified thousands of large between-twin differences at CpG sites in five monozygotic twin pairs. Young et al. conducted biological pathway analysis, which revealed that impairments in GABA signaling were common to all ALS individuals. Other altered pathways were also identified, including some relevant to ALS such as glutamate metabolism and the Golgi apparatus [108]. Importantly, Young et al. applied to their 450K data the Horvath algorithm of epigenetic age [166]—an aging clock of chromatin states derived from the characterization of 353 CpG sites—and found that ALS-affected twins were epigenetically older than their unaffected co-twins, confirming previous findings that ALS is characterized by accelerated brain aging [108, 167]. In all cases, the other twin may develop disease in time, but the question would still remain regarding why a significant time gap in onset exists.

Stress is also an environmental variable that has received increased attention in recent years. Both histone methylation and acetylation modifications have been observed in rodents because of stress after social defeat [168]—acute and chronic stress has been shown to activate and repress genes through histone modifications [169]. Interestingly, transposable elements are repressed during acute stress, as are hippocampal coding and noncoding RNA as a result of stress-induced histone modifications. These expression changes have been suggested to impair genomic stability and give rise to cognitive impairments [109–114].

ROS, a species of free radical, can be induced through environmental signals, causing oxidative stress and, ultimately, cause a range of epigenetic modifications altering gene expression [162, 170–172]. Heavy metals are believed to cause cellular stress and toxicity through ROS, driving protein denaturation and aggregation and preventing proteasomes from eliminating dysfunctional proteins [120]. One study used carboxy-DCFDA (5-(and-6)-carboxy-2′,7′-dichlorofluorescin diacetate) to quantify stress-induced ROS production from metal sulfates in human neurons [121] and found aluminum sulfate induced the most ROS.

Repetitive electromagnetic field exposure is also believed to trigger epigenetic changes, including DNA methylation and histone modifications, as was suggested by a study of a large cohort of workers [122]. Resistance welders had a higher incidence of Alzheimer's Disease and ALS, potentially because they are exposed regularly to low frequency magnetic fields [122]. Other studies have suggested that exposure to other heavy metals such as mercury, lead, and selenium, plus pesticides and herbicides containing organophosphate may increase risk for ALS, though no clear association has been found [115–119, 123–130].

As science continues to demonstrate the environmental effects of some high-contact sports on mental health, additional studies to explore the increased ALS incidence in athletes that play American football and soccer [131, 173], and in war veterans [132, 133] are needed. These data further suggest that some ALS cases arise from environmental exposures, potentially from epigenetic consequences of violent jarring in the brain [134]. While various methods, including illicit drug use and ischemia from head injuries, have been proposed to increase ROS production and drive dementia [135], the exact molecular mechanism leading to ALS and FTD needs to be clearly mapped through future rigorous studies.

1.3.4   Therapeutic Potential

The ultimate goal in ALS and FTD research is to develop therapeutic interventions for these diseases, allowing those who are affected to live a long and healthy life. Here, we discuss potential effects of epigenetic therapeutics.

Targeted epigenetic modifiers capable of regulating expression of both the normal and mutant alleles are an exciting possibility. Although significant work must be done in this area, this concept was successfully demonstrated in 1994, suggesting epigenetic therapy is feasible [174]. Since that discovery, other research has been performed in neurodegeneration and cancer in an effort to translate this for clinical use [175, 176]. A recent ALS-specific study utilized bromodomain small molecule inhibitors to increase mRNA and pre-mRNA expression for the normal *C9orf72* allele without destroying the epigenetic markers that repress expression of the expanded allele [177]. Increasing expression of the normal allele without increasing expression for the disease-causing allele is a significant achievement and demonstrates the reality of epigenetic therapy.

DNA methylation effects have been extensively evaluated across many diseases. One study found DNA methylation changes may be good ALS biomarkers for disease and potentially future therapeutic targets [79]. DNMTS have been shown to promote apoptosis and increase 5mC levels in motor neurons, and administering Dnmt inhibitors in a motor neuron-degenerative mouse model mitigated both apoptosis and 5mC levels in the motor neurons [36].

The therapeutic potential of oligonucleotides targeting miRNAs has also been evaluated by researchers using mouse models of ALS. Two groups showed that oligonucleotides able to inhibit either miR-155 or miR-29a extended the lifespan of *SOD1* p.G93A transgenic mice [178, 179]. Similarly, Morel et al. found that injecting oligonucleotides targeting miR-124a in the same transgenic mouse model prevented the pathological loss of EAAT2/GLT1 (encoded by human *SLC1A2*), an astroglial glutamate transporter known to be implicated in ALS [180].

Researchers have shown that histone marks at the *C9orf72* locus are associated with reduced gene expression in ALS and FTD patients carrying a repeat expansion when compared to controls [98]. They were then able to increase *C9orf72* mRNA expression by treating patient-derived fibroblasts with 5-aza-2-deoxycytidine (a demethylating agent). While this study demonstrates a proof of concept, increasing the mutated allele might not be a good therapeutic approach for c9ALS/FTD, as results from others suggest that *C9orf72*-associated hypermethylation may actually be neuroprotective in patients [92, 93]. Nonetheless, similar epigenetic strategies have been developed for cancer therapy, where HDAC inhibitors have reversed the effects of cancer-induced epigenetic changes [181]. These techniques have also been applied in ALS both in vitro and in an animal model, and later proceeded to clinical trials. Specifically, sodium phenylbutyrate (NaPB) prolonged *SOD1* p.G93A mouse survival [182], and was subsequently tested in a phase 2 clinical trial. The participants of this clinical trial tolerated this treatment well, and increased histone acetylation in participant blood samples [183]. These results present an exciting and realistic opportunity to treat ALS and FTD using targeted epigenetic therapeutics.

Given the epigenome's dynamic and targetable nature combined with its apparent involvement in disease, it is a primary target for additional therapeutic efforts. As a field, researchers studying epigenetics in neurodegenerative diseases have made significant progress characterizing their involvement, but it is unclear whether the observed epigenetic modifications are driving disease or whether they are just a consequence. For example, researchers recently observed clear transcriptomic and epigenetic differences between c9ALS and sALS brains [99, 184], but it has not been shown whether reversing them would rescue neuronal health. If researchers can establish that the epigenetic dysregulation is driving these changes and that reversing them rescues neuronal behavior, epigenetic therapeutics would revolutionize ALS and FTD treatment.

Acknowledgments We would like to thank Dr. Tamas Ordog, Director of the Epigenomics Translational Program at Mayo Clinic Center for Individualized Medicine for reviewing and providing critical input for this manuscript.

References

1. Cronin S, Hardiman O, Traynor BJ. Ethnic variation in the incidence of ALS: a systematic review. Neurology. 2007;68:1002–7. https://doi.org/10.1212/01.wnl.0000258551.96893.6f.
2. Marin B, Boumediene F, Logroscino G, Couratier P, Babron MC, Leutenegger AL, et al. Variation in worldwide incidence of amyotrophic lateral sclerosis: a meta-analysis. Int J Epidemiol. 2017;46:57–74. https://doi.org/10.1093/ije/dyw061.
3. Bradley WG. Neurology in clinical practice. 3rd ed. Boston, MA: Butterworth-Heinemann; 2000.
4. Chio A, Logroscino G, Hardiman O, Swingler R, Mitchell D, Beghi E, et al. Prognostic factors in ALS: a critical review. Amyotroph Lateral Scler. 2009;10:310–23. https://doi.org/10.3109/17482960802566824.
5. Burrell JR, Kiernan MC, Vucic S, Hodges JR. Motor neuron dysfunction in frontotemporal dementia. Brain. 2011;134:2582–94. https://doi.org/10.1093/brain/awr195. awr195 [pii].
6. Giordana MT, Ferrero P, Grifoni S, Pellerino A, Naldi A, Montuschi A. Dementia and cognitive impairment in amyotrophic lateral sclerosis: a review. Neurol Sci. 2011;32:9–16. https://doi.org/10.1007/s10072-010-0439-6.
7. Gordon PH, Delgadillo D, Piquard A, Bruneteau G, Pradat PF, Salachas F, et al. The range and clinical impact of cognitive impairment in French patients with ALS: a cross-sectional study of neuropsychological test performance. Amyotroph Lateral Scler. 2011;12:372–8. https://doi.org/10.3109/17482968.2011.580847.
8. Lomen-Hoerth C, Anderson T, Miller B. The overlap of amyotrophic lateral sclerosis and frontotemporal dementia. Neurology. 2002;59:1077–9.
9. Ringholz GM, Appel SH, Bradshaw M, Cooke NA, Mosnik DM, Schulz PE. Prevalence and patterns of cognitive impairment in sporadic ALS. Neurology. 2005;65:586–90.
10. Knopman DS, Roberts RO. Estimating the number of persons with frontotemporal lobar degeneration in the US population. J Mol Neurosci. 2011;45:330–5. https://doi.org/10.1007/s12031-011-9538-y.
11. Hodges JR, Davies R, Xuereb J, Kril J, Halliday G. Survival in frontotemporal dementia. Neurology. 2003;61:349–54.

12. Pal S, Tyler JK. Epigenetics and aging. Sci Adv. 2016;2:e1600584. https://doi.org/10.1126/sciadv.1600584.
13. Luger K, Mader AW, Richmond RK, Sargent DF, Richmond TJ. Crystal structure of the nucleosome core particle at 2.8 A resolution. Nature. 1997;389:251–60. https://doi.org/10.1038/38444.
14. Illingworth RS, Gruenewald-Schneider U, Webb S, Kerr AR, James KD, Turner DJ, et al. Orphan CpG islands identify numerous conserved promoters in the mammalian genome. PLoS Genet. 2010;6:e1001134. https://doi.org/10.1371/journal.pgen.1001134.
15. Maunakea AK, Nagarajan RP, Bilenky M, Ballinger TJ, D'Souza C, Fouse SD, et al. Conserved role of intragenic DNA methylation in regulating alternative promoters. Nature. 2010;466:253–7. https://doi.org/10.1038/nature09165.
16. Saxonov S, Berg P, Brutlag DL. A genome-wide analysis of CpG dinucleotides in the human genome distinguishes two distinct classes of promoters. Proc Natl Acad Sci U S A. 2006;103:1412–7. https://doi.org/10.1073/pnas.0510310103.
17. He Y, Ecker JR. Non-CG methylation in the human genome. Annu Rev Genomics Hum Genet. 2015;16:55–77. https://doi.org/10.1146/annurev-genom-090413-025437.
18. Stroud H, Su SC, Hrvatin S, Greben AW, Renthal W, Boxer LD, et al. Early-life gene expression in neurons modulates lasting epigenetic states. Cell. 2017;171:1151–1164.e16. https://doi.org/10.1016/j.cell.2017.09.047.
19. Kinney SR, Pradhan S. Regulation of expression and activity of DNA (cytosine-5) methyltransferases in mammalian cells. Prog Mol Biol Transl Sci. 2011;101:311–33. https://doi.org/10.1016/B978-0-12-387685-0.00009-3.
20. Gabel HW, Kinde B, Stroud H, Gilbert CS, Harmin DA, Kastan NR, et al. Disruption of DNA-methylation-dependent long gene repression in Rett syndrome. Nature. 2015;522:89–93. https://doi.org/10.1038/nature14319.
21. Guo JU, Su Y, Shin JH, Shin J, Li H, Xie B, et al. Distribution, recognition and regulation of non-CpG methylation in the adult mammalian brain. Nat Neurosci. 2014;17:215–22. https://doi.org/10.1038/nn.3607.
22. Lister R, Mukamel EA, Nery JR, Urich M, Puddifoot CA, Johnson ND, et al. Global epigenomic reconfiguration during mammalian brain development. Science. 2013;341:1237905. https://doi.org/10.1126/science.1237905.
23. Darst RP, Pardo CE, Ai L, Brown KD, Kladde MP (2010) Bisulfite sequencing of DNA. Curr Protoc Mol Biol Chapter 7:Unit 7 9 1-17. doi:https://doi.org/10.1002/0471142727.mb0709s91
24. Hahn MA, Szabo PE, Pfeifer GP. 5-Hydroxymethylcytosine: a stable or transient DNA modification? Genomics. 2014;104:314–23. https://doi.org/10.1016/j.ygeno.2014.08.015.
25. Wu X, Zhang Y. TET-mediated active DNA demethylation: mechanism, function and beyond. Nat Rev Genet. 2017;18:517–34. https://doi.org/10.1038/nrg.2017.33.
26. Lu X, Han D, Zhao BS, Song CX, Zhang LS, Dore LC, et al. Base-resolution maps of 5-formylcytosine and 5-carboxylcytosine reveal genome-wide DNA demethylation dynamics. Cell Res. 2015;25:386–9. https://doi.org/10.1038/cr.2015.5.
27. Shmookler Reis RJ, Goldstein S. Mitochondrial DNA in mortal and immortal human cells. Genome number, integrity, and methylation. J Biol Chem. 1983;258:9078–85.
28. Infantino V, Castegna A, Iacobazzi F, Spera I, Scala I, Andria G, et al. Impairment of methyl cycle affects mitochondrial methyl availability and glutathione level in Down's syndrome. Mol Genet Metab. 2011;102:378–82. https://doi.org/10.1016/j.ymgme.2010.11.166.
29. Bellizzi D, D'Aquila P, Scafone T, Giordano M, Riso V, Riccio A, et al. The control region of mitochondrial DNA shows an unusual CpG and non-CpG methylation pattern. DNA Res. 2013;20:537–47. https://doi.org/10.1093/dnares/dst029.
30. Shock LS, Thakkar PV, Peterson EJ, Moran RG, Taylor SM. DNA methyltransferase 1, cytosine methylation, and cytosine hydroxymethylation in mammalian mitochondria. Proc Natl Acad Sci U S A. 2011;108:3630–5. https://doi.org/10.1073/pnas.1012311108.

31. Byun HM, Barrow TM. Analysis of pollutant-induced changes in mitochondrial DNA methylation. Methods Mol Biol. 2015;1265:271–83. https://doi.org/10.1007/978-1-4939-2288-8_19.

32. Byun HM, Panni T, Motta V, Hou L, Nordio F, Apostoli P, et al. Effects of airborne pollutants on mitochondrial DNA methylation. Part Fibre Toxicol. 2013;10:18. https://doi.org/10.1186/1743-8977-10-18.

33. Hong EE, Okitsu CY, Smith AD, Hsieh CL. Regionally specific and genome-wide analyses conclusively demonstrate the absence of CpG methylation in human mitochondrial DNA. Mol Cell Biol. 2013;33:2683–90. https://doi.org/10.1128/MCB.00220-13.

34. Liu B, Du Q, Chen L, Fu G, Li S, Fu L, et al. CpG methylation patterns of human mitochondrial DNA. Sci Rep. 2016;6:23421. https://doi.org/10.1038/srep23421.

35. Mechta M, Ingerslev LR, Fabre O, Picard M, Barres R. Evidence suggesting absence of mitochondrial DNA methylation. Front Genet. 2017;8:166. https://doi.org/10.3389/fgene.2017.00166.

36. Chestnut BA, Chang Q, Price A, Lesuisse C, Wong M, Martin LJ. Epigenetic regulation of motor neuron cell death through DNA methylation. J Neurosci. 2011;31:16619–36. https://doi.org/10.1523/JNEUROSCI.1639-11.2011.

37. Doynova MD, Berretta A, Jones MB, Jasoni CL, Vickers MH, O'Sullivan JM. Interactions between mitochondrial and nuclear DNA in mammalian cells are non-random. Mitochondrion. 2016;30:187–96. https://doi.org/10.1016/j.mito.2016.08.003.

38. Rodley CD, Grand RS, Gehlen LR, Greyling G, Jones MB, O'Sullivan JM. Mitochondrial-nuclear DNA interactions contribute to the regulation of nuclear transcript levels as part of the inter-organelle communication system. PLoS One. 2012;7:e30943. https://doi.org/10.1371/journal.pone.0030943.

39. Glover-Cutter K, Kim S, Espinosa J, Bentley DL. RNA polymerase II pauses and associates with pre-mRNA processing factors at both ends of genes. Nat Struct Mol Biol. 2008;15:71–8. https://doi.org/10.1038/nsmb1352.

40. Liu J, Jia G. Methylation modifications in eukaryotic messenger RNA. J Genet Genomics. 2014;41:21–33. https://doi.org/10.1016/j.jgg.2013.10.002.

41. Liu H, Kiledjian M. Decapping the message: a beginning or an end. Biochem Soc Trans. 2006;34:35–8. https://doi.org/10.1042/BST20060035.

42. Kiani J, Grandjean V, Liebers R, Tuorto F, Ghanbarian H, Lyko F, et al. RNA-mediated epigenetic heredity requires the cytosine methyltransferase Dnmt2. PLoS Genet. 2013;9:e1003498. https://doi.org/10.1371/journal.pgen.1003498.

43. Amort T, Souliere MF, Wille A, Jia XY, Fiegl H, Worle H, et al. Long noncoding RNAs as targets for cytosine methylation. RNA Biol. 2013;10:1003–8. https://doi.org/10.4161/rna.24454.

44. Dominissini D, Moshitch-Moshkovitz S, Schwartz S, Salmon-Divon M, Ungar L, Osenberg S, et al. Topology of the human and mouse m6A RNA methylomes revealed by m6A-seq. Nature. 2012;485:201–6. https://doi.org/10.1038/nature11112.

45. Roundtree IA, Luo GZ, Zhang Z, Wang X, Zhou T, Cui Y, et al. YTHDC1 mediates nuclear export of N(6)-methyladenosine methylated mRNAs. Elife. 2017;6:pii: e31311. https://doi.org/10.7554/eLife.31311.

46. Jia G, Fu Y, He C. Reversible RNA adenosine methylation in biological regulation. Trends Genet. 2013;29:108–15. https://doi.org/10.1016/j.tig.2012.11.003.

47. Jia G, Fu Y, Zhao X, Dai Q, Zheng G, Yang Y, et al. N6-methyladenosine in nuclear RNA is a major substrate of the obesity-associated FTO. Nat Chem Biol. 2011;7:885–7. https://doi.org/10.1038/nchembio.687.

48. Zheng G, Dahl JA, Niu Y, Fedorcsak P, Huang CM, Li CJ, et al. ALKBH5 is a mammalian RNA demethylase that impacts RNA metabolism and mouse fertility. Mol Cell. 2013;49:18–29. https://doi.org/10.1016/j.molcel.2012.10.015.

49. Consortium EP. An integrated encyclopedia of DNA elements in the human genome. Nature. 2012;489:57–74. https://doi.org/10.1038/nature11247.

50. Knowling S, Morris KV. Non-coding RNA and antisense RNA. Nature's trash or treasure? Biochimie. 2011;93:1922–7. https://doi.org/10.1016/j.biochi.2011.07.031.
51. Baek D, Villen J, Shin C, Camargo FD, Gygi SP, Bartel DP. The impact of microRNAs on protein output. Nature. 2008;455:64–71. https://doi.org/10.1038/nature07242.
52. Guo H, Ingolia NT, Weissman JS, Bartel DP. Mammalian microRNAs predominantly act to decrease target mRNA levels. Nature. 2010;466:835–40. https://doi.org/10.1038/nature09267.
53. Gosline SJ, Gurtan AM, JnBaptiste CK, Bosson A, Milani P, Dalin S, et al. Elucidating microRNA regulatory networks using transcriptional, post-transcriptional, and histone modification measurements. Cell Rep. 2016;14:310–9. https://doi.org/10.1016/j.celrep.2015.12.031.
54. Lewis BP, Shih IH, Jones-Rhoades MW, Bartel DP, Burge CB. Prediction of mammalian microRNA targets. Cell. 2003;115:787–98.
55. Saito Y, Saito H, Liang G, Friedman JM. Epigenetic alterations and microRNA misexpression in cancer and autoimmune diseases: a critical review. Clin Rev Allergy Immunol. 2014;47:128–35. https://doi.org/10.1007/s12016-013-8401-z.
56. Saito Y, Liang G, Egger G, Friedman JM, Chuang JC, Coetzee GA, et al. Specific activation of microRNA-127 with downregulation of the proto-oncogene BCL6 by chromatin-modifying drugs in human cancer cells. Cancer Cell. 2006;9:435–43. https://doi.org/10.1016/j.ccr.2006.04.020.
57. Carmell MA, Girard A, van de Kant HJ, Bourc'his D, Bestor TH, de Rooij DG, et al. MIWI2 is essential for spermatogenesis and repression of transposons in the mouse male germline. Dev Cell. 2007;12:503–14. https://doi.org/10.1016/j.devcel.2007.03.001.
58. Rouget C, Papin C, Boureux A, Meunier AC, Franco B, Robine N, et al. Maternal mRNA deadenylation and decay by the piRNA pathway in the early Drosophila embryo. Nature. 2010;467:1128–32. https://doi.org/10.1038/nature09465.
59. Watanabe T, Takeda A, Tsukiyama T, Mise K, Okuno T, Sasaki H, et al. Identification and characterization of two novel classes of small RNAs in the mouse germline: retrotransposon-derived siRNAs in oocytes and germline small RNAs in testes. Genes Dev. 2006;20:1732–43. https://doi.org/10.1101/gad.1425706.
60. Yang N, Kazazian HH Jr. L1 retrotransposition is suppressed by endogenously encoded small interfering RNAs in human cultured cells. Nat Struct Mol Biol. 2006;13:763–71. https://doi.org/10.1038/nsmb1141.
61. Guttman M, Amit I, Garber M, French C, Lin MF, Feldser D, et al. Chromatin signature reveals over a thousand highly conserved large non-coding RNAs in mammals. Nature. 2009;458:223–7. https://doi.org/10.1038/nature07672.
62. Khalil AM, Guttman M, Huarte M, Garber M, Raj A, Rivea Morales D, et al. Many human large intergenic noncoding RNAs associate with chromatin-modifying complexes and affect gene expression. Proc Natl Acad Sci U S A. 2009;106:11667–72. https://doi.org/10.1073/pnas.0904715106.
63. De Santa F, Barozzi I, Mietton F, Ghisletti S, Polletti S, Tusi BK, et al. A large fraction of extragenic RNA pol II transcription sites overlap enhancers. PLoS Biol. 2010;8:e1000384. https://doi.org/10.1371/journal.pbio.1000384.
64. Kim TK, Hemberg M, Gray JM, Costa AM, Bear DM, Wu J, et al. Widespread transcription at neuronal activity-regulated enhancers. Nature. 2010;465:182–7. https://doi.org/10.1038/nature09033.
65. Preker P, Nielsen J, Kammler S, Lykke-Andersen S, Christensen MS, Mapendano CK, et al. RNA exosome depletion reveals transcription upstream of active human promoters. Science. 2008;322:1851–4. https://doi.org/10.1126/science.1164096.
66. Affymetrix ETP, Cold Spring Harbor Laboratory ETP. Post-transcriptional processing generates a diversity of 5′-modified long and short RNAs. Nature. 2009;457:1028–32. https://doi.org/10.1038/nature07759.

67. Han J, Kim D, Morris KV. Promoter-associated RNA is required for RNA-directed transcriptional gene silencing in human cells. Proc Natl Acad Sci U S A. 2007;104:12422–7. https://doi.org/10.1073/pnas.0701635104.
68. Morris KV, Santoso S, Turner AM, Pastori C, Hawkins PG. Bidirectional transcription directs both transcriptional gene activation and suppression in human cells. PLoS Genet. 2008;4:e1000258. https://doi.org/10.1371/journal.pgen.1000258.
69. Wang X, Arai S, Song X, Reichart D, Du K, Pascual G, et al. Induced ncRNAs allosterically modify RNA-binding proteins in cis to inhibit transcription. Nature. 2008;454:126–30. https://doi.org/10.1038/nature06992.
70. Jenuwein T, Allis CD. Translating the histone code. Science. 2001;293:1074–80. https://doi.org/10.1126/science.1063127.
71. Ernst J, Kellis M. Chromatin-state discovery and genome annotation with ChromHMM. Nat Protoc. 2017;12:2478–92. https://doi.org/10.1038/nprot.2017.124.
72. Hake SB, Allis CD. Histone H3 variants and their potential role in indexing mammalian genomes: the "H3 barcode hypothesis". Proc Natl Acad Sci U S A. 2006;103:6428–35. https://doi.org/10.1073/pnas.0600803103.
73. DeJesus-Hernandez M, Mackenzie IR, Boeve BF, Boxer AL, Baker M, Rutherford NJ, et al. Expanded GGGGCC hexanucleotide repeat in noncoding region of C9ORF72 causes chromosome 9p-linked FTD and ALS. Neuron. 2011;72:245–56. https://doi.org/10.1016/j.neuron.2011.09.011. S0896-6273(11)00828-2 [pii].
74. Rademakers R, van Blitterswijk M. Motor neuron disease in 2012: novel causal genes and disease modifiers. Nat Rev Neurol. 2013;9:63–4. https://doi.org/10.1038/nrneurol.2012.276. nrneurol.2012.276 [pii].
75. Renton AE, Majounie E, Waite A, Simon-Sanchez J, Rollinson S, Gibbs JR, et al. A hexanucleotide repeat expansion in C9ORF72 is the cause of chromosome 9p21-linked ALS-FTD. Neuron. 2011;72:257–68. https://doi.org/10.1016/j.neuron.2011.09.010. S0896-6273(11)00797-5 [pii].
76. Renton AE, Chio A, Traynor BJ. State of play in amyotrophic lateral sclerosis genetics. Nat Neurosci. 2014;17:17–23. https://doi.org/10.1038/nn.3584. nn.3584 [pii].
77. Coppede F, Stoccoro A, Mosca L, Gallo R, Tarlarini C, Lunetta C, et al. Increase in DNA methylation in patients with amyotrophic lateral sclerosis carriers of not fully penetrant SOD1 mutations. Amyotroph Lateral Scler Frontotemporal Degener. 2017;19(1-2):93. https://doi.org/10.1080/21678421.2017.1367401.
78. Morahan JM, Yu B, Trent RJ, Pamphlett R. A genome-wide analysis of brain DNA methylation identifies new candidate genes for sporadic amyotrophic lateral sclerosis. Amyotroph Lateral Scler. 2009;10:418–29. https://doi.org/10.3109/17482960802635397.
79. Figueroa-Romero C, Hur J, Bender DE, Delaney CE, Cataldo MD, Smith AL, et al. Identification of epigenetically altered genes in sporadic amyotrophic lateral sclerosis. PLoS One. 2012;7:e52672. https://doi.org/10.1371/journal.pone.0052672.
80. Tremolizzo L, Messina P, Conti E, Sala G, Cecchi M, Airoldi L, et al. Whole-blood global DNA methylation is increased in amyotrophic lateral sclerosis independently of age of onset. Amyotroph Lateral Scler Frontotemporal Degener. 2014;15:98–105. https://doi.org/10.3109/21678421.2013.851247.
81. Banzhaf-Strathmann J, Claus R, Mucke O, Rentzsch K, van der Zee J, Engelborghs S, et al. Promoter DNA methylation regulates progranulin expression and is altered in FTLD. Acta Neuropathol Commun. 2013;1:16. https://doi.org/10.1186/2051-5960-1-16.
82. Galimberti D, D'Addario C, Dell'osso B, Fenoglio C, Marcone A, Cerami C, et al. Progranulin gene (GRN) promoter methylation is increased in patients with sporadic frontotemporal lobar degeneration. Neurol Sci. 2013;34:899–903. https://doi.org/10.1007/s10072-012-1151-5.
83. Li Y, Chen JA, Sears RL, Gao F, Klein ED, Karydas A, et al. An epigenetic signature in peripheral blood associated with the haplotype on 17q21.31, a risk factor for neurodegenerative tauopathy. PLoS Genet. 2014;10:e1004211. https://doi.org/10.1371/journal.pgen.1004211.

84. Wong M, Gertz B, Chestnut BA, Martin LJ. Mitochondrial DNMT3A and DNA methylation in skeletal muscle and CNS of transgenic mouse models of ALS. Front Cell Neurosci. 2013;7:279. https://doi.org/10.3389/fncel.2013.00279.

85. Dobrowolny G, Bernardini C, Martini M, Baranzini M, Barba M, Musaro A. Muscle Expression of SOD1(G93A) Modulates microRNA and mRNA Transcription Pattern Associated with the Myelination Process in the Spinal Cord of Transgenic Mice. Front Cell Neurosci. 2015;9:463. https://doi.org/10.3389/fncel.2015.00463.

86. Figueroa-Romero C, Hur J, Lunn JS, Paez-Colasante X, Bender DE, Yung R, et al. Expression of microRNAs in human post-mortem amyotrophic lateral sclerosis spinal cords provides insight into disease mechanisms. Mol Cell Neurosci. 2016;71:34–45. https://doi.org/10.1016/j.mcn.2015.12.008.

87. Marcuzzo S, Bonanno S, Kapetis D, Barzago C, Cavalcante P, D'Alessandro S, et al. Up-regulation of neural and cell cycle-related microRNAs in brain of amyotrophic lateral sclerosis mice at late disease stage. Mol Brain. 2015;8:5. https://doi.org/10.1186/s13041-015-0095-0.

88. Toivonen JM, Manzano R, Olivan S, Zaragoza P, Garcia-Redondo A, Osta R. MicroRNA-206: a potential circulating biomarker candidate for amyotrophic lateral sclerosis. PLoS One. 2014;9:e89065. https://doi.org/10.1371/journal.pone.0089065.

89. Zhang Z, Almeida S, Lu Y, Nishimura AL, Peng L, Sun D, et al. Downregulation of microRNA-9 in iPSC-derived neurons of FTD/ALS patients with TDP-43 mutations. PLoS One. 2013;8:e76055. https://doi.org/10.1371/journal.pone.0076055.

90. Jiao J, Herl LD, Farese RV, Gao FB. MicroRNA-29b regulates the expression level of human progranulin, a secreted glycoprotein implicated in frontotemporal dementia. PLoS One. 2010;5:e10551. https://doi.org/10.1371/journal.pone.0010551.

91. Gascon E, Lynch K, Ruan H, Almeida S, Verheyden JM, Seeley WW, et al. Alterations in microRNA-124 and AMPA receptors contribute to social behavioral deficits in frontotemporal dementia. Nat Med. 2014;20:1444–51. https://doi.org/10.1038/nm.3717.

92. Bauer PO. Methylation of C9orf72 expansion reduces RNA foci formation and dipeptide-repeat proteins expression in cells. Neurosci Lett. 2016;612:204–9. https://doi.org/10.1016/j.neulet.2015.12.018.

93. Liu EY, Russ J, Wu K, Neal D, Suh E, McNally AG, et al. C9orf72 hypermethylation protects against repeat expansion-associated pathology in ALS/FTD. Acta Neuropathol. 2014;128:525–41. https://doi.org/10.1007/s00401-014-1286-y.

94. Xi Z, Rainero I, Rubino E, Pinessi L, Bruni AC, Maletta RG, et al. Hypermethylation of the CpG-island near the C9orf72 G(4)C(2)-repeat expansion in FTLD patients. Hum Mol Genet. 2014;23:5630–7. https://doi.org/10.1093/hmg/ddu279.

95. Xi Z, Zhang M, Bruni AC, Maletta RG, Colao R, Fratta P, et al. The C9orf72 repeat expansion itself is methylated in ALS and FTLD patients. Acta Neuropathol. 2015;129:715–27. https://doi.org/10.1007/s00401-015-1401-8.

96. Xi Z, Zinman L, Moreno D, Schymick J, Liang Y, Sato C, et al. Hypermethylation of the CpG island near the G4C2 repeat in ALS with a C9orf72 expansion. Am J Hum Genet. 2013;92:981–9. https://doi.org/10.1016/j.ajhg.2013.04.017.

97. Esanov R, Belle KC, van Blitterswijk M, Belzil VV, Rademakers R, Dickson DW, et al. C9orf72 promoter hypermethylation is reduced while hydroxymethylation is acquired during reprogramming of ALS patient cells. Exp Neurol. 2015;277:171–7. https://doi.org/10.1016/j.expneurol.2015.12.022.

98. Belzil VV, Bauer PO, Prudencio M, Gendron TF, Stetler CT, Yan IK, et al. Reduced C9orf72 gene expression in c9FTD/ALS is caused by histone trimethylation, an epigenetic event detectable in blood. Acta Neuropathol. 2013;126:895–905. https://doi.org/10.1007/s00401-013-1199-1.

99. Ebbert MTW, Ross CA, Pregent LJ, Lank RJ, Zhang C, Katzman RB, et al. Conserved DNA methylation combined with differential frontal cortex and cerebellar expression

distinguishes C9orf72-associated and sporadic ALS, and implicates SERPINA1 in disease. Acta Neuropathol. 2017;134:715–28. https://doi.org/10.1007/s00401-017-1760-4.

100. Banack SA, Cox PA. Biomagnification of cycad neurotoxins in flying foxes: implications for ALS-PDC in Guam. Neurology. 2003;61:387–9.

101. Bradley WG, Mash DC. Beyond Guam: the cyanobacteria/BMAA hypothesis of the cause of ALS and other neurodegenerative diseases. Amyotroph Lateral Scler. 2009;10(Suppl 2):7–20. https://doi.org/10.3109/17482960903286009.

102. Chiu AS, Gehringer MM, Welch JH, Neilan BA. Does alpha-amino-beta-methylaminopropionic acid (BMAA) play a role in neurodegeneration? Int J Environ Res Public Health. 2011;8:3728–46. https://doi.org/10.3390/ijerph8093728.

103. Dastur DK. Cycad toxicity in monkeys: clinical, pathological, and biochemical aspects. Fed Proc. 1964;23:1368–9.

104. Polsky FI, Nunn PB, Bell EA. Distribution and toxicity of alpha-amino-beta-methylaminopropionic acid. Fed Proc. 1972;31:1473–5.

105. Remely M, Stefanska B, Lovrecic L, Magnet U, Haslberger AG. Nutriepigenomics: the role of nutrition in epigenetic control of human diseases. Curr Opin Clin Nutr Metab Care. 2015;18:328–33. https://doi.org/10.1097/MCO.0000000000000180.

106. Meltz Steinberg K, Nicholas TJ, Koboldt DC, Yu B, Mardis E, Pamphlett R. Whole genome analyses reveal no pathogenetic single nucleotide or structural differences between monozygotic twins discordant for amyotrophic lateral sclerosis. Amyotroph Lateral Scler Frontotemporal Degener. 2015;16:385–92. https://doi.org/10.3109/21678421.2015.1040029.

107. Xi Z, Yunusova Y, van Blitterswijk M, Dib S, Ghani M, Moreno D, et al. Identical twins with the C9orf72 repeat expansion are discordant for ALS. Neurology. 2014;83:1476–8. https://doi.org/10.1212/WNL.0000000000000886.

108. Young PE, Kum Jew S, Buckland ME, Pamphlett R, Suter CM. Epigenetic differences between monozygotic twins discordant for amyotrophic lateral sclerosis (ALS) provide clues to disease pathogenesis. PLoS One. 2017;12:e0182638. https://doi.org/10.1371/journal.pone.0182638.

109. Erwin JA, Marchetto MC, Gage FH. Mobile DNA elements in the generation of diversity and complexity in the brain. Nat Rev Neurosci. 2014;15:497–506. https://doi.org/10.1038/nrn3730.

110. Hunter RG, Gagnidze K, McEwen BS, Pfaff DW. Stress and the dynamic genome: steroids, epigenetics, and the transposome. Proc Natl Acad Sci U S A. 2015;112:6828–33. https://doi.org/10.1073/pnas.1411260111.

111. Hunter RG, McEwen BS, Pfaff DW. Environmental stress and transposon transcription in the mammalian brain. Mob Genet Elements. 2013;3:e24555. https://doi.org/10.4161/mge.24555.

112. Johnson R, Guigo R. The RIDL hypothesis: transposable elements as functional domains of long noncoding RNAs. RNA. 2014;20:959–76. https://doi.org/10.1261/rna.044560.114.

113. McEwen BS, Bowles NP, Gray JD, Hill MN, Hunter RG, Karatsoreos IN, et al. Mechanisms of stress in the brain. Nat Neurosci. 2015;18:1353–63. https://doi.org/10.1038/nn.4086.

114. Reilly MT, Faulkner GJ, Dubnau J, Ponomarev I, Gage FH. The role of transposable elements in health and diseases of the central nervous system. J Neurosci. 2013;33:17577–86.https://doi.org/10.1523/JNEUROSCI.3369-13.2013.

115. Bakir F, Damluji SF, Amin-Zaki L, Murtadha M, Khalidi A, al-Rawi NY, et al. Methylmercury poisoning in Iraq. Science. 1973;181:230–41.

116. Cicero CE, Mostile G, Vasta R, Rapisarda V, Signorelli SS, Ferrante M, et al. Metals and neurodegenerative diseases. A systematic review. Environ Res. 2017;159:82–94. https://doi.org/10.1016/j.envres.2017.07.048.

117. Combs GF Jr. Selenium in global food systems. Br J Nutr. 2001;85:517–47.

118. Fang F, Peters TL, Beard JD, Umbach DM, Keller J, Mariosa D, et al. Blood Lead, Bone Turnover, and Survival in Amyotrophic Lateral Sclerosis. Am J Epidemiol. 2017;186:1057–64. https://doi.org/10.1093/aje/kwx176.

119. Johnson FO, Atchison WD. The role of environmental mercury, lead and pesticide exposure in development of amyotrophic lateral sclerosis. Neurotoxicology. 2009;30:761–5. https://doi.org/10.1016/j.neuro.2009.07.010.

120. Migliore L, Coppede F. Environmental-induced oxidative stress in neurodegenerative disorders and aging. Mutat Res. 2009;674:73–84. https://doi.org/10.1016/j.mrgentox.2008.09.013.

121. Pogue AI, Jones BM, Bhattacharjee S, Percy ME, Zhao Y, Lukiw WJ. Metal-sulfate induced generation of ROS in human brain cells: detection using an isomeric mixture of 5- and 6-carboxy-2′,7′-dichlorofluorescein diacetate (carboxy-DCFDA) as a cell permeant tracer. Int J Mol Sci. 2012;13:9615–26. https://doi.org/10.3390/ijms13089615.

122. Hakansson N, Gustavsson P, Johansen C, Floderus B. Neurodegenerative diseases in welders and other workers exposed to high levels of magnetic fields. Epidemiology. 2003;14:420–6; . discussion 427–428. https://doi.org/10.1097/01.EDE.0000078446.76859.c9.

123. Cronin S, Greenway MJ, Prehn JH, Hardiman O. Paraoxonase promoter and intronic variants modify risk of sporadic amyotrophic lateral sclerosis. J Neurol Neurosurg Psychiatry. 2007;78:984–6. https://doi.org/10.1136/jnnp.2006.112581.

124. Diekstra FP, Beleza-Meireles A, Leigh NP, Shaw CE, Al-Chalabi A. Interaction between PON1 and population density in amyotrophic lateral sclerosis. Neuroreport. 2009;20:186–90. https://doi.org/10.1097/WNR.0b013e32831af220.

125. Matin MA, Hussain K. Striatal neurochemical changes and motor dysfunction in mipafox-treated animals. Methods Find Exp Clin Pharmacol. 1985;7:79–81.

126. Merwin SJ, Obis T, Nunez Y, Re DB. Organophosphate neurotoxicity to the voluntary motor system on the trail of environment-caused amyotrophic lateral sclerosis: the known, the misknown, and the unknown. Arch Toxicol. 2017;91:2939–52. https://doi.org/10.1007/s00204-016-1926-1.

127. Morahan JM, Yu B, Trent RJ, Pamphlett R. A gene-environment study of the paraoxonase 1 gene and pesticides in amyotrophic lateral sclerosis. Neurotoxicology. 2007;28:532–40. https://doi.org/10.1016/j.neuro.2006.11.007.

128. Saeed M, Siddique N, Hung WY, Usacheva E, Liu E, Sufit RL, et al. Paraoxonase cluster polymorphisms are associated with sporadic ALS. Neurology. 2006;67:771–6. https://doi.org/10.1212/01.wnl.0000227187.52002.88.

129. Sanchez-Santed F, Colomina MT, Herrero Hernandez E. Organophosphate pesticide exposure and neurodegeneration. Cortex. 2016;74:417–26. https://doi.org/10.1016/j.cortex.2015.10.003.

130. Valdmanis PN, Kabashi E, Dyck A, Hince P, Lee J, Dion P, et al. Association of paraoxonase gene cluster polymorphisms with ALS in France, Quebec, and Sweden. Neurology. 2008;71:514–20. https://doi.org/10.1212/01.wnl.0000324997.21272.0c. 71/7/514 [pii].

131. Chio A, Benzi G, Dossena M, Mutani R, Mora G. Severely increased risk of amyotrophic lateral sclerosis among Italian professional football players. Brain. 2005;128:472–6. https://doi.org/10.1093/brain/awh373.

132. Horner RD, Grambow SC, Coffman CJ, Lindquist JH, Oddone EZ, Allen KD, et al. Amyotrophic lateral sclerosis among 1991 Gulf War veterans: evidence for a time-limited outbreak. Neuroepidemiology. 2008;31:28–32. https://doi.org/10.1159/000136648.

133. Miranda ML, Alicia Overstreet Galeano M, Tassone E, Allen KD, Horner RD. Spatial analysis of the etiology of amyotrophic lateral sclerosis among 1991 Gulf War veterans. Neurotoxicology. 2008;29:964–70. https://doi.org/10.1016/j.neuro.2008.05.005.

134. Pupillo E, Poloni M, Bianchi E, Giussani G, Logroscino G, Zoccolella S, et al. Trauma and amyotrophic lateral sclerosis: a european population-based case-control study from the EURALS consortium. Amyotroph Lateral Scler Frontotemporal Degener. 2018;19(1-2):118. https://doi.org/10.1080/21678421.2017.1386687.

135. Szczygielski J, Mautes A, Steudel WI, Falkai P, Bayer TA, Wirths O. Traumatic brain injury: cause or risk of Alzheimer's disease? A review of experimental studies. J Neural Transm (Vienna). 2005;112:1547–64. https://doi.org/10.1007/s00702-005-0326-0.

136. Oates N, Pamphlett R. An epigenetic analysis of SOD1 and VEGF in ALS. Amyotroph Lateral Scler. 2007;8:83–6. https://doi.org/10.1080/17482960601149160.
137. Yang Y, Gozen O, Vidensky S, Robinson MB, Rothstein JD. Epigenetic regulation of neuron-dependent induction of astroglial synaptic protein GLT1. Glia. 2010;58:277–86. https://doi.org/10.1002/glia.20922.
138. Baker M, Mackenzie IR, Pickering-Brown SM, Gass J, Rademakers R, Lindholm C, et al. Mutations in progranulin cause tau-negative frontotemporal dementia linked to chromosome 17. Nature. 2006;442:916–9. https://doi.org/10.1038/nature05016.
139. Gass J, Cannon A, Mackenzie IR, Boeve B, Baker M, Adamson J, et al. Mutations in progranulin are a major cause of ubiquitin-positive frontotemporal lobar degeneration. Hum Mol Genet. 2006;15:2988–3001. https://doi.org/10.1093/hmg/ddl241.
140. Rademakers R, Neumann M, Mackenzie IR. Advances in understanding the molecular basis of frontotemporal dementia. Nat Rev Neurol. 2012;8:423–34. https://doi.org/10.1038/nrneurol.2012.117.
141. Finch N, Baker M, Crook R, Swanson K, Kuntz K, Surtees R, et al. Plasma progranulin levels predict progranulin mutation status in frontotemporal dementia patients and asymptomatic family members. Brain. 2009;132:583–91. https://doi.org/10.1093/brain/awn352.
142. Baker M, Litvan I, Houlden H, Adamson J, Dickson D, Perez-Tur J, et al. Association of an extended haplotype in the tau gene with progressive supranuclear palsy. Hum Mol Genet. 1999;8:711–5.
143. Caffrey TM, Wade-Martins R. The role of MAPT sequence variation in mechanisms of disease susceptibility. Biochem Soc Trans. 2012;40:687–92. https://doi.org/10.1042/BST20120063.
144. Hoglinger GU, Melhem NM, Dickson DW, Sleiman PM, Wang LS, Klei L, et al. Identification of common variants influencing risk of the tauopathy progressive supranuclear palsy. Nat Genet. 2011;43:699–705. https://doi.org/10.1038/ng.859.
145. Stefansson H, Helgason A, Thorleifsson G, Steinthorsdottir V, Masson G, Barnard J, et al. A common inversion under selection in Europeans. Nat Genet. 2005;37:129–37.https://doi.org/10.1038/ng1508.
146. Arai T, Hasegawa M, Akiyama H, Ikeda K, Nonaka T, Mori H, et al. TDP-43 is a component of ubiquitin-positive tau-negative inclusions in frontotemporal lobar degeneration and amyotrophic lateral sclerosis. Biochem Biophys Res Commun. 2006;351:602–11.https://doi.org/10.1016/j.bbrc.2006.10.093.
147. Neumann M, Sampathu DM, Kwong LK, Truax AC, Micsenyi MC, Chou TT, et al. Ubiquitinated TDP-43 in frontotemporal lobar degeneration and amyotrophic lateral sclerosis. Science. 2006;314:130–3.
148. Buratti E, De Conti L, Stuani C, Romano M, Baralle M, Baralle F. Nuclear factor TDP-43 can affect selected microRNA levels. FEBS J. 2010;277:2268–81.https://doi.org/10.1111/j.1742-4658.2010.07643.x.
149. Emde A, Eitan C, Liou LL, Libby RT, Rivkin N, Magen I, et al. Dysregulated miRNA biogenesis downstream of cellular stress and ALS-causing mutations: a new mechanism for ALS. EMBO J. 2015;34:2633–51. https://doi.org/10.15252/embj.201490493.
150. Kawahara Y, Mieda-Sato A. TDP-43 promotes microRNA biogenesis as a component of the Drosha and Dicer complexes. Proc Natl Acad Sci U S A. 2012;109:3347–52. https://doi.org/10.1073/pnas.1112427109.
151. King IN, Yartseva V, Salas D, Kumar A, Heidersbach A, Ando DM, et al. The RNA-binding protein TDP-43 selectively disrupts microRNA-1/206 incorporation into the RNA-induced silencing complex. J Biol Chem. 2014;289:14263–71. https://doi.org/10.1074/jbc.M114.561902.
152. Li Z, Lu Y, Xu XL, Gao FB. The FTD/ALS-associated RNA-binding protein TDP-43 regulates the robustness of neuronal specification through microRNA-9a in Drosophila. Hum Mol Genet. 2013;22:218–25. https://doi.org/10.1093/hmg/dds420.
153. Almeida S, Gascon E, Tran H, Chou HJ, Gendron TF, Degroot S, et al. Modeling key pathological features of frontotemporal dementia with C9ORF72 repeat expansion in

iPSC-derived human neurons. Acta Neuropathol. 2013;126:385–99. https://doi.org/10.1007/s00401-013-1149-y.

154. Ciura S, Lattante S, Le Ber I, Latouche M, Tostivint H, Brice A, et al. Loss of function of C9orf72 causes motor deficits in a zebrafish model of Amyotrophic Lateral Sclerosis. Ann Neurol. 2013;74:180. https://doi.org/10.1002/ana.23946.

155. Donnelly CJ, Zhang PW, Pham JT, Haeusler AR, Mistry NA, Vidensky S, et al. RNA toxicity from the ALS/FTD C9ORF72 expansion is mitigated by antisense intervention. Neuron. 2013;80:415–28. https://doi.org/10.1016/j.neuron.2013.10.015.

156. Fratta P, Poulter M, Lashley T, Rohrer JD, Polke JM, Beck J, et al. Homozygosity for the C9orf72 GGGGCC repeat expansion in frontotemporal dementia. Acta Neuropathol. 2013;126:401–9. https://doi.org/10.1007/s00401-013-1147-0.

157. Gendron TF, van Blitterswijk M, Bieniek KF, Daughrity LM, Jiang J, Rush BK, et al. Cerebellar c9RAN proteins associate with clinical and neuropathological characteristics of C9ORF72 repeat expansion carriers. Acta Neuropathol. 2015;130:559–73. https://doi.org/10.1007/s00401-015-1474-4. s00401-015-1474-4 [pii].

158. Gijselinck I, Van Langenhove T, van der Zee J, Sleegers K, Philtjens S, Kleinberger G, et al. A C9orf72 promoter repeat expansion in a Flanders-Belgian cohort with disorders of the frontotemporal lobar degeneration-amyotrophic lateral sclerosis spectrum: a gene identification study. Lancet Neurol. 2012;11:54–65. https://doi.org/10.1016/S1474-4422(11)70261-7. S1474-4422(11)70261-7 [pii].

159. Mori K, Weng SM, Arzberger T, May S, Rentzsch K, Kremmer E, et al. The C9orf72 GGGGCC repeat is translated into aggregating dipeptide-repeat proteins in FTLD/ALS. Science. 2013;339:1335–8. https://doi.org/10.1126/science.1232927. science.1232927 [pii].

160. Russ J, Liu EY, Wu K, Neal D, Suh E, Irwin DJ, et al. Hypermethylation of repeat expanded C9orf72 is a clinical and molecular disease modifier. Acta Neuropathol. 2015;129:39–52. https://doi.org/10.1007/s00401-014-1365-0.

161. Ritossa F. Discovery of the heat shock response. Cell Stress Chaperones. 1996;1:97–8.

162. Meaney MJ, Szyf M. Environmental programming of stress responses through DNA methylation: life at the interface between a dynamic environment and a fixed genome. Dialogues Clin Neurosci. 2005;7:103–23.

163. Caller TA, Doolin JW, Haney JF, Murby AJ, West KG, Farrar HE, et al. A cluster of amyotrophic lateral sclerosis in New Hampshire: a possible role for toxic cyanobacteria blooms. Amyotroph Lateral Scler. 2009;10(Suppl 2):101–8. https://doi.org/10.3109/17482960903278485.

164. Karlsson O, Roman E, Berg AL, Brittebo EB. Early hippocampal cell death, and late learning and memory deficits in rats exposed to the environmental toxin BMAA (beta-N-methylamino-L-alanine) during the neonatal period. Behav Brain Res. 2011;219:310–20. https://doi.org/10.1016/j.bbr.2011.01.056.

165. Purdie EL, Samsudin S, Eddy FB, Codd GA. Effects of the cyanobacterial neurotoxin beta-N-methylamino-L-alanine on the early-life stage development of zebrafish (Danio rerio). Aquat Toxicol. 2009;95:279–84. https://doi.org/10.1016/j.aquatox.2009.02.009.

166. Horvath S. DNA methylation age of human tissues and cell types. Genome Biol. 2013;14:R115. https://doi.org/10.1186/gb-2013-14-10-r115.

167. Zhang M, Tartaglia MC, Moreno D, Sato C, McKeever P, Weichert A, et al. DNA methylation age-acceleration is associated with disease duration and age at onset in C9orf72 patients. Acta Neuropathol. 2017;134:271–9. https://doi.org/10.1007/s00401-017-1713-y.

168. Tsankova NM, Berton O, Renthal W, Kumar A, Neve RL, Nestler EJ. Sustained hippocampal chromatin regulation in a mouse model of depression and antidepressant action. Nat Neurosci. 2006;9:519–25. https://doi.org/10.1038/nn1659.

169. Reul JM, Chandramohan Y. Epigenetic mechanisms in stress-related memory formation. Psychoneuroendocrinology. 2007;32(Suppl 1):S21–5. https://doi.org/10.1016/j.psyneuen.2007.03.016.

170. Griffiths BB, Hunter RG. Neuroepigenetics of stress. Neuroscience. 2014;275:420–35. https://doi.org/10.1016/j.neuroscience.2014.06.041.
171. Hunter RG, McEwen BS. Stress and anxiety across the lifespan: structural plasticity and epigenetic regulation. Epigenomics. 2013;5:177–94. https://doi.org/10.2217/epi.13.8.
172. Reul JM. Making memories of stressful events: a journey along epigenetic, gene transcription, and signaling pathways. Front Psych. 2014;5:00005. https://doi.org/10.3389/fpsyt.2014.00005.
173. Abel EL. Football increases the risk for Lou Gehrig's disease, amyotrophic lateral sclerosis. Percept Mot Skills. 2007;104:1251–4. https://doi.org/10.2466/pms.104.4.1251-1254.
174. Wassenegger M, Heimes S, Riedel L, Sanger HL. RNA-directed de novo methylation of genomic sequences in plants. Cell. 1994;76:567–76.
175. Ho AS, Turcan S, Chan TA. Epigenetic therapy: use of agents targeting deacetylation and methylation in cancer management. Onco Targets Ther. 2013;6:223–32. https://doi.org/10.2147/OTT.S34680.
176. Veerappan CS, Sleiman S, Coppola G. Epigenetics of Alzheimer's disease and frontotemporal dementia. Neurotherapeutics. 2013;10:709–21. https://doi.org/10.1007/s13311-013-0219-0.
177. Zeier Z, Esanov R, Belle KC, Volmar CH, Johnstone AL, Halley P, et al. Bromodomain inhibitors regulate the C9ORF72 locus in ALS. Exp Neurol. 2015;271:241–50. https://doi.org/10.1016/j.expneurol.2015.06.017.
178. Koval ED, Shaner C, Zhang P, du Maine X, Fischer K, Tay J, et al. Method for widespread microRNA-155 inhibition prolongs survival in ALS-model mice. Hum Mol Genet. 2013;22:4127–35. https://doi.org/10.1093/hmg/ddt261.
179. Nolan K, Mitchem MR, Jimenez-Mateos EM, Henshall DC, Concannon CG, Prehn JH. Increased expression of microRNA-29a in ALS mice: functional analysis of its inhibition. J Mol Neurosci. 2014;53:231–41. https://doi.org/10.1007/s12031-014-0290-y.
180. Morel L, Regan M, Higashimori H, Ng SK, Esau C, Vidensky S, et al. Neuronal exosomal miRNA-dependent translational regulation of astroglial glutamate transporter GLT1. J Biol Chem. 2013;288:7105–16. https://doi.org/10.1074/jbc.M112.410944.
181. Lakshmaiah KC, Jacob LA, Aparna S, Lokanatha D, Saldanha SC. Epigenetic therapy of cancer with histone deacetylase inhibitors. J Cancer Res Ther. 2014;10:469–78. https://doi.org/10.4103/0973-1482.137937.
182. Ryu H, Smith K, Camelo SI, Carreras I, Lee J, Iglesias AH, et al. Sodium phenylbutyrate prolongs survival and regulates expression of anti-apoptotic genes in transgenic amyotrophic lateral sclerosis mice. J Neurochem. 2005;93:1087–98. https://doi.org/10.1111/j.1471-4159.2005.03077.x.
183. Cudkowicz ME, Andres PL, Macdonald SA, Bedlack RS, Choudry R, Brown RH Jr, et al. Phase 2 study of sodium phenylbutyrate in ALS. Amyotroph Lateral Scler. 2009;10:99–106. https://doi.org/10.1080/17482960802320487.
184. Prudencio M, Belzil VV, Batra R, Ross CA, Gendron TF, Pregent LJ, et al. Distinct brain transcriptome profiles in C9orf72-associated and sporadic ALS. Nat Neurosci. 2015;18:1175–82. https://doi.org/10.1038/nn.4065.

Chapter 2
Mechanism of Splicing Regulation of Spinal Muscular Atrophy Genes

Ravindra N. Singh and Natalia N. Singh

Abstract Spinal muscular atrophy (SMA) is one of the major genetic disorders associated with infant mortality. More than 90% cases of SMA result from deletions or mutations of *Survival Motor Neuron 1* (*SMN1*) gene. *SMN2*, a nearly identical copy of *SMN1*, does not compensate for the loss of *SMN1* due to predominant skipping of exon 7. However, correction of *SMN2* exon 7 splicing has proven to confer therapeutic benefits in SMA patients. The only approved drug for SMA is an antisense oligonucleotide (Spinraza™/Nusinersen), which corrects *SMN2* exon 7 splicing by blocking intronic splicing silencer N1 (ISS-N1) located immediately downstream of exon 7. ISS-N1 is a complex regulatory element encompassing overlapping negative motifs and sequestering a cryptic splice site. More than 40 protein factors have been implicated in the regulation of *SMN* exon 7 splicing. There is evidence to support that multiple exons of *SMN* are alternatively spliced during oxidative stress, which is associated with a growing number of pathological conditions. Here, we provide the most up to date account of the mechanism of splicing regulation of the *SMN* genes.

Keywords SMN · SMA · Splicing · ISS-N1 · ISS-N2 · Cryptic splice site · U1 snRNA

2.1 Introduction

Pre-mRNA splicing is an essential process in eukaryotic cells during which noncoding (intronic) sequences are removed and coding (exonic) sequences are joined together to generate mRNA. The complex reaction of splicing is catalyzed by a spliceosome, a macromolecular machinery [1]. The most critical step of a splicing reaction is the accurate determination of the 5′ and 3′ splice sites (5′ss and 3′ss) that

R. N. Singh (✉) · N. N. Singh
Department of Biomedical Sciences, Iowa State University, Ames, IA, USA
e-mail: singhr@iastate.edu

© Springer International Publishing AG, part of Springer Nature 2018   31
R. Sattler, C. J. Donnelly (eds.), *RNA Metabolism in Neurodegenerative Diseases*, Advances in Neurobiology 20,
https://doi.org/10.1007/978-3-319-89689-2_2

mark the beginning and end of an intron, respectively [2]. All intron-containing human genes have potential to be alternatively spliced, generating multiple mRNA isoforms from a single gene [3]. Decision to include or exclude an exon during pre-mRNA splicing is dictated by a combinatorial control of *cis*-elements and transacting factors. The same *cis*-element when presented in a different context may have different effects on splicing [4, 5]. Hence, the relative impact of a *cis*-element cannot be accurately predicted, it [the impact] requires experimental validation. Interpreting the consequences of a splicing-associated mutation remains a puzzle, since a single mutation can cause at least one of the following changes: loss of a positive element, gain of a negative element, change of a structural context, and nonsense-mediated decay (NMD) due to creation of an in-frame premature termination codon (PTC) [6, 7]. Rules of splicing are quite flexible and are heavily influenced by the relative abundance of various splicing factors in different tissues [8]. Further, splicing is coupled to other events including transcription, the 5′ capping, and the 3′ polyadenylation [6, 9]. Therefore, deciphering the mechanism by which a given exon is alternatively spliced remains a daunting task. A growing number of disorders are linked to aberrant splicing [10, 11]. Each case of aberrant splicing calls for an in-depth analysis of the context-specific rules so that strategies to manipulate splicing could be devised in a gene-specific manner.

Humans carry two near identical copies of the *Survival Motor Neuron* gene: *SMN1* and *SMN2* [12]. Both *SMN* genes code for SMN, a multifunction protein essential for the survival of all animal cells. The ability of SMN to interact with nucleic acids and proteins allows it to participate in various cellular processes, including but not limited to transcription, splicing, translation, macromolecular trafficking, and signal transduction [13]. The critical difference between *SMN1* and *SMN2* is the splicing of exon 7. Unlike *SMN1* exon 7, *SMN2* exon 7 is predominantly skipped in most tissues, except in testis [14]. The exon 7-skipped transcript generated by *SMN2* codes for SMNΔ7, a partially functional and unstable protein [15–17]. Loss of *SMN1* creates SMN deficit, leading to spinal muscular atrophy (SMA), a major genetic disease of children and infants [18, 19]. Aberrant expression and/or localization of SMN have been associated with several other diseases, including amyotrophic lateral sclerosis (ALS), metabolic disorders, male infertility, and stress-associated disorders [14, 20–22]. Correction of *SMN2* exon 7 splicing has proven to confer therapeutic benefits in mouse models of SMA [23, 24]. The first approved drug for SMA, Nusinersen (Spinraza™), is an antisense oligonucleotide (ASO) that promotes inclusion of *SMN2* exon 7 by sequestering an inhibitory *cis*-element called Intronic Splicing Silencer N1 or ISS-N1 [25, 26]. In this review, we describe studies that culminated in the discovery of ISS-N1 and analyze how the characterization of ISS-N1 paved the way for a better understanding of pre-mRNA splicing in the context of a human disease. We summarize the role of various cis-elements and transacting factors that regulate *SMN* exon 7 splicing. We also discuss how lessons learnt from the *SMN* genes will help find effective therapies for genetic diseases associated with aberrant splicing.

2.2   Organization of Human *SMN* Genes

The presence of two *SMN* genes in humans is attributed to the intrachromosomal duplication of ~500 kb segment at the 5q13.3 locus on chromosome 5 (Fig. 2.1a; [12, 35, 36]). Despite conservation of the coding region of *SMN* between human and rodents, there are substantial differences in the promoter, intronic, and the untranslated regions (UTRs). The abundance of Alu elements in human *SMN* genes suggests a distinct regulation of transcription and splicing of *SMN* in primates. Both *SMN* genes are ~34 kb long including ~6 kb long promoter sequence. Several mutations within the promoter region distinguish *SMN1* from *SMN2*, suggesting that transcription of these genes might be differentially regulated under certain conditions, such as stress (Fig. 2.1b). Each *SMN* gene is comprised of 10 exons, that is, 1, 2A, 2B, 3, 4, 5, 6, 6B, 7 and 8 (Fig. 2.1c). About 2/3rd of exon 1 serves as the 5′UTR, whereas the remaining 1/3rd serves as the coding sequence. Exon 8 is the longest exon that encodes the 3′UTR. *SMN2* intronic sequences flanking exon 7 contain several substitutions and a 5-nt deletion (Fig. 2.1c). A C-to-T substitution at the sixth position (C6U) of exon 7, a G-to-A substitution at the -44th position (G-44A) of intron 6, and an A-to-G substitution at the 100th position (A100G) of intron 7 are associated with skipping of *SMN2* exon 7 [37–40]. Recently discovered exon 6B is generated by exonization of an Alu element within intron 6 [33]. Another alternative transcript is generated by intron 3 retention. It codes for a short protein called axonal SMN or aSMN [34]. Considering intron 3 is conserved between human and mouse, expression of aSMN has been detected in mice as well. SMN contains several functional domains and interacts with various proteins. All isoforms of SMN possess identical N-terminus that is involved in interactions with both proteins and nucleic acids (Fig. 2.1d; [13]).

Recent reports reveal that two antisense transcripts, which function as long noncoding RNAs (lncRNAs), are generated from *SMN* locus. One of these lncRNAs termed *SMN-AS1* is ~1.6 kb long; it starts and finishes within intron 1 (Fig. 2.1a; [27]). Other one termed *SMN-AS1** is ~10 kb long; it starts within intergenic region downstream of exon 8 and extends till intron 5 (Fig. 2.1a; [28]). These lncRNAs are specific to humans and their expressions appear to downregulate SMN levels through transcriptional control. The significance of fine-tuning of SMN levels within cells is underscored by a recent study that showed the pathogenesis of osteoarthritis caused by aberrantly high expression of SMN [41]. Factors that regulate *SMN* transcription and splicing modulate SMN levels in a cell-specific manner. Testis happens to be one of the tissues with a very high SMN demand. This demand is met by an entirely different set of rules that govern transcription and splicing of the *SMN* genes in testis. Here, we describe a critical role of the context-specific *cis*-elements in *SMN* splicing and outline the emerging rules that are likely to be applicable in most cell types.

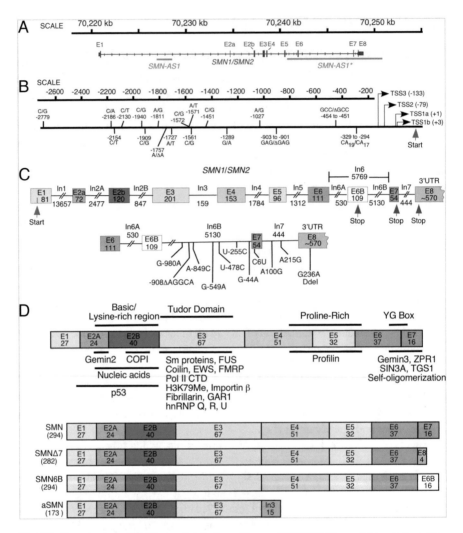

Fig. 2.1 Organization of *SMN* gene. (**a**) A view of human *SMN1/SMN2* gene(s) located on chromosome 5. Exons and introns are shown as boxes and lines, respectively. Loci of antisense RNAs, SMN-AS1 [27], and SMN-AS1* [28] are marked with bars. (**b**) Diagrammatic representation of human SMN promoter region. Multiple transcription start sites (TSS) identified so far are indicated using arrows. Numbers in brackets correspond to their position relative to TSS1a (+1). TSS1a and TSS2 were identified in [29] as transcription start sites preferentially used in adult and fetal tissues, respectively. TSS1b was mapped in Echaniz-Laguna et al. [30], and TSS3 was identified in Monani et al., [31]. Nucleotide differences between the *SMN1* and *SMN2* promoters are indicated based on Monani et al., [31]; [29, 32]), where nucleotide positions were calculated from TSS1a. Translation initiation site is marked as Start. (**c**) Diagrammatic representation of the *SMN1/SMN2* pre-mRNA. Exons and introns are shown as boxes and lines, respectively. Sizes of exons and introns are indicated in nucleotides (nts). The translation initiation and termination sites are marked as Start and Stop, respectively. Exon 8 is mostly used as the 3′ untranslated region (UTR). The bottom panel indicates nucleotides differences between *SMN1* and *SMN2* in the region located downstream of exon 6B. The last position of intron 6B is designated as −1. For exons 7 and 8, as

2.3 Regulation of *SMN* Exon 7 Splicing

Our understanding of *SMN* exon 7 skipping is continuing to evolve as more and more regulatory elements are being discovered within this relatively short exon and its flanking intronic sequences. Early studies established that the C6U substitution is the primary cause of *SMN2* exon 7 skipping [38, 39]. It was also shown that the 3'ss of *SMN2* exon 7 is weakened by the C6U substitution; but the usage of this 3'ss was enhanced when the downstream 3'ss of exon 8 was blocked [42]. Based on bioinformatics predictions and in vitro studies, it was proposed that C6U abrogates an enhancer associated with SRSF1 (ASF/SF2), a member of the highly conserved family of serine/arginine (SR)-rich proteins (Fig. 2.2a; [46]). However, this simple "SRSF1 abrogation" hypothesis did not hold true in a subsequent cell-based study, where the depletion of SRSF1 did not cause the expected enhancement of *SMN1* exon 7 skipping (Fig. 2.2a; [47]). A more recent study suggests a surprising dual role of *SRSF1* in regulation of *SMN2* exon 7 splicing, as both overexpression and depletion of *SRSF1* caused enhanced skipping of *SMN2* exon 7 [48]. An alternative hypothesis that C6U creates a silencer associated with hnRNP A1/A2 was proposed to explain the skipping of *SMN2* exon 7 [47]. Supporting this hypothesis, depletion of hnRNP A1/A2 promoted *SMN2* exon 7 inclusion [47, 49, 50]. Subsequent studies implicated the role of multiple hnRNP A1/A2 sites in the regulation of *SMN* exon 7 splicing [37, 51–53]. These findings brought additional complexity to the interpretations of the hnRNP A1/A2 depletion experiments, since the observed effect could be attributed to abrogation of hnRNP A1/A2 binding to any/all of these sites within *SMN2* pre-mRNA. Interestingly, *hnRNP A1* knockout mice show muscle-specific developmental defects [54]. Hence, depletion of hnRNP A1 cannot be exploited for a potential therapy of SMA.

The hnRNP A1/A2 model has been subsequently modified to include Sam68 as an additional factor associated with the inhibitory effect of C6U (Fig. 2.2a; [55]). Consistent with the role of hnRNP A1 and Sam68 in *SMN2* exon 7 splicing, low extracellular pH that increased the nuclear concentrations of hnRNP A1 and Sam68 was found to enhance *SMN2* exon 7 skipping [56]. Another mechanism by which C6U might affect *SMN2* exon 7 splicing is through creation of an extended inhibitory context (Exinct) that consists of overlapping negative motifs [57]. Interestingly, C6U also strengthens a predicted terminal stem-loop structure, TSL1 (Fig. 2.2b). Supporting the distinct inhibitory role of TSL1, mutations that disrupted TSL1 without abrogating C6U-associated hnRNP A1/A2 motif promoted *SMN2* exon 7

Fig. 2.1 (coninued) well as intron 7, counting starts with the first position of the respective exon or intron. (**d**). Diagrammatic representation of SMN protein isoforms. Protein regions encoded by each exon are shown as colored boxes with the number of amino acids given. In the top panel, protein domains are indicted above, while SMN interacting partners are shown below the diagrammatic representation of the full-length SMN. For further details see Singh et al. [13]. The bottom panel shows the known SMN isoforms as compared to the full-length SMN protein. These isoforms are generated either due to exon 7 skipping or exonization of a region within intron 6 [33] or intron 3 retention [34]. The size of each isoform (in amino acids) is given in brackets. Abbreviations are given in Table 2.2

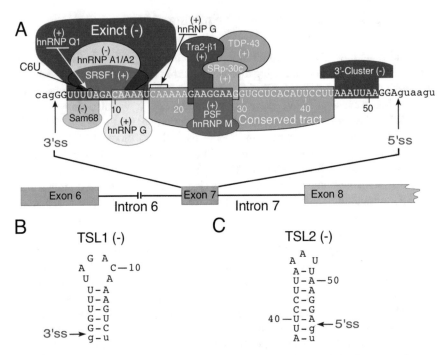

Fig. 2.2 Exon 7 splicing regulation. (**a**) Diagrammatic representation of cis-elements and trans-acting factors that modulate *SMN* exon 7 splicing. Positive and negative elements are indicated by (+) and (−), respectively (For further details see [43]). Numbering of nucleotides starts with the first position of exon 7. Exonic and intronic sequences are shown in upper- and lower-case letters, respectively. The 3′ and 5′ss are indicated by arrows. (**b**) Terminal stem-loop structure, TSL1, formed at the beginning of *SMN2* exon 7 as determined by enzymatic structure probing [44, 45]. Both TSL1 and TSL2 are marked by (−) because they contribute toward exon 7 skipping. Numbering of nucleotides starts with the first position of the exon. Exonic and intronic sequences are shown in upper- and lower-case letters, respectively. The 3′ ss is indicated by an arrow. (**c**) Terminal stem-loop structure, TSL2, formed at the end of *SMN2* exon 7 as determined by enzymatic structure probing [44, 45]. (−) indicates that TSL1 contributes to exon 7 skipping. Numbering of nucleotides starts with the first position of the exon. Exonic and intronic sequences are shown in upper- and lower-case letters, respectively. The 5′ss is indicated by an arrow. Abbreviations are given in Table 2.2

inclusion [57]. It should be noted that the proposed hypotheses associated with the inhibitory effect of C6U are not mutually exclusive. Recent years have witnessed a shift in the debate as critical roles of several other negative elements located away from the C6U site have been discovered.

As per the exon definition model, positive factors bridge cross-exon interactions before splicing takes place [58]. An early study implicated SFRS10 (Tra2-beta1) as one of the factors that interacts directly with a GA-rich sequence located in the middle of exon 7 (Fig. 2.2a; [59]). Several other proteins, including TDP43, SRSF9 (SRp30c), PSF and hnRNP M, were subsequently shown to stimulate exon 7 inclusion through a direct or indirect interaction with exon 7 (Fig. 2.2a; [60–64]). Surprisingly, a follow-up study in a mouse model of SMA established that SFRS10 is dispensable for *SMN* exon 7 splicing [65]. This finding underscored the complexity of splicing regulation when the loss of a positive factor could be tolerated due to the presence of other factors with redundant/overlapping functions. Thus far, studies suggest that skipping of *SMN2* exon 7 is driven largely by the occurrence of negative interactions. The list of factors that regulate *SMN2* exon 7 is large and continues to grow (Table 2.1). However, interaction sites for most of the identified transacting factors remain unknown. There have been very limited attempts to correlate the effect of the naturally occurring mutation within a given factor and splicing of *SMN* exon 7.

Table 2.1 Factors tested for an effect on *SMN2* exon 7 splicing

Factor (Gene)	Effect on exon 7 splicing	Binding Location	Effect of overexpression		Effect of depletion		References
			$SMN2^m$	$SMN2^g$	$SMN2^m$	$SMN2^g$	
ASF/SF2 (*SRSF1*)	Positive Neutral	Exon 7	No	Yes	No	Yes No(	[46, 47, 66, 37, 53, 67, 68, 48]
SC35 (*SRSF2*)	Negative Neutral	–	No	Yes	ND	Yes	[64, 47, 48]
SRp20 (*SRSF3*)	Negative Neutral	–	No	Yes	ND	Yes	[64, 48]
SRp75 (*SRSF4*)	Negative	–	ND	No	ND	Yes	[48]
SRp40 (*SRSF5*)	Negative Neutral	–	No	Yes	ND	Yes	[64, 48]
SRp55 (*SRSF6*)	Negative Neutral	–	No	No	ND	Yes	[64, 48]
9G8 (*SRSF7*)	Negative Neutral	–	No	Yes	Yes	Yes	[47, 64, 69, 48]
SRp30c (*SRSF9*)	Positive Neutral	–	Yes	No	ND	No	[64, 67, 48]
SRp38 (*SRSF10*)	Neutral	–	ND	No	ND	No	[48]
SRSF11 (*SRSF11*)	Negative	–	ND	Yes	ND	Yes	[48]
Tra2-β1 (*TRA2B*)	Positive	Exon 7	Yes	Yes	No	ND	[59, 64, 60, 70]
ZIS/ZNF265 (*ZRANB2*)	Negative	–	Yes	ND	ND	ND	[71]
hnRNP A1 (*HNRNPA1*)	Negative	Exon 7, Intron 7, 3′ss	ND	Yes	Yes	Yes	[47, 66, 37, 53, 60, 67, 52, 51, 69, 50]
hnRNPA2/B1 (*HNRNPA2/B1*)	Negative	–	ND	ND	Yes	Yes	[47, 37, 67, 52, 69, 50, 48]
hnRNPC (*HNRNPC*)	Positive Negative Neutral	I6-E7 junction	ND	ND	No; Yes	Yes	[68, 69, 48]

(continued)

Table 2.1 (continued)

			Effect of overexpression		Effect of depletion		
hnRNP D (*HNRNPD*)	Neutral	–	ND	ND	ND	No	[48]
hnRNP F (*HNRNPF*)	Neutral	–	ND	ND	No	No	[69, 48]
hnRNP G (*RBMX*)	Positive	Exon 7	Yes	ND	ND	ND	[63, 72, 70, 60, 73]
hnRNP H (*HNRNPH1*)	Neutral	–	ND	ND	Yes; No	No	[51, 69, 48]
hnRNP K (*HNRNPK*)	Neutral	–	ND	ND	No	ND	[69]
hnRNP L (*HNRNPL*)	Neutral	–	ND	ND	No	ND	[69]
hnRNP M (*HNRNPM*)	Positive	–	Yes	ND	Yes	Yes	[67, 69, 62]
RALY (*RALY*)	Neutral	–	ND	ND	No	ND	[69]
hnRNP Q (*SYNCRIP*)	Positive	Exon 7	Yes	Yes	Yes	ND	[67]
hnRNP U (*HNRNPU*)	Negative	–	ND	Yes	ND	Yes	[69, 48]
CHERP (*CHERP*)	Negative	–	ND	ND	Yes	ND	[69]
HuR (*ELAVL1*)	Negative	3′-UTR	ND	ND	ND	Yes	[48]
PSF (*SFPQ*)	Positive	Exon 7	Yes	Yes	ND	Yes	[67, 62]
PUF60 (*PUF60*)	Negative	3′ss	ND	ND	Yes	Yes	[69, 74]
TDP-43 (*TARDBP*)	Positive	–	Yes	ND	No	ND	[60]
TIA1 (*TIA1*)	Positive	Intron 7	Yes	Yes	Yes	Yes	[75]
RBM10 (*RBM10*)	Negative	–	ND	ND	Yes	Yes	[69, 76]
Sam68 (*KHDRBS1*)	Negative	Exon 7	Yes	ND	Yes	ND	[55]
SF1 (*SF1*)	Negative	Branch Point	ND	ND	Yes	ND	[69]
SmD3 (*SNRPD3*)	Positive	–	ND	ND	ND	Yes	[77]
SON (*SON*)	Negative	–	ND	ND	Yes	ND	[69]
U1-70K (*SNRNP70*)	Positive	–	ND	ND	ND	Yes	[77]
U2AF35 (*U2AF1*)	Negative	–	ND	ND	Yes	Yes	[69]
U2AF65 (*U2AF2*)	Negative	3′ss	ND	ND	Yes	Yes	[67, 69, 74]
U2B″ (*U2B″*)	Positive	–	ND	ND	ND	Yes	[77]

Abbreviations: positive, positive effect on exon 7 splicing; negative, negative effect on exon 7 splicing; neutral, neutral effect on exon 7 splicing; 3′ss, 3′ splice site; 3′-UTR, 3′-untranslated region; ND, not performed or assayed; Yes, observed; No, not observed; *SMN2^m*, *SMN2* minigene; *SMN2^g*, Endogenous *SMN2* gene

Table 2.2 Abbreviations and terminology used in this study

Abbreviation	Full name	Relevant figures
3′ss	3′ splice site	2
3′-UTR	3′ untranslated region	
5′ss	5′ splice site	2
5′-UTR	5′ untranslated region	
ALS	amyotrophic lateral sclerosis	
ASO	Antisense oligonucleotide	5
bp	Base pair	
C6U	A C-to-U substitution at the sixth position of *SMN2* exon 7	2
Element 1	Negative *cis*-element located within *SMN* intron 6	3
Element 2	Positive *cis*-element located within *SMN* intron 7	3
FTD	Frontotemporal dementia	
eU1	Engineered U1 snRNA	6
hnRNP	Hetero-nuclear ribonucleoprotein	2
ISS-N1	Intronic splicing silencer N1 (located within *SMN* intron 7)	3, 5
ISS-N2	Intronic splicing silencer N2 (located within *SMN* intron 7)	4, 5
ISTL1	Internal stem formed by LDI-1 (located within *SMN* intron 7)	4, 5
ISTL2	Internal stem formed by LDI-2 (located within *SMN* intron 7)	4
ISTL3	Internal stem formed by LDI-3 (located within *SMN* intron 7)	4
ISTL4	Internal stem formed by LDI-4 (located within *SMN* intron 7)	4
nt	Nucleotide	
LDI	Long-distance interaction (located within *SMN* intron 7)	3, 4, 5
lncRNA	Long non-coding RNA	1
Nusinersen	An ASO drug that targets ISS-N1 sequence (synonym of Spinraza™)	5
PMD	Pelizaeus–Merzbacher disease	
PLP1	Proteolipid protein 1	
SMA	Spinal Muscular Atrophy	
SMN (Italics)	Survival motor neuron gene or transcript	
SMN-AS1 (Italics)	Antisense transcript (lncRNA) generated from *SMN* locus	1
SMN2^m	*SMN2* minigene	
SMN2^g	Endogenous *SMN2* gene	
*SMN-AS1** (Italics)	Antisense transcript (lncRNA) generated from *SMN* locus	1
SMN	Survival motor neuron protein	
SMN6B	SMN6B protein	7
Spinraza™	An ASO drug that targets ISS-N1 sequence (synonym of Nusinersen)	5
TSL1	Terminal stem-loop 1 located within *SMN* exon 7	2

(continued)

Table 2.2 (continued)

Abbreviation	Full name	Relevant figures
TSL2	Terminal stem-loop 2 located within *SMN* exon 7	2, 4, 5
TSS	Transcription start site	1
U1 or U1 snRNA	U1 small nuclear RNA	5, 6
U1 snRNP	U1 small nuclear ribonucleoprotein	5, 6
URC1	U-rich cluster 1 located within intron 7	3, 5
URC2	U-rich cluster 2 located within intron 7	3, 5
URC3	U-rich cluster 3 located within intron 7	3, 5
UTR	Untranslated region	
WDM	Welander distal myopathy	
wt	Wild-type	

2.3.1   In Vivo Selection of Exon 7

In vivo selection is a powerful method to determine the position-specific role of every exonic residue on splicing of a given exon. The feasibility of in vivo selection for an entire exon was first demonstrated in the context of *SMN1* exon 7 [78]. The method employed a partially randomized exon 7 and repeated rounds of selection for sequences that promoted exon 7 inclusion [78]. The approach was modeled on in vitro selection of a large sequence used for the simultaneous identification of *cis*-elements and structural motifs critical for RNA-protein interaction [45, 79]. The results of in vivo selection confirmed the presence of "Exinct" in the beginning of exon 7 (Fig. 2.2a; [78]). The findings of in vivo selection also uncovered the role of a "conserved tract," a long stretch of nucleotides in the middle of exon 7 that constituted a number of overlapping positive *cis*-elements (Fig. 2.2a; [78]). In addition, the results of in vivo selection revealed the existence of a negative *cis*-element, the "3'-cluster," located toward the end of exon 7 (Fig. 2.2a; [78]). Of note, the "3'-cluster" overlaps with the exonic region that is not conserved between human and rodents, suggesting that human *SMN* exon 7 acquired this negative regulator of splicing after the divergence from the common rodent ancestor ~80 million years ago. Major findings of in vivo selection were independently confirmed by an antisense microwalk as well as by a machine-learning-based simulation study [80, 81].

The most surprising finding of in vivo selection was the overwhelming selection of a non-wild type G residue (A54G) at the last position of exon 7 [78]. Validating experiments confirmed the strong stimulatory effect of A54G substitution on *SMN2* exon 7 splicing. For instance, substitutions abrogating various positive *cis*-elements of exon 7 were fully tolerated in the presence of 54G. Numerous mechanisms by which 54G imparts such a strong stimulatory effect on *SMN2* exon 7 splicing could be envisioned. For example, 54G is predicted to disrupt an inhibitory structure (terminal stem loop 2 or TSL2) that sequesters the 5'ss of exon 7 (Fig. 2.2c). In addition, 54G increases the base pairing between U1 snRNP and the 5'ss of exon 7. Indeed, both of these predictions turned out to be true [82]. Hence, findings of in vivo selection had a transformative effect on our understanding of *SMN* exon 7 splicing. In particular, they revealed that the 5′ss of exon 7 is weak in both *SMN1* and *SMN2*. Subsequent studies focused on the mechanism that defines the 5′ss of exon 7 [43, 44, 83, 84]. These studies culminated in discoveries that led to the first therapy for SMA.

2.3.2 Effect of Terminal Stem Loop 2

In order to demonstrate the role of an RNA structure in pre-mRNA splicing, one must first perform structure probing to definitively confirm the existence of a specific RNA structure. In addition, using site-specific mutagenesis one must then show a correlation between disruption of the structure and altered splicing. Validating experiments must also demonstrate that the splicing pattern is restored when the structure is reinstated. Thus far only a handful of studies have fulfilled the above-mentioned requirements to conclusively establish the role of an RNA structure in pre-mRNA splicing. Inspired by the results of in vivo selection, we performed a systematic study uncovering the role of the terminal stem-loop 2 (TSL2) predicted to partially sequester the 5′ss of exon 7 in splicing regulation of this exon (Fig. 2.2c). Enzymatic structure probing confirmed the existence of both TSL1 and TSL2 [82]. Supporting the inhibitory role of TSL2, U40G or A54C substitution that disrupted TSL2 was found to promote *SMN2* exon 7 inclusion. As expected, when U40G and A54C substitutions were combined to reinstate the TSL2 structure, a strong inhibitory effect on *SMN2* exon 7 splicing was restored [82]. These results unequivocally confirmed that TSL2 plays the inhibitory role in the regulation of *SMN* exon 7 splicing. One of the mechanisms by which TSL2 prevents *SMN2* exon 7 inclusion is through poor recruitment of U1 snRNP at the 5′ss of exon 7. Consistent with this argument, a mutated U1 snRNA with extended complementarity to the 5′ss of exon 7 was found to restore *SMN2* exon 7 inclusion [82]. Independently validating these findings, an ASO-mediated depletion of endogenous U1 snRNP was found to promote skipping of exon 7 from both *SMN1* and *SMN2* [49]. However, the effect of U1 snRNP depletion was less pronounced in case of *SMN1* exon 7 than *SMN2* exon 7. This could be due to C6U substitution strengthening TSL1 and as a consequence stabilizing TSL2. It is also possible that the stimulatory factor(s) interacting with *SMN1* exon 7 disrupt TSL2.

2.3.3 Effect of Intronic Splicing Silencer N1

In an effort to identify additional *cis*-elements that might suppress the recognition of the 5′ss of *SMN2* exon 7, we analyzed the intronic sequences immediately downstream of the 5′ss of exon 7. Using the *SMN2* minigene we generated a set of mutants with overlapping deletions and tested their splicing pattern. Our results

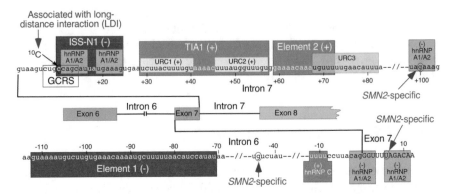

Fig. 2.3 Diagrammatic representation of intronic cis-elements and transacting factors that modulate *SMN* exon 7 splicing. Positive and negative elements are indicated by (+) and (−), respectively. Positive and neutral numbers indicate nucleotide positions within intron 7 and exon 7, respectively, starting with the first intronic/exonic position. Negative numbers indicate nucleotide positions within intron 6, starting with the last intronic position. Exonic and intronic sequences are shown in upper- and lower-case letters, respectively. Exons and introns are also shown as colored boxes and lines. *SMN2*-spesific single nucleotide substitutions are indicated. Intron 7-located ISS-N1, the overlapping GC-rich sequence (GCRS) and ¹⁰C contribute to skipping of exon 7 [43]. ISS-N1 harbors two hnRNP A1/A2B1-binding sites that are highlighted in pink. An *SMN2*-specific C6U substitution in exon 7 and A100G substitution in intron 7 create additional binding sites for hnRNP A1 [37, 47]. Another hnRNP A1-binding site is located at the junction of intron 6 and exon 7 [51]. Element 2 and U-rich clusters (URC1 and URC2) are positive *cis*-elements [75, 86]. TIA1 interacts with URC1 and URC2 and promotes exon 7 inclusion [75]. Intron 6-located Element 1 is highlighted in red [87]. It serves as a binding site for PTB and FUSE-BP [88]. A binding site for the stimulatory hnRNP C1/C2 within intron 6 is highlighted in green [68]

revealed that the sequence spanning from the 10th to 24th positions of intron 7 is highly inhibitory for exon 7 inclusion [85]. We termed this sequence as intronic splicing silencer N1 or ISS-N1 (Fig. 2.3; [85]). ISS-N1 deletion obviated the requirement for several positive *cis*-elements responsible for *SMN* exon 7 inclusion. We next employed type 1 SMA patient fibroblasts (GM03813) to validate the inhibitory effect of ISS-N1 in the context of the endogenous *SMN2*. Of note, GM03813 cells carry only *SMN2* and offer an invaluable tool to examine the effect of compounds on spicing of *SMN2* exon 7. As expected, an ASO that blocked ISS-N1 fully restored *SMN2* exon 7 inclusion in GM03813 cells [85]. Importantly, ISS-N1-targeting ASO had a pronounced stimulatory effect on *SMN2* exon 7 splicing even at a low concentration of 5 nM. This could be due to strong inhibitory nature of ISS-N1 combined with its high accessibility for an ASO that targets it.

Among several hundred targets examined thus far, ISS-N1 remains the most effective target for an ASO-mediated stimulation of *SMN2* exon 7 inclusion [89]. Numerous studies employing various mouse models have independently validated the in vivo efficacy of ISS-N1-targeting ASOs [23]. The recently approved ISS-N1-targeting drug for SMA, Nusinersen (synonyms: ISIS-SMNRx, IONIS-SMNRx and Spinraza™), is a modified oligonucleotide that carries phosphorothioate backbone and encompasses methoxyethyl modification at the 2′-hydroxyl position of the sugar moiety [23]. The above-mentioned modifications are known to enhance the in vivo stability of oligonucleotides. Multiple reports published recently discuss different aspects of the drug development process that led to the FDA approval of Nusinersen [25, 26, 49, 90–93]. More than a dozen independent studies employing ASOs with different chemistries have validated the stimulatory effect of ISS-N1 sequestration on *SMN2* exon 7 splicing [89, 94]. An in-depth analysis of these studies for an improved future ASO-based therapy is beyond the scope of this review.

Several studies have been performed to uncover the mechanism of ISS-N1 function. The inhibitory effect of ISS-N1 was only partially maintained in a heterologous background, suggesting that the context of *SMN2* makes ISS-N1 a strong negative regulator of splicing [85]. An early report implicated two putative-binding sites of hnRNP A/A2 within ISS-N1 as the major cause of the inhibitory effect of this *cis*-element (Fig. 2.3; [52]). This model has been recently revised to suggest that two RNA-recognition motifs (RRMs) of a single hnRNP A1 molecule interact with two putative sites within ISS-N1 [95]. Noticeably, the cytosine residue at the first position (^{10}C) of ISS-N1 does not fall within the putative hnRNP A1/A2-binding site. Yet, sequestration of ^{10}C was found to be absolutely critical for an ASO-mediated splicing correction of *SMN2* exon 7 (Fig. 2.3; [96]). It has been also confirmed that ASO-mediated sequestration of two putative hnRNP A1/A2-binding sites within ISS-N1 is not enough to produce a stimulatory effect on *SMN2* exon 7 splicing [50, 96]. Overall, several studies suggest a more complex mode of ISS-N1 action. Furthermore, motifs upstream and downstream of ISS-N1 appear to be involved in it as well [49, 50, 96, 97].

In search for the shortest ASO that effectively restores *SMN2* exon 7 inclusion, we performed an ultra-refined antisense microwalk within and around ISS-N1 sequence [97]. Of note, ASO sizes and their respective targets in our ultra-refined antisense microwalk differed by single nucleotides. Such approach unequivocally guarantees success for the identification of the shortest therapeutic ASO [98]. Our results showed that sequestration of a GC-rich sequence (GCRS) by an 8-mer ASO fully restored *SMN2* exon 7 inclusion (Fig. 2.3; [97]). Interestingly, GCRS-targeting ASO was found to be more specific than an ISS-N1-targeting ASO, particularly at higher concentrations [97]. This is not entirely surprising, since long ASOs can tolerate mismatched base pairs, whereas as shorter ASOs require total complementarity. Subsequent studies confirmed the therapeutic efficacy of a GCRS-targeting ASO in both mild and severe mouse models of SMA [99]. Although GCRS partially overlaps with ISS-N1, it may represent a distinct negative element. Future studies will determine if a specific factor associates with GCRS.

2.3.4   Effect of U-Rich Clusters Within Intron 7

SMN intron 7 contains multiple U-rich clusters (URCs). URC1 and URC2 are located next to each other immediately downstream of ISS-N1 (Fig. 2.3). Element 2, the very first intronic *cis*-element shown to promote exon 7 inclusion, is located downstream of URC2 [86]. It partially overlaps with the third U-rich cluster, URC3 (Fig. 2.3). Overlapping deletions in the *SMN2* minigene confirmed the strong stimulatory nature of the above URCs and Element 2. Subsequent experiments linked the stimulatory effect of URC1 and URC2 with TIA1, a glutamine-rich RNA-binding protein [75]. TIA1 and its related protein TIAR generally interact with URCs immediately downstream of a 5′ss and stimulate exon inclusion by promoting recruitment of U1 snRNP to suboptimal 5′ss [100]. However, the context of TIA1/TIAR interactions in *SMN2* intron 7 is somewhat different due to the presence of ISS-N1 between the 5′ss of exon 7 and URC1/URC2 sites to which TIA1 binds. Overexpression of TIA1 fully restored *SMN2* exon 7 inclusion, suggesting that factors that interact with ISS-N1 interfere with recruitment of TIA1 to URC1/URC2 [75]. Supporting the role TIA1 in *SMN* exon 7 splicing in the context of a human disease, Welander distal myopathy (WDM) patients carrying a TIA1 mutation display an elevated level of *SMN* exon 7 skipping [101]. Recently, mutations in TIA1 have been also linked to frontotemporal dementia (FTD) and ALS [102]. However, it is not known if FTD/ALS patients carrying TIA1 mutations display *SMN* exon 7 skipping in any of their tissues. Notably, nervous tissue of *Tia1* knockout mouse shows dysregulated expression of lipid storage and membrane dynamics factors [103]. However, effect of *Tia1* deletion on *SMN2* exon 7 splicing cannot be evaluated because mice lack *SMN2*. To obviate this problem, we generated a *Tia1* knockout mouse in the context of a mild SMA model harboring *SMN2* alleles [104]. Interestingly, loss of *Tia1* in this mouse model did not show changes in *SMN2* exon 7 splicing, although the severity of the SMA disease was affected in a gender-specific manner [104]. Several reasons may account for the discrepancy between the effects of *Tia1* deletion (in mouse) and *TIA1* mutation (in human). For instance, TIA1 is involved in various types of protein-protein and RNA-protein interactions during pre-mRNA splicing, stress granule formation, and mRNA trafficking [105, 106]. It is likely that a mutant TIA1 protein perturbs protein-protein and RNA-protein interactions in the above-mentioned processes. On the other hand, the complete loss of *Tia1* in the mouse model is tolerated due to the presence of its related protein Tiar and/or other glutamine-rich RNA-binding protein.

2.3.5   *Effect of Long-Distance Interactions Within Intron 7*

Splicing of *SMN* exon 7 is modulated by a unique RNA structure formed by long-distance interactions (LDI) within intron 7 [43, 50, 96]. This structure is termed as "Internal-Stem formed by LDI 1" or ISTL1 (Fig. 2.4; [50]). Chemical structure probing confirmed the formation of ISTL1 along with several other structures within intron 7 (Fig. 2.4). Two strands of ISTL1 are separated from each other by 279-nts, of which 189 residues are located within the independently folded modules. The 5′ strand of ISTL1 overlaps with the 5′ss of exon 7 as well as ^{10}C, which occupies the first position of ISS-N1. It appears that the formation of ISTL1 strengthens TSL2. Consistently, F14, a 14-mer ASO that sequesters the first 14 residues of ISS-N1,

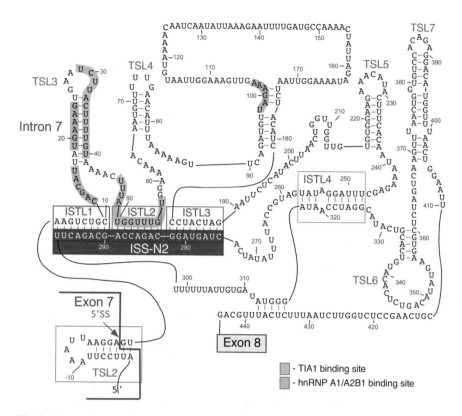

Fig. 2.4 Secondary structure of *SMN2* intron 7 derived from chemical probing. Numbering starts from the first position of intron 7. Negative numbers represent upstream sequences within exon 7. TSLs, ISTLs and binding sites for TIA1 and hnRNP A1/A2B1 are shown and highlighted. ISS-N2 is composed of the 3′ strands of ISTL1, ISTL2 and ISTL3 [43, 50]. The 5′ss of exon 7 is indicated by a red arrow. Abbreviations are given in Table 2.2

including ¹⁰C, destabilizes both ISTL1 and TSL2 [50, 96]. On the contrary, L14, a 14-mer ASO that sequesters the last 14 residues of ISS-N1, but not ¹⁰C, strengthens both ISTL1 and TSL2. Consequently, F14 and L14 have opposite effects on *SMN2* exon 7 splicing: F14 promotes *SMN2* exon 7 inclusion, while L14 causes skipping of this exon [50, 96]. The opposite effects of F14 or L14 were found to be independent of the oligonucleotide chemistry, suggesting that ASO-induced structural rearrangement at the 5′ss of exon 7 was the driving force behind the splicing outcomes [96]. This is a rare example in which two ASOs of identical size annealing to sequences differing only by a single nucleotide produce opposite effects on premRNA splicing.

The 3′ strand of ISTL1 overlaps with ISS-N2, a negative element located deep within intron 7 (Fig. 2.4; [50]). ISS-N2 also participates in the formation of ISTL2 and ISTL3, other intra-intronic structures formed by LDIs (Fig. 2.4). Formation of ISTL2 sequesters URC2, one of the binding sites of TIA1. Similar to ISS-N1, deletion or an ASO-mediated sequestration of ISS-N2 restores *SMN2* exon 7 inclusion.

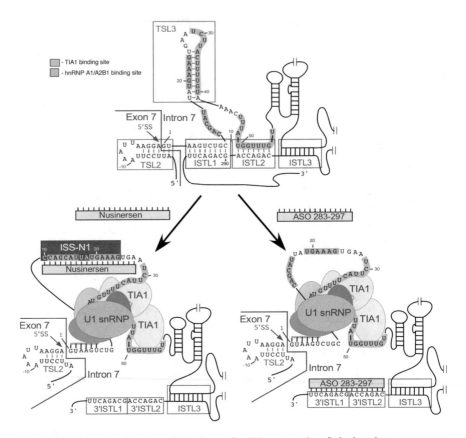

Fig. 2.5 ASO-based mechanism of *SMN2* exon 7 splicing correction. Only the relevant sequences of exon 7/intron 7 are given. Nucleotide numbering starts from the first position of intron 7. ISS-N1 and the binding sites for TIA1 and hnRNP A1/A2B1 are marked by colored boxes. The 5′ ss of exon 7 is indicated by a red arrow. The annealing positions of U1 snRNA to this 5′ ss are shown. TSL2 and 3 are local RNA secondary structures, while ISTL1, 2 and 3 are the structures formed by long-distance interactions. These structures are boxed. Nusinersen and ASO 283–297 are shown as yellow bars [25, 107]. Their annealing positions within intron 7 are indicated. Targeting of the corresponding intronic sequences by Nusinersen and ASO 283–297 causes massive structural rearrangements, including disruption of TSL3 and ISTL1. As the results TIA1-binding sites become accessible, the recruitment of U1 snRNP to the 5′ ss of exon 7 is increased and, in case of Nusinersen, the binding of hnRNP A1/A2 to ISS-N1 is blocked. Abbreviations are given in Table 2.2

Interestingly, ASO-mediated sequestration of ISS-N1 and ISS-N2 brings the similar structural changes at the 5′ss of *SMN2* exon 7, suggesting a common mechanism of action. It appears that both ISS-N1- and ISS-N2-targeting ASOs promote inclusion of *SMN2* exon 7 through abrogation of ISTL1 and an improved recruitment of TIA1 (Fig. 2.5). In vivo study with an ISS-N2 targeting ASO was recently shown to confer gender-specific therapeutic benefits in a mild mouse model of SMA [107].

2.3.6   Extension of Exon 7 by the Activation of a Cryptic 5'ss

Various instances of SMA caused by enhanced exon 7 skipping triggered by mutations at the 3' or the 5'ss of *SMN1* exon 7 have been reported [12, 108, 109]. Such patients cannot benefit from Nusinersen or any other therapeutic approach requiring the fully functional splice sites of exon 7. However, these patients can take advantage of an engineered U1 snRNA (eU1)-based approach aimed at the activation of a cryptic 5'ss located downstream of the natural 5'ss of exon 7. The proof of principle has recently been established in the context of a pathogenic G-to-C mutation at the first position (G1C) of *SMN1* intron 7 (Fig. 2.6; [49]). As expected, *SMN1* exon 7 carrying G1C substitution undergoes complete skipping of exon 7 with or without an ISS-N1-targeting ASO. However, eU1s targeting ISS-N1 or sequences upstream or downstream of this *cis*-element activate a cryptic 5'ss (Cr1) leading to the inclusion of an "extended" exon 7. Of note, another cryptic 5'ss, Cr2, located within URC2 could also be activated by a different set eU1s, albeit with less efficiency [49]. Cr1 and Cr2 usage increases the length of exon 7 by 23 and 51 nts, respectively (Fig. 2.6). Since the stop codon of SMN is located within exon 7, activation of Cr1 or Cr2 will have no consequences for the protein. Indeed, the activation of Cr1 in *SMN1* construct carrying pathogenic G1C mutation led to the production of SMN, confirming that transcripts generated by Cr1 activation are stable and translation competent (Fig. 2.6; [49]).

The discovery of Cr1 and Cr2 brings new perspective to our understanding of *SMN* exon 7 splicing regulation. Cr1 partially overlaps with ISS-N1, suggesting that the factors interacting with ISS-N1 are likely to suppress the activation of Cr1 as well. Interestingly, Cr1 is efficiently activated even by those eU1s that did not anneal to Cr1 directly [49]. Also, activation of Cr1 does not require assistance of the endogenous U1 snRNP, suggesting that usage of Cr1 can occur in the absence of the typical RNA:RNA duplex formed between the 5'ss and the U1 snRNA. This finding has broad implications as it suggests that the U1 snRNP can affect selection of a 5'ss from distance. It appears that positive *cis*-elements required for inclusion of *SMN* exon 7 are dispensable for Cr1 activation. For instance, point mutations that activated Cr1 in *SMN2* tolerated the loss of the enhancer associated with Tra2-beta1. Further, eU1s targeting Cr1 prevented skipping of exon 7 associated with the pathogenic mutation at the 3'ss of *SMN1* exon 7. Overall, these findings suggest that the activation of Cr1 might employ an entirely different set of rules.

2.3.7   Role of cis-Elements Within Intron 6

Various mutations at the 3'ss of *SMN1* intron 6 have been found to be associated with SMA pathogenesis [12, 110, 111]. However, very limited studies have been done to uncover the role of cis-elements within *SMN* intron 6. Element 1, an extended inhibitory sequence situated immediately upstream of the 3'ss of exon 7, was the first cis-element to be reported within intron 6 (Fig. 2.3; [87]). Deletion or

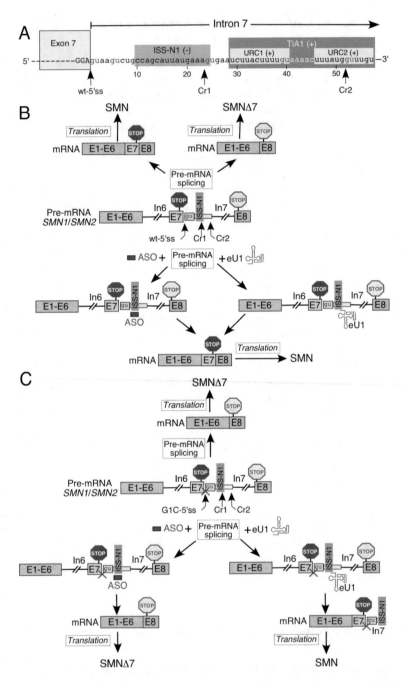

Fig. 2.6 Effect of an ASO and eU1 on splicing of exon 7. (**a**) Diagrammatic representation of exon 7/intron 7 junction. Exonic and intronic sequences are shown in upper- and lower-case letters, respectively. Exon7 is also shown as a blue box. Nucleotide numbering starts from the first position of intron 7. ISS-N1 and URC1 and URC2 are marked by colored boxes. The wild type and the cryptic

an ASO-mediated sequestration of Element 1 promoted *SMN2* exon 7 inclusion [87, 112]. A recent report demonstrated an in vivo efficacy of an Element 1-targeting ASO in a severe mouse model of SMA [112]. Another negative *cis*-element at the junction of intron 6 and exon 7 has been suggested to constitute a binding site for hnRNP A1 (Fig. 2.3; [51]). The location of this site right next to the other hnRNP A1-binding site created by the C6U mutation within exon 7 strikingly resembles the arrangement of two putative hnRNP A1 sites within ISS-N1. As recently proposed, close proximity of the two hnRNP A1 sites is conducive for a tight interaction involving two RRMs of a single hnRNP A1 molecule [95]. The polypyrimidine tract (PPT) at the 3′ss of exon 7 has been suggested to harbor a positive element associated with hnRNP C (Fig. 2.3; [68]). However, the role of hnRNP C in *SMN* exon 7 splicing could not be independently validated by depletion experiments [48, 113]. Interestingly, an A-to-G substitution at the -44th position (A-44G) of intron 6 has been found to promote *SMN2* exon 7 inclusion (Fig. 2.3; [40]). The A-44G substitution is naturally present in human population and SMA patients carrying A-44G substitution show mild phenotype [40].

2.4 Exonization of an Intronic Alu-Element

Alu elements are primate-specific transposable elements encompassing ~300 bp bipartite motifs derived from the 7SL RNA, an essential component of the protein signal recognition complex [114]. Insertion of Alu elements has played a significant role in primate evolution due to their drastic effect on chromatin remodeling, transcription and generation of novel exons [115, 116]. Multi-exon skipping detection assay (MESDA) is a powerful technique that simultaneously detects most *SMN* splice isoforms in a single reaction [117]. Employing MESDA, we have recently reported a novel exon, exon 6B, generated by the exonization of an Alu element located within intron 6 [33]. Expression of exon 6B-containing transcripts has been confirmed in various tissues of a mouse model of SMA as well as in human tissues examined [33]. Both *SMN1* and *SMN2* produce exon 6B-containing transcripts. Generally, the right arm of an antisense sequence of an Alu is used for exonization

Fig. 2.6 (continued) 5′ ss of exon 7 (Cr1 and 2) are indicated by arrows. GU dinucleotides are highlighted in red. (**b**) Model of how in the context of the intact 5′ ss of exon 7 an ASO and eU1 promote production of the full-length SMN protein (Adapted from [49]). The ASO block ISS-N1 and eU1 activates usage of the wild-type 5′ ss of exon 7. Exons and introns are indicated by the colored boxes and lines, respectively. The ASO is shown as a red bar, and eU1 as a blue structure. ISS-N1, stop codons in exon 7 and 8 and the 5′ ss of exon 7, wild type and cryptic, are indicated. (**c**) Model of how in the context of the mutated 5′ ss of exon 7 only eU1 promotes production of the full-length SMN protein (Adapted from [49]). The G to C mutation at the first position of intron 7 is shown in red. The inactivation of the 5′ ss is signified by a red cross. The ASO blocks ISS-N1 and eU1 activate usage of the cryptic 5′ ss of exon 7, Cr1. Exons and introns are indicated by the colored boxes and lines, respectively. The ASO is shown as a red bar, and eU1 as a blue structure. ISS-N1, stop codons in exon 7 and 8 and the 5′ ss of exon 7, wild type and cryptic, are indicated. Abbreviations are given in Table 2.2

[118]. However, the 109-nt long exon 6B originated from the left antisense arm of an Alu element. The low expression of exon 6B-containing transcripts is attributed to various factors, including suppression by hnRNP C and degradation by Nonsense Mediated Decay (NMD). An overwhelming 39% of *SMN* sequence is occupied by >40 Alu elements located within introns. Exon 6B is the first and only known example of *SMN* exon derived from the exonization of an intronic Alu element. Due to its location upstream of exons 7, it is likely that splicing of exon 6B is influenced by exon 7 and vice versa. However, the mechanism of exon 6B splicing regulation remains to be determined.

Amino acids coded by exon 7 define the critical C-terminus of SMN and confer protein stability. The loss of amino acids coded by exon 7 is the primary reasons why SMNΔ7 is less stable than SMN (Fig. 2.7; [16, 119]). Irrespective of exon 7 inclusion or skipping, the exon 6B-containing transcripts code for SMN6B protein in which the last 16 amino acids are coded by exon 6B. The altered C-terminus makes SMN6B less stable than SMN. However, SMN6B was found to be more stable than SMNΔ7, suggesting that the altered C-terminus of SMN6B is not deleterious as observed in case of SMNΔ7 (Fig. 2.7; [33]). As expected, SMN6B retains the ability to interact with Gemin2, a key protein required for most SMN functions. Similar to SMN, SMN6B localizes to both, nuclear and cytosolic compartments. Hence, it is likely that SMN6B will be able to ameliorate SMA pathology if expressed at sufficient levels.

2.5   Alternative Splicing of Other *SMN* Exons

The diversity of *SMN* splice isoforms is best demonstrated by MESDA, which captures susceptibility of various *SMN* exons to skipping under normal and stress-associated conditions [117]. Low levels of exon 3 and exon 5-skipped transcripts are generated under normal conditions in most tissues from both *SMN1* and *SMN2* [117]. *SMN2* exons 5 and 7 become highly susceptible to skipping under the conditions of oxidative stress, although skipping of *SMN1* exon 5 is also enhanced by oxidative stress. A recent study examined the effect of paraquat, an oxidative-stress-causing agent, on splicing of various *SMN2* exons in different tissues of a transgenic mouse model harboring *SMN2* [120]. Findings of this study revealed tissue-specific effect of oxidative stress on splicing of various *SMN2* exons. For instance, skipping of *SMN2* exons 3, 5, and 7 was found to be substantially increased under oxidative stress in lung as compared to brain and spinal cord, which instead showed significant enhancement of *SMN2* exons 5 and 7 skipping. The study also captured individual differences of the effect of oxidative stress on splicing of various *SMN2* exons. For example, one of the four animals examined showed enhanced co-skipping of exons 3, 4, 5, 6 and 7 in liver at 8 h post paraquat treatment. Another animal showed enhanced co-skipping of exons 3, 5, 6 and 7 in liver at 12 h post paraquat treatment. While reasons for these individual differences remain unknown, findings underscore that the rules of stress-associated splicing regulation should be interpreted with caution.

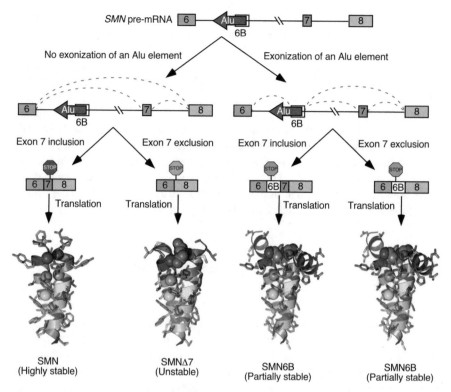

Fig. 2.7 A model showing skipping and inclusion of *SMN* exon 6B. Exon 6B is derived from an Alu element located within *SMN* intron 6 [33]. Transcripts that include exon 7 but exclude exon 6B produce full-length SMN, a highly stable protein. Transcripts that lack both exons 6B and exon 7 produce SMNΔ7, an unstable and partially functional protein. Transcripts that include exon 6B produce SMN6B protein irrespective of inclusion or exclusion of exon 7. SMN6B protein is more stable than SMNΔ7 [33]

Depletion of U1 snRNP creates a stress on the splicing machinery as well as on other co-transcriptional events dependent upon the availability of U1 snRNP [121]. A diverse set of *SMN* transcripts is generated upon depletion of U1 snRNP by an ASO that sequesters the 5′ end of endogenous U1 snRNA [49]. MESDA profile of *SMN* transcripts generated under U1 snRNP depletion condition is distinct from those observed under the conditions of oxidative stress. For example, splicing of all exons was affected under U1 snRNP depletion, whereas splicing of *SMN2* exons 5 and 7 was the most affected under oxidative stress condition [49, 117]. Interestingly, skipping of exon 6 was the least among all other internal exons of *SMN* under both U1 snRNP depletion and oxidative stress conditions [49, 117]. This could be attributed to relatively high accessibility of the 5′ss of exon 6 coupled with a strong duplex between U1 snRNP and the 5′ss of exon 6.

It is likely that the energy (ATP) deficit created by oxidative stress downregulates the biogenesis of snRNPs, particularly U1 snRNP, which is generally maintained at a higher level than other snRNPs. It has been recently shown that the depletion of DHX9, an RNA helicase that resolves the double-stranded RNA structures, enhances the Alu-induced RNA processing defects, including aberrant pre-mRNA splicing and circRNA production from transcripts harboring Alu repeats [122]. Similar to snRNP biogenesis, RNA helicases require ATP for their function. Therefore, it is likely that large RNA:RNA duplexes formed by Alu elements positioned in opposite orientations in *SMN* pre-mRNA are not appropriately resolved by RNA helicases under the conditions of oxidative stress. Preliminary analysis of the publicly available circRNA database suggests production of circRNAs by *SMN* [123]. However, it is not known what fraction of *SMN* transcripts make circRNAs and which of the circRNAs are predominantly expressed in most cell types. Future studies will determine how Alu elements might impact generation of *SMN* circRNAs under normal and stress-associated conditions in a cell-specific manner.

2.6 Effect of Transcription on Splicing of Various *SMN* Exons

Transcription requires opening of chromatin structure followed by recruitment of transcription initiation factors [9]. Transcription in vivo is coupled to splicing through two likely mechanisms: "recruitment coupling" and "kinetic coupling" [124]. These two mechanisms are not mutually exclusive and it is often difficult to conclusively distinguish one mechanism from the other. In case of recruitment coupling, RNA polymerase II (pol II) recruits splicing factors at the promoter site and then transports it to the splice sites. In case of kinetic coupling, the rate of transcription elongation influences the outcome of splicing. The evidence that transcription affects splicing of *SMN* exon 7 comes from a promoter-swapping experiment performed in minigene systems. In particular, the replacement of the wild-type *SMN* promoter with CMV or TK promoter caused enhanced skipping of exon 7 in both *SMN1* and *SMN2* minigenes [117]. These results suggested that wild-type promoter harbors sequences that are stimulatory for exon 7 splicing.

Additional evidence that transcription affects *SMN* splicing comes from small molecules that affect the activity of histone acetylases (HATs) and histone deacetylases (HDACs). The former and the latter enzymes activate and suppress transcription, respectively. Various HDAC inhibitorsm, including trichostatin A (TSA), suberoylanilide hydroxamic acid (SAHA), and benzamide M344, have been shown to modulate splicing of *SMN* exon 7 [125]. Another mechanism by which transcription could modulate splicing of *SMN* exons is through the regulation of the formation of loops within pre-mRNA. PTB and hnRNP A1/A2 have been implicated in deciding splicing outcomes through looping out specific sequences [126, 127]. In particular, looping out of an exon promotes its skipping, whereas looping out of an intra-intronic sequence promotes exon inclusion. Furthermore, a slow elongating pol II might delay the formation of a specific loop. Considering that *SMN* pre-mRNA contains binding sites for the loop-forming hnRNP A1/A2 protein, it is highly likely that splicing of various *SMN* exons is regulated by transcription.

2.7 Conclusions

SMA is one of the leading genetic diseases associated with infant mortality. As soon as the association of SMA with *SMN1* deletion/mutations was established in 1995, attempts began to find a potential cure/therapy for this disorder. Since *SMN2* is almost universally present in SMA patients, it offers an obvious therapeutic target for exon 7 splicing correction. The major breakthrough came when the critical role of the context-specific *cis*-elements located away from the pathogenic mutations, such as C6U, was beginning to be established. In particular, the discovery of the intronic *cis*-element, ISS-N1, reported in 2006 produced an effective target, sequestration of which fully corrected *SMN2* exon 7 splicing and restored SMN levels in SMA patient cells. General interest in ISS-N1 combined with subsequent independent validations of its therapeutic potential paved a way to the first FDA-approved drug for SMA. In addition, the detailed characterization of ISS-N1 led to the discovery of a unique RNA structure formed by long-distance intra-intronic interactions that contributes to exon 7 skipping. Interestingly, abrogation of a similar structure within intron 3 of the proteolipid protein 1 (*PLP1*) gene has been recently suggested to cause X-linked Pelizaeus–Merzbacher disease or PMD [128]. Growing evidence suggests that splicing of various exons is differentially regulated under the normal and stress-associated conditions. It is also becoming obvious that the intronic Alu elements are capable of increasing the diversity of *SMN* splice isoforms and may play an important role in the generation of circRNAs [123]. Furthermore, new findings that two antisense transcripts are produced from the *SMN* locus highlight the existence of an addition layer of *SMN* transcription and potentially splicing control. The development of novel tools and reliable assays that accurately capture transcription-coupled splicing events would tremendously advance our understanding of how expression of the *SMN* gene is regulated, including the pre-mRNA splicing step. This advancement would also uncover the likely mechanisms of the tissue-specific modulation of splicing of various *SMN* exons under the normal and stress-associated conditions. A better understanding of *SMN* splicing has implications for several diseases impacted by the low levels of the SMN protein. Lessons learnt from *SMN* would also provide unique insights into our understanding of a growing number of human diseases associated with aberrant splicing.

Acknowledgements This work was supported by grants from the National Institutes of Health (R01 NS055925 and R21 NS101312), Iowa Center for Advanced Neurotoxicology (ICAN), and Salsbury Endowment (Iowa State University, Ames, IA, USA) to RNS. The authors acknowledge and regret not being able to include several references due to lack of space.

Disclosures and Competing Interests
    The ISS-N1 target (US Patent# US7838657) was discovered in the Singh laboratory at UMass Medical School (MA, USA). Inventors, including RN Singh, NN Singh and UMASS Medical School, are currently benefiting from licensing of the ISS-N1 target to Ionis Pharmaceuticals and Biogen. Iowa State University holds intellectual property rights on GC-rich and ISS-N2 targets. Therefore, inventors including RN Singh, NN Singh and Iowa State University could potentially benefit from any future commercial exploitation of GC-rich and ISS-N2 targets.

References

1. Wahl MC, Will CL, Luhrmann R. The spliceosome: design principles of a dynamic RNP machine. Cell. 2009;136:701–18. https://doi.org/10.1016/j.cell.2009.02.009.
2. Hertel KJ. Combinatorial control of exon recognition. J Biol Chem. 2008;283(3):1211–5. https://doi.org/10.1074/jbc.R700035200.
3. Raj B, Blencowe BJ. Alternative splicing in the mammalian nervous system: recent insights into mechanisms and functional roles. Neuron. 2015;87:14–27. https://doi.org/10.1016/j.neuron.2015.05.004.
4. Erkelenz S, Mueller WF, Evans MS, Busch A, Schöneweis K, Hertel KJ, Schaal H. Position-dependent splicing activation and repression by SR and hnRNP proteins rely on common mechanisms. RNA. 2013;19(1):96–102. https://doi.org/10.1261/rna.037044.112.
5. Huelga SC, Vu AQ, Arnold JD, Liang TY, Liu PP, Yan BY, Donohue JP, Shiue L, Hoon S, Brenner S, Ares M Jr, Yeo GW. Integrative genome-wide analysis reveals cooperative regulation of alternative splicing by hnRNP proteins. Cell Rep. 2012;1(2):167–78. https://doi.org/10.1016/j.celrep.2012.02.001.
6. Lee Y, Rio DC. Mechanisms and regulation of alternative pre-mRNA splicing. Annu Rev Biochem. 2015;84:291–323. https://doi.org/10.1146/annurev-biochem-060614-034316.
7. Shepard PJ, Hertel KJ. Conserved RNA secondary structures promote alternative splicing. RNA. 2008;14(8):1463–9. https://doi.org/10.1261/rna.1069408.
8. Fu XD, Ares M Jr. Context-dependent control of alternative splicing by RNA-binding proteins. Nat Rev Genet. 2014;15(10):689–701. https://doi.org/10.1038/nrg3778.
9. Saldi T, Cortazar MA, Sheridan RM, Bentley DL. Coupling of RNA polymerase II transcription elongation with pre-mRNA splicing. J Mol Biol. 2016;428(12):2623–35. https://doi.org/10.1016/j.jmb.2016.04.017.
10. Cooper TA, Wan L, Dreyfuss G. RNA and disease. Cell. 2009;136(4):777–93. https://doi.org/10.1016/j.cell.2009.02.011.
11. Deschênes M, Chabot B. The emerging role of alternative splicing in senescence and aging. Aging Cell. 2017;16(5):918–33. https://doi.org/10.1111/acel.12646.
12. Lefebvre S, Bürglen L, Reboullet S, Clermont O, Burlet P, Viollet L, Benichou B, Cruaud C, Millasseau P, Zeviani M, Le Paslier D, Frézal J, Cohen D, Weissenbach J, Munnich A, Melki J. Identification and characterization of a spinal muscular atrophy-determining gene. Cell. 1995;80(1):155–65.
13. Singh RN, Howell MD, Ottesen EW, Singh NN. Diverse role of survival motor neuron protein. Biochim Biophys Acta. 2017;1860(3):299–315. https://doi.org/10.1016/j.bbagrm.2016.12.008.
14. Ottesen EW, Howell MD, Singh NN, Seo J, Whitley EM, Singh RN. Severe impairment of male reproductive organ development in a low SMN expressing mouse model of spinal muscular atrophy. Sci Rep. 2016;6:17. https://doi.org/10.1038/srep20193.
15. Burnett BG, Muñoz E, Tandon A, Kwon DY, Sumner CJ, Fischbeck KH. Regulation of SMN protein stability. Mol Cell Biol. 2009;29(5):1107–15. https://doi.org/10.1128/MCB.01262-08.
16. Cho SC, Dreyfuss G. A degron created by SMN2 exon 7 skipping is a principal contributor to spinal muscular atrophy severity. Genes Dev. 2010;24(5):438–42. https://doi.org/10.1101/gad.1884910.
17. Vitte J, Fassier C, Tiziano FD, Dalard C, Soave S, Roblot N, Brahe C, Saugier-Veber P, Bonnefont JP, Melki J. Refined characterization of the expression and stability of the SMN gene products. Am J Pathol. 2007;171(4):1269–80. https://doi.org/10.2353/ajpath.2007.070399.
18. Ahmad S, Bhatia K, Kannan A, Gangwani L. Molecular Mechanisms of Neurodegeneration in Spinal Muscular Atrophy. J Exp Neuro. 2016;10:39–49. https://doi.org/10.4137/jen.s33122.
19. Nash LA, Burns JK, Chardon JW, Kothary R, Parks RJ. Spinal muscular atrophy: more than a disease of motor neurons? Curr Mol Med. 2016;16(9):779–92. https://doi.org/10.2174/1566524016666161128113338.

20. Bowerman M, Michalski JP, Beauvais A, Murray LM, DeRepentigny Y, Kothary R. Defects in pancreatic development and glucose metabolism in SMN-depleted mice independent of canonical spinal muscular atrophy neuromuscular pathology. Hum Mol Genet. 2014;23(13):3432–44. https://doi.org/10.1093/hmg/ddu052.

21. Dominguez CE, Cunningham D, Chandler DS. SMN regulation in SMA and in response to stress: new paradigms and therapeutic possibilities. Hum Genet. 2017;136:1173. https://doi.org/10.1007/s00439-017-1835-2.

22. Rodriguez-Muela N, Litterman NK, Norabuena EM, Mull JL, Galazo MJ, Sun C, Ng SY, Makhortova NR, White A, Lynes MM, Chung WK, Davidow LS, Macklis JD, Rubin LL. Single-cell analysis of SMN reveals its broader role in neuromuscular disease. Cell Rep. 2017;18(6):1484–98. https://doi.org/10.1016/j.celrep.2017.01.035.

23. Howell MD, Singh NN, Singh RN. Advances in therapeutic development for spinal muscular atrophy. Future Med Chem. 2014;6(9):1081–99. https://doi.org/10.4155/fmc.14.63.

24. Seo J, Howell MD, Singh NN, Singh RN. Spinal muscular atrophy: an update on therapeutic progress. Biochim Biophys Acta. 2013;1832(12):2180–90. https://doi.org/10.1016/j.bbadis.2013.08.005.

25. Ottesen EW. ISS-N1 makes the first FDA-approved drug for spinal muscular atrophy. Transl Neurosci. 2017;8:1–6. https://doi.org/10.1515/tnsci-2017-0001.

26. Singh NN, Howell MD, Androphy EJ, Singh RN. How the discovery of ISS-N1 led to the first medical therapy for spinal muscular atrophy. Gene Ther. 2017b;24:520–6. https://doi.org/10.1038/gt.2017.34.

27. d'Ydewalle C, Ramos DM, Pyles NJ, Ng SY, Gorz M, Pilato CM, Ling K, Kong L, Ward AJ, Rubin LL, Rigo F, Bennett CF, Sumner CJ. Antisense Transcript SMN-AS1 Regulates SMN Expression and Is a Novel Therapeutic Target for Spinal Muscular Atrophy. Neuron. 2017;93(1):66–79. https://doi.org/10.1016/j.neuron.2016.11.033.

28. Woo CJ, Maier VK, Davey R, Brennan J, Li G, Brothers J 2nd, Schwartz B, Gordo S, Kasper A, Okamoto TR, Johansson HE, Mandefro B, Sareen D, Bialek P, Chau BN, Bhat B, Bullough D, Barsoum J. Gene activation of SMN by selective disruption of lncRNA-mediated recruitment of PRC2 for the treatment of spinal muscular atrophy. Proc Natl Acad Sci U S A. 2017;114(8):E1509–18. https://doi.org/10.1073/pnas.1616521114.

29. Germain-Desprez D, Brun T, Rochette C, Semionov A, Rouget R, Simard LR. The SMN genes are subject to transcriptional regulation during cellular differentiation. Gene. 2001;279(2):109–17. https://doi.org/10.1016/S0378-1119(01)00758-2.

30. Echaniz-Laguna A, Miniou P, Bartholdi D, Melki J. The promoters of the survival motor neuron gene (SMN) and its copy (SMNc) share common regulatory elements. Am J Hum Genet. 1999;64(5):1365–70. https://doi.org/10.1086/302372.

31. Monani UR, McPherson JD, Burghes AH. Promoter analysis of the human centromeric and telomeric survival motor neuron genes (SMNC and SMNT). Biochim Biophys Acta. 1999b;1445(3):330–6.

32. Boda B, Mas C, Giudicelli C, Nepote V, Guimiot F, Levacher B, Zvara A, Santha M, LeGall I, Simonneau M. Survival motor neuron SMN1 and SMN2 gene promoters: identical sequences and differential expression in neurons and non-neuronal cells. Eur J Hum Genet. 2004;12(9):729–37. https://doi.org/10.1038/sj.ejhg.5201217.

33. Seo J, Singh NN, Ottesen EW, Lee BM, Singh RN. A novel human-specific splice isoform alters the critical C-terminus of Survival Motor Neuron protein. Sci Rep. 2016a;6:14. https://doi.org/10.1038/srep30778.

34. Setola V, Terao M, Locatelli D, Bassanini S, Garattini E, Battaglia G. Axonal-SMN (a-SMN), a protein isoform of the survival motor neuron gene, is specifically involved in axonogenesis. Proc Nat Acad Sci U S A. 2007;104(6):1959–64. https://doi.org/10.1073/pnas.0610660104.

35. Rochette CF, Gilbert N, Simard LR. SMN gene duplication and the emergence of the SMN2 gene occurred in distinct hominids: SMN2 is unique to Homo sapiens. Hum Genet. 2001;108(3):255–66. https://doi.org/10.1007/s004390100473.

36. Schmutz J, Martin J, Terry A, Couronne O, Grimwood J, Lowry S, Gordon LA, Scott D, Xie G, Huang W, Hellsten U, Tran-Gyamfi M, She X, Prabhakar S, Aerts A, et al. The DNA

sequence and comparative analysis of human chromosome 5. Nature. 2004;431(7006):268–74. https://doi.org/10.1038/nature02919.

37. Kashima T, Rao N, Manley JL. An intronic element contributes to splicing repression in spinal muscular atrophy. Proc Natl Acad Sci U S A. 2007b;104(9):3426–31. https://doi.org/10.1073/pnas.0700343104.

38. Lorson CL, Hahnen E, Androphy EJ, Wirth B. A single nucleotide in the SMN gene regulates splicing and is responsible for spinal muscular atrophy. Proc Natl Acad Sci U S A. 1999;96(11):6307–11. https://doi.org/10.1073/pnas.96.11.6307.

39. Monani UR, Lorson CL, Parsons DW, Prior TW, Androphy EJ, Burghes AH, McPherson JD. A single nucleotide difference that alters splicing patterns distinguishes the SMA gene SMN1 from the copy gene SMN2. Hum Mol Genet. 1999a;8(7):1177–83. https://doi.org/10.1093/hmg/8.7.1177.

40. Wu X, Wang SH, Sun J, Krainer AR, Hua Y, Prior TW. A-44G transition in SMN2 intron 6 protects patients with spinal muscular atrophy. Hum Mol Genet. 2017;26(14):2768–80. https://doi.org/10.1093/hmg/ddx166.

41. Cucchiarini M, Madry H, Terwilliger EF. Enhanced expression of the central survival of motor neuron (SMN) protein during the pathogenesis of osteoarthritis. J Cell Mol Med. 2014;18(1):115–24. https://doi.org/10.1111/jcmm.12170.

42. Lim SR, Hertel KJ. Modulation of survival motor neuron pre-mRNA splicing by inhibition of alternative 3′ splice site pairing. J Biol Chem. 2001;276(48):45476–83. https://doi.org/10.1074/jbc.M107632200.

43. Singh NN, Lee BM, Singh RN. Splicing regulation in spinal muscular atrophy by a RNA structure formed by long distance interactions. Ann N Y Acad Sci. 2015b;1341:176–87. https://doi.org/10.1111/nyas.12727.

44. Singh RN. Evolving concepts on human SMN Pre-mRNA splicing. RNA Biol. 2007a;4(1):7–10. https://doi.org/10.4161/rna.4.1.4535.

45. Singh RN. Unfolding the mystery of alternative splicing through a unique method of in vivo selection. Front Biosci. 2007b;12:3263–72. https://doi.org/10.2741/2310.

46. Cartegni L, Krainer AR. Disruption of an SF2/ASF-dependent exonic splicing enhancer in SMN2 causes spinal muscular atrophy in the absence of SMN1. Nat Genet. 2002;30(4):377–84. https://doi.org/10.1038/ng854.

47. Kashima T, Manley JL. (2003). A negative element in SMN2 exon 7 inhibits splicing in spinal muscular atrophy. Nat Genet. 2003;34(4):460–3. https://doi.org/10.1038/ng1207.

48. Wee CD, Havens MA, Jodelka FM, Hastings ML. Targeting SR proteins improves SMN expression in spinal muscular atrophy cells. PLoS One. 2014;9(12):e115205. https://doi.org/10.1371/journal.pone.0115205.

49. Singh NN, Del Rio-Malewski JB, Luo D, Ottesen EW, Howell MD, Singh RN. Activation of a cryptic 5′ splice site reverses the impact of pathogenic splice site mutations in the spinal muscular atrophy gene. Nucleic Acids Res. 2017a;45:12214. https://doi.org/10.1093/nar/gkx824.

50. Singh NN, Lawler MN, Ottesen EW, Upreti D, Kaczynski JR, Singh RN. An intronic structure enabled by a long-distance interaction serves as a novel target for splicing correction in spinal muscular atrophy. Nucleic Acids Res. 2013;41(17):8144–65. https://doi.org/10.1093/nar/gkt609.

51. Doktor TKd, Schroeder LD, Vested A, Palmfeldt J, Andersen HS, Gregersen N, Andresen BS. SMN2 exon 7 splicing is inhibited by binding of hnRNP A1 to a common ESS motif that spans the 3' splice site. Hum Mutat. 2011;32(2):220–30. https://doi.org/10.1002/humu.21419.

52. Hua Y, Vickers TA, Okunola HL, Bennett CF, Krainer AR. Antisense masking of an hnRNP A1/A2 intronic splicing silencer corrects SMN2 splicing in transgenic mice. Am J Hum Genet. 2008;82(4):834–48. https://doi.org/10.1016/j.ajhg.2008.01.014.

53. Kashima T, Rao N, David CJ, Manley JL. hnRNP A1 functions with specificity in repression of SMN2 exon 7 splicing. Hum Mol Genet. 2007a;16(24):3149–59. https://doi.org/10.1093/hmg/ddm276.

54. Liu TY, Chen YC, Jong YJ, Tsai HJ, Lee CC, Chang YS, Chang JG, Chang YF. Muscle developmental defects in heterogeneous nuclear Ribonucleoprotein A1 knockout mice. Open Biol. 2017;7(1):pii: 160303. https://doi.org/10.1098/rsob.160303.
55. Pedrotti S, Bielli P, Paronetto MP, Ciccosanti F, Fimia GM, Stamm S, Manley JL, Sette C. The splicing regulator Sam68 binds to a novel exonic splicing silencer and functions in SMN2 alternative splicing in spinal muscular atrophy. EMBO J. 2010;29(7):1235–47. https://doi.org/10.1038/emboj.2010.19.
56. Chen YC, Yuo CY, Yang WK, Jong YJ, Lin HH, Chang YS, Chang JG. Extracellular pH change modulates the exon 7 splicing in SMN2 mRNA. Mol Cell Neurosci. 2008b;39(2):268–72. https://doi.org/10.1016/j.mcn.2008.07.002.
57. Singh NN, Androphy EJ, Singh RN. An extended inhibitory context causes skipping of exon 7 of SMN2 in spinal muscular atrophy. Biochem Biophys Res Commun. 2004a;315(2):381–8. https://doi.org/10.1016/j.bbrc.2004.01.067.
58. De Conti L, Baralle M, Buratti E. Exon and intron definition in pre-mRNA splicing. Wiley Interdiscip Rev RNA. 2013;4(1):49–60. https://doi.org/10.1002/wrna.1140.
59. Hofmann Y, Lorson CL, Stamm S, Androphy EJ, Wirth B. Htra2-beta 1 stimulates an exonic splicing enhancer and can restore full-length SMN expression to survival motor neuron 2 (SMN2). Proc Natl Acad Sci U S A. 2000;97(17):9618–23. https://doi.org/10.1073/pnas.160181697.
60. Bose JK, Wang I-F, Hung L, Tarn W-Y. Shen C-KJ (2008). TDP-43 overexpression enhances exon 7 inclusion during the survival of motor neuron pre-mRNA splicing. J Biol Chem. 2008;283(43):28852–9. https://doi.org/10.1074/jbc.M805376200.
61. Cho S, Moon H, Loh TJ, Oh HK, Cho S, Choy HE, Song WK, Chun J-S, Zheng X, Shen H. hnRNP M facilitates exon 7 inclusion of SMN2 pre-mRNA in spinal muscular atrophy by targeting an enhancer on exon 7. Biochim Biophys Acta. 2014a;1839(4):306–15. https://doi.org/10.1016/j.bbagrm.2014.02.006.88.
62. Cho S, Moon H, Loh TJ, Oh HK, Williams DR, Liao DJ, Zhou J, Green MR, Zheng X, Shen H. PSF contacts exon 7 of SMN2 pre-mRNA to promote exon 7 inclusion. Biochim Biophys Acta. 2014b;1839(6):517–25. https://doi.org/10.1016/j.bbagrm.2014.03.003.
63. Hofmann Y, Wirth B. hnRNP-G promotes exon 7 inclusion of survival motor neuron (SMN) via direct interaction with Htra2-beta1. Hum Mol Genet. 2002;11(17):2037–49. https://doi.org/10.1093/hmg/11.17.2037.
64. Young PJ, DiDonato CJ, Hu D, Kothary R, Androphy EJ, Lorson CL. SRp30c-dependent stimulation of survival motor neuron (SMN) exon 7 inclusion is facilitated by a direct interaction with hTra2 beta 1. Hum Mol Genet. 2002;11(5):577–87. https://doi.org/10.1093/hmg/11.5.577.
65. Mende Y, Jakubik M, Riessland M, Schoenen F, Rossbach K, Kleinridders A, Köhler C, Buch T, Wirth B. (2010). Deficiency of the splicing factor Sfrs10 results in early embryonic lethality in mice and has no impact on full-length SMN/Smn splicing. Hum Mol Genet. 2010;19(11):2154–67. https://doi.org/10.1093/hmg/ddq094.
66. Cartegni L, Hastings ML, Calarco JA, de Stanchina E, Krainer AR. Determinants of exon 7 splicing in the spinal muscular atrophy genes, SMN1 and SMN2. Am J Hum Genet. 2006;78(1):63–77. https://doi.org/10.1086/498853.
67. Chen H-H, Chang J-G, Lu R-M, Peng T-Y, Tarn W-Y. The RNA Binding Protein hnRNP Q Modulates the Utilization of Exon 7 in the Survival Motor Neuron 2 (SMN2) Gene. Mol Cell Biol. 2008a;28(22):6929–38. https://doi.org/10.1128/MCB.01332-08.
68. Irimura S, Kitamura K, Kato N, Saiki K, Takeuchi A, Gunadi, Matsuo M, Nishio H, Lee MJ. HnRNP C1/C2 may regulate exon 7 splicing in the spinal muscular atrophy gene SMN1. Kobe J Med Sci. 2009;54(5):E227–36.
69. Xiao R, Tang P, Yang B, Huang J, Zhou Y, Shao C, Li H, Sun H, Zhang Y, Fu X-D. Nuclear matrix factor hnRNP U/SAF-A exerts a global control of alternative splicing by regulating U2 snRNP maturation. Mol Cell. 2012;45(5):656–68. https://doi.org/10.1016/j.molcel.2012.01.009.

70. Cléry A, Jayne S, Benderska N, Dominguez C, Stamm S, Allain FH-T. Molecular basis of purine-rich RNA recognition by the human SR-like protein Tra2-β1. Nat Struct Mol Biol. 2011;18(4):443–50. https://doi.org/10.1038/nsmb.2001.

71. Li J, Chen X, Xiao P, Li L, Lin W, Huang J, Xu P. Expression pattern and splicing function of mouse ZNF265. Neurochem Res. 2008;33(3):483–9. https://doi.org/10.1007/s11064-007-9461-3.

72. Heinrich B, Zhang Z, Raitskin O, Hiller M, Benderska N, Hartmann AM, Bracco L, Elliott D, Ben-Ari S, Soreq H, Sperling J, Sperling R, Stamm S. Heterogeneous nuclear ribonucleoprotein G regulates splice site selection by binding to CC(A/C)-rich regions in pre-mRNA. J Biol Chem. 2009;284(21):14303–15. https://doi.org/10.1074/jbc.M901026200.

73. Moursy A, Allain FH-T, Cléry A. Characterization of the RNA recognition mode of hnRNP G extends its role in SMN2 splicing regulation. Nucleic Acids Res. 2014;42(10):6659–72. https://doi.org/10.1093/nar/gku244.

74. Hastings ML, Allemand E, Duelli DM, Myers MP, Krainer AR. Control of pre-mRNA splicing by the general splicing factors PUF60 and U2AF(65). PLoS One. 2007;2(6):e538. https://doi.org/10.1371/journal.pone.0000538.

75. Singh NN, Seo JB, Ottesen EW, Shishimorova M, Bhattacharya D, Singh RN. TIA1 prevents skipping of a critical exon associated with spinal muscular atrophy. Mol Cell Biol. 2011;31(5):935–54. https://doi.org/10.1128/mcb.00945-10.

76. Sutherland LC, Thibault P, Durand M, Lapointe E, Knee JM, Beauvais A, Kalatskaya I, Hunt SC, Loiselle JJ, Roy JG, Tessier SJ, Ybazeta G, Stein L, Kothary R, Klinck R, Chabot B. Splicing arrays reveal novel RBM10 targets, including SMN2 pre-mRNA. BMC Mol Biol. 2017;18(1):19. https://doi.org/10.1186/s12867-017-0096-x.

77. Jodelka FM, Ebert AD, Duelli DM, Hastings ML. A feedback loop regulates splicing of the spinal muscular atrophy-modifying gene, SMN2. Hum Mol Genet. 2010;19(24):4906–17. https://doi.org/10.1093/hmg/ddq425.

78. Singh NN, Androphy EJ, Singh RN. In vivo selection reveals combinatorial controls that define a critical exon in the spinal muscular atrophy genes. RNA. 2004b;10(8):1291–305. https://doi.org/10.1261/rna.7580704.

79. Singh RN, Saldanha RJ, D'Souza LM, Lambowitz AM. Binding of a group II intron-encoded reverse transcriptase/maturase to its high affinity intron RNA binding site involves sequence-specific recognition and autoregulates translation. J Mol Biol. 2002;318(2):287–303. https://doi.org/10.1016/S0022-2836(02)00054-2.

80. Hua Y, Vickers TA, Baker BF, Bennett CF, Krainer AR. Enhancement of SMN2 exon 7 inclusion by antisense oligonucleotides targeting the exon. PLoS Biol. 2007;5(4):e73. https://doi.org/10.1371/journal.pbio.0050073.

81. Xiong HY, Alipanahi B, Lee LJ, Bretschneider H, Merico D, Yuen RKC, Hua Y, Gueroussov S, Najafabadi HS, Hughes TR, Morris Q, Barash Y, Krainer AR, Jojic N, Scherer SW, Blencowe BJ, Frey BJ. RNA splicing. The human splicing code reveals new insights into the genetic determinants of disease. Science. 2015;347(6218):1254806. https://doi.org/10.1126/science.1254806.

82. Singh NN, Singh RN, Androphy EJ. Modulating role of RNA structure in alternative splicing of a critical exon in the spinal muscular atrophy genes. Nucleic Acids Res. 2007;35(2):371–89. https://doi.org/10.1093/nar/gkl1050.

83. Singh NN, Singh RN. Alternative splicing in spinal muscular atrophy underscores the role of an intron definition model. RNA Biol. 2011;8(4):600–6. https://doi.org/10.4161/rna.8.4.16224.

84. Singh NN, Androphy EJ, Singh RN. The regulation and regulatory activities of alternative splicing of the SMN gene. Crit Rev Eukaryot Gene Expr. 2004c;14(4):271–85. https://doi.org/10.1615/CritRevEukaryotGeneExpr.v14.i4.30.

85. Singh NK, Singh NN, Androphy EJ, Singh RN. Splicing of a critical exon of human survival motor neuron is regulated by a unique silencer element located in the last intron. Mol Cell Biol. 2006;26(4):1333–46. https://doi.org/10.1128/mcb.26.4.1333-1346.2006.

86. Miyaso H, Okumura M, Kondo S, Higashide S, Miyajima H, Imaizumi K. An intronic splicing enhancer element in survival motor neuron (SMN) pre-mRNA. J Biol Chem. 2003;278(18):15825–31. https://doi.org/10.1074/jbc.M209271200.
87. Miyajima H, Miyaso H, Okumura M, Kurisu J, Imaizumi K. Identification of a cis-acting element for the regulation of SMN exon 7 splicing. J Biol Chem. 2002;277(26):23271–7. https://doi.org/10.1074/jbc.M200851200.
88. Baughan TD, Dickson A, Osman EY, Lorson CL. Delivery of bifunctional RNAs that target an intronic repressor and increase SMN levels in an animal model of spinal muscular atrophy. Hum Mol Genet. 2009;18(9):1600–11. https://doi.org/10.1093/hmg/ddp076.
89. Singh NN, Lee BM, DiDonato CJ, Singh RN. Mechanistic principles of antisense targets for the treatment of spinal muscular atrophy. Future Med Chem. 2015a;7:1793–808. https://doi.org/10.4155/fmc.15.101.
90. Aartsma-Rus A. FDA approval of nusinersen for spinal muscular atrophy makes 2016 the year of splice modulating oligonucleotides. Nucleic Acid Ther. 2017;27(2):67–9. https://doi.org/10.1089/nat.2017.0665.
91. Glascock J, Lenz M, Hobby K, Jarecki J. Cure SMA and our patient community celebrate the first approved drug for SMA. Gene Ther. 2017;24(9):498–500. https://doi.org/10.1038/gt.2017.39.
92. Wan L, Dreyfuss G. Splicing-correcting therapy for SMA. Cell. 2017;170(1):5. https://doi.org/10.1016/j.cell.2017.06.028.
93. Wood MJA, Talbot K, Bowerman M. Spinal muscular atrophy: antisense oligonucleotide therapy opens the door to an integrated therapeutic landscape. Hum Mol Genet. 2017;26(R2):R151–9. https://doi.org/10.1093/hmg/ddx215.
94. Sivanesan S, Howell MD, DiDonato CJ, Singh RN. Antisense oligonucleotide mediated therapy of spinal muscular atrophy. Transl Neurosci. 2013;4:1–7. https://doi.org/10.2478/s13380-013-0109-2.
95. Beusch I, Barraud P, Moursy A, Cléry A, Allain FH. Tandem hnRNP A1 RNA recognition motifs act in concert to repress the splicing of survival motor neuron exon 7. Elife. 2017;6:pii: e25736. https://doi.org/10.7554/eLife.25736.
96. Singh NN, Hollinger K, Bhattacharya D, Singh RN. An antisense microwalk reveals critical role of an intronic position linked to a unique long-distance interaction in pre-mRNA splicing. RNA. 2010;16:1167–81. https://doi.org/10.1261/rna.2154310.
97. Singh NN, Shishimorova M, Cao LC, Gangwani L, Singh RN. A short antisense oligonucleotide masking a unique intronic motif prevents skipping of a critical exon in spinal muscular atrophy. RNA Biol. 2009;6:341–50. https://doi.org/10.4161/rna.6.3.8723.
98. Seo J, Ottesen EW, Singh RN. Antisense methods to modulate pre-mRNA splicing. Methods Mol Biol. 2014;1126:271–83. https://doi.org/10.1007/978-1-62703-980-2_20.
99. Kiel JM, Seo J, Howell MD, Hsu WH, Singh RN, DiDonato CJ. A short antisense oligonucleotide ameliorates symptoms of severe mouse models of spinal muscular atrophy. Mol Ther Nucleic Acids. 2014;3:e174. https://doi.org/10.1038/mtna.2014.23.
100. Förch P, Puig O, Martínez C, Séraphin B, Valcárcel J. The splicing regulator TIA-1 interacts with U1-C to promote U1 snRNP recruitment to 5′ splice sites. EMBO J. 2002;21(24):6882–92.
101. Klar J, Sobol M, Melberg A, Mäbert K, Ameur A, Johansson ACV, Feuk L, Entesarian M, Orlén H, Casar-Borota O, Dahl N. Welander distal myopathy caused by an ancient founder mutation in TIA1 associated with perturbed splicing. Hum Mutat. 2013;34(4):572–7. https://doi.org/10.1002/humu.22282.
102. Hirsch-Reinshagen V, Pottier C, Nicholson AM, Baker M, Hsiung GR, Krieger C, Sengdy P, Boylan KB, Dickson DW, Mesulam M, Weintraub S, Bigio E, Zinman L, Keith J, Rogaeva E, Zivkovic SA, Lacomis D, Taylor JP, Rademakers R, Mackenzie IRA. Clinical and neuropathological features of ALS/FTD with TIA1 mutations. Acta Neuropathol Commun. 2017;5(1):96. https://doi.org/10.1186/s40478-017-0493-x.
103. Heck MV, Azizov M, Stehning T, Walter M, Kedersha N, Auburger G. Dysregulated expression of lipid storage and membrane dynamics factors in Tia1 knockout mouse nervous tissue. Neurogenetics. 2014;15(2):135–44. https://doi.org/10.1007/s10048-014-0397-x.

104. Howell MD, Ottesen EW, Singh NN, Anderson RL, Seo J, Sivanesan S, Whitley EM, Singh RN. TIA1 is a gender-specific disease modifier of a mild mouse model of spinal muscular atrophy. Sci Rep. 2017a;7:18. https://doi.org/10.1038/s41598-017-07468-2.

105. Díaz-Muñoz MD, Kiselev VY, Novère NL, Curk T, Ule J, Turner M. Tia1 dependent regulation of mRNA subcellular location and translation controls p53 expression in B cells. Nat Commun. 2017;8(1):530. https://doi.org/10.1038/s41467-017-00454-2.

106. Vanderweyde T, Apicco DJ, Youmans-Kidder K, Ash PEA, Cook C, Lummertz da Rocha E, Jansen-West K, Frame AA, Citro A, Leszyk JD, Ivanov P, Abisambra JF, Steffen M, Li H, Petrucelli L, Wolozin B. Interaction of tau with the RNA-binding Protein TIA1 regulates tau pathophysiology and toxicity. Cell Rep. 2016;15(7):1455–66. https://doi.org/10.1016/j.celrep.2016.04.045.

107. Howell MD, Ottesen EW, Singh NN, Anderson RL, Singh RN. Gender-specific amelioration of SMA phenotype upon disruption of a deep intronic structure by an oligonucleotide. Mol Ther. 2017b;25(6):1328–41. https://doi.org/10.1016/j.ymthe.2017.03.036.

108. Ronchi D, Previtali SC, Sora MGN, Barera G, Del Menico B, Corti S, Bresolin N, Comi GP. Novel splice-site mutation in SMN1 associated with a very severe SMA-I phenotype. J Mol Neurosci. 2015;56:212–5. https://doi.org/10.1007/s12031-014-0483-4.

109. Wirth B, Herz M, Wetter A, Moskau S, Hahnen E, Rudnik-Schöneborn S, Wienker T, Zerres K. Quantitative analysis of survival motor neuron copies: identification of subtle SMN1 mutations in patients with spinal muscular atrophy, genotype-phenotype correlation, and implications for genetic counseling. Am J Hum Genet. 1999;64(5):1340–56. https://doi.org/10.1086/302369.

110. Sheng-Yuan Z, Xiong F, Chen YJ, Yan TZ, Zeng J, Li L, Zhang YN, Chen WQ, Bao XH, Zhang C, Xu XM. Molecular characterization of SMN copy number derived from carrier screening and from core families with SMA in a Chinese population. Eur J Hum Genet. 2010;18(9):978–84. https://doi.org/10.1038/ejhg.2010.54.

111. Vezain M, Gérard B, Drunat S, Funalot B, Fehrenbach S, N'Guyen-Viet V, Vallat JM, Frébourg T, Tosi M, Martins A, Saugier-Veber P. A leaky splicing mutation affecting SMN1 exon 7 inclusion explains an unexpected mild case of spinal muscular atrophy. Hum Mutat. 2011;32(9):989–94. https://doi.org/10.1002/humu.21528.

112. Osman EY, Washington CW 3rd, Kaifer KA, Mazzasette C, Patitucci TN, Florea KM, Simon ME, Ko CP, Ebert AD, Lorson CL. Optimization of morpholino antisense oligonucleotides targeting the intronic repressor element1 in spinal muscular atrophy. Mol Ther. 2016;24(9):1592–601. https://doi.org/10.1038/mt.2016.145.

113. Zarnack K, Konig J, Tajnik M, Martincorena I, Eustermann S, Stevant I, Reyes A, Anders S, Luscombe NM, Ule J. Direct Competition between hnRNP C and U2AF65 Protects the Transcriptome from the Exonization of Alu Elements. Cell. 2013;152(3):453–66. https://doi.org/10.1016/j.cell.2012.12.023.

114. Deininger P. Alu elements: know the SINEs. Genome Biol. 2011;12(12):12. https://doi.org/10.1186/gb-2011-12-12-236.

115. Bouttier M, Laperriere D, Memari B, Mangiapane J, Fiore A, Mitchell E, Verway M, Behr MA, Sladek R, Barreiro LB, Mader S, White JH. Alu repeats as transcriptional regulatory platforms in macrophage responses to M-tuberculosis infection. Nucleic Acids Res. 2016;44(22):10571–87. https://doi.org/10.1093/nar/gkw782.

116. Daniel C, Silberberg G, Behm M, Ohman M. Alu elements shape the primate transcriptome by cis-regulation of RNA editing. Genome Biol. 2014;15(2):17. https://doi.org/10.1186/gb-2014-15-2-r28.

117. Singh NN, Seo J, Singh RN. A multi-exon-skipping detection assay reveals surprising diversity of splice isoforms of spinal muscular atrophy genes. Plos One. 2012;7(11):17. https://doi.org/10.1371/journal.pone.0049595.

118. Sorek R, Ast G, Graur D. Alu-containing exons are alternatively spliced. Genome Res. 2002;12(7):1060–7. https://doi.org/10.1101/gr.229302.

119. Lorson CL, Strasswimmer J, Yao JM, Baleja JD, Hahnen E, Wirth B, Le T, Burghes AH. AndrophyEJ (1998). SMN oligomerization defect correlates with spinal muscular atrophy severity. Nat Genet. 1998;19(1):63–6. https://doi.org/10.1038/ng0598-63.
120. Seo J, Singh NN, Ottesen EW, Sivanesan S, Shishimorova M, Singh RN. Oxidative stress triggers body-wide skipping of multiple exons of the spinal muscular atrophy gene. PLoS One. 2016b;11(4):31. https://doi.org/10.1371/journal.pone.0154390.
121. Oh JM, Di C, Venters CC, Guo J, Arai C, So BR, Pinto AM, Zhang Z, Wan L, Younis I, Dreyfuss G. U1 snRNP telescripting regulates a size-function-stratified human genome. Nat Struct Mol Biol. 2017;24:993. https://doi.org/10.1038/nsmb.3473.
122. Aktaş T, Ilik IA, Maticzka D, Bhardwaj V, Rodrigues CP, Mittler G, Manke T, Backofen R, Akhtar A. DHX9 suppresses RNA processing defects originating from the Alu invasion of the human genome. Nature. 2017;544(7648):115–9. https://doi.org/10.1038/nature21715.
123. Ottesen EW, Seo J, Singh NN, Singh RN. A multilayered control of the human Survival Motor Neuron gene expression by Alu elements. Front Microbiol. 2017;8:2252. https://doi.org/10.3389/fmicb.2017.02252.
124. Acuña LIG, Kornblihtt AR. Long range chromatin organization: a new layer in splicing regulation? Transcription. 2014;5(3):e28726. https://doi.org/10.4161/trns.28726.
125. Singh NN, Howell MD, Singh RN. Transcriptional and splicing regulation of spinal muscular atrophy genes. In: Charlotte SJ, Paushkin S, Ko C-P, editors. Spinal muscular atrophy: disease mechanisms and therapy. Amsterdam: Elsevier Inc.; 2016.
126. Martinez-Contreras R, Fisette JF, Nasim FU, Madden R, Cordeau M, Chabot B. Intronic binding sites for hnRNP A/B and hnRNP F/H proteins stimulate pre-mRNA splicing. PLoS Biol. 2006;4(2):e21. https://doi.org/10.1371/journal.pbio.0040021.
127. Spellman R, Smith CW. Novel modes of splicing repression by PTB. Trends Biochem Sci. 2006;31(2):73–6. https://doi.org/10.1016/j.tibs.2005.12.003.
128. Taube JR, Sperle K, Banser L, Seeman P, Cavan BC, Garbern JY, Hobson GM. PMD patient mutations reveal a long-distance intronic interaction that regulates PLP1/DM20 alternative splicing. Hum Mol Genet. 2014;23(20):5464–78. https://doi.org/10.1093/hmg/ddu271.
129. Liang DM, Wilusz JE. Short intronic repeat sequences facilitate circular RNA production. Genes Dev. 2014;28(20):2233–47. https://doi.org/10.1101/gad.251926.114.

Chapter 3
RNA Editing Deficiency in Neurodegeneration

Ileana Lorenzini, Stephen Moore, and Rita Sattler

Abstract The molecular process of RNA editing allows changes in RNA transcripts that increase genomic diversity. These highly conserved RNA editing events are catalyzed by a group of enzymes known as adenosine deaminases acting on double-stranded RNA (ADARs). ADARs are necessary for normal development, they bind to over thousands of genes, impact millions of editing sites, and target critical components of the central nervous system (CNS) such as glutamate receptors, serotonin receptors, and potassium channels. Dysfunctional ADARs are known to cause alterations in CNS protein products and therefore play a role in chronic or acute neurodegenerative and psychiatric diseases as well as CNS cancer. Here, we review how RNA editing deficiency impacts CNS function and summarize its role during disease pathogenesis.

Keywords RNA editing · MARCH · AMPA · GluA2 · 5HT receptors · K channels · Excitotoxicity · Neurodegeneration · Psychiatric diseases · Cancer

Ileana Lorenzini and Stephen Moore contributed equally to this work.

I. Lorenzini
Barrow Neurological Institute, Department of Neurobiology, Dignity Health, St. Joseph's Hospital and Medical Center, Phoenix, AZ, USA

S. Moore
Barrow Neurological Institute, Department of Neurobiology, Dignity Health, St. Joseph's Hospital and Medical Center, Phoenix, AZ, USA

Interdisciplinary Graduate Program in Neuroscience, Arizona State University, Tempe, AZ, USA

R. Sattler (✉)
Department of Neurobiology and Neurology, Dignityhealth St. Joseph's Hospital, Barrow Neurological Institute, Phoenix, AZ, USA
e-mail: rita.sattler@dignityhealth.org

© Springer International Publishing AG, part of Springer Nature 2018
R. Sattler, C. J. Donnelly (eds.), *RNA Metabolism in Neurodegenerative Diseases*, Advances in Neurobiology 20,
https://doi.org/10.1007/978-3-319-89689-2_3

3.1   Introduction: RNA Editing Overview

RNA editing is a molecular process that allows changes in the sequence of specific RNA transcripts to increase the diversity of different RNAs that can be generated from the genome. This can result in translation of different protein variants, but can also alter alternative splicing events and micro RNA-binding efficiencies [1]. RNA editing occurs during or after transcription through two distinct mechanisms: (1) chemically modifying a nucleotide, and therefore, altering the nucleotide sequence; (2) inserting or deleting nucleotides and changing the length of the mRNA. This chapter will focus on the most common form of RNA editing, the Adenosine to Inosine (A/I) nucleotide modification of RNA catalyzed by a family of enzymes known as adenosine deaminase acting on double-stranded RNA (ADARs). We will summarize how the dysfunction of these RNA editing enzymes and the subsequent substrate alterations contributes to central nervous system (CNS) diseases [2–5].

While other RNA editing events such as C-to-U or G-to-A exist, the catalytic deamination of Adenosine into Inosine is the most prevalent [6, 7]. There are three ADAR gene family members in mammals: ADAR1, ADAR2, and ADAR3 (These enzymes have also been referred to as ADAR, ADARB1, and ADARB2 respectively). All ADAR proteins contain double-stranded RNA-binding domains (dsRBDs), a nuclear localization sequence (NLS), and a C terminal deaminase domain [8] (see Fig. 3.1). There are two ADAR1 isoforms, ADAR1 p150 containing two additional Z DNA-binding domains and an NES and ADAR1 p110 a truncation isoform maintaining one Z DNA-binding domain and no NES [9]. ADAR1 is widely expressed throughout the body and to a lesser extent in the CNS. It has been shown to bind to over 10,000 genes and is necessary for normal development [10, 11].

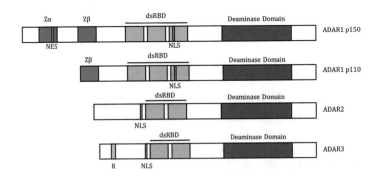

Fig. 3.1 ADAR domain structures. ADAR family members do share certain domain structures, including a C-terminal Deaminase Domain, dsRNA-binding domains (dsRBD), and a nuclear localization domain (NLS). ADAR1 comes in two isoforms, ADAR1 p150 and p110. ADAR1 p150 has two Z-DNA-binding domains, Zα and Zβ, in addition to a nuclear export sequence, which explains why ADAR1 p150 can be found both in the nucleus and in the cytoplasm. ADAR1 p110 only has a Zβ domain, and is only expressed in the nucleus. ADAR3 differs from ADAR1 and ADAR2 by the existence of a N-terminal RG-rich region. ADAR1 and ADAR2 are ubiquitously expressed throughout the body, while ADAR3 is CNS specific. The chromosomal locations of the *ADAR1–3* genes are 1q21.3, 21q22.3, and 10p15.3, respectively

ADAR1$^{-/-}$ mice are embryonic lethal and die around day E11.5 [10]. The mouse embryos undergo widespread apoptosis and show severe liver disintegration due to the loss of ADAR1 [10, 11]. ADAR2 is highly expressed in the CNS, and to a lesser extent in peripheral tissues [12]. ADAR2 has been shown to be responsible for the A/I editing of transcripts that are most actively edited. Knockout mouse models have shown that ADAR2 is required for normal development and ADAR2$^{-/-}$ mice die by P20 and become progressively seizure prone [13]. The third and final member of the family, ADAR3, is thought to have no RNA editing activity [14]. ADAR3 contains an additional arginine-rich domain [14]. Unlike its family members, ADAR3 is expressed exclusively in the brain [15]. Because no ADAR3 editing activity has been reported, the function of the enzyme is still an area of debate. There is a growing amount of evidence to suggest that ADAR3 acts as a negative regulator of overall RNA editing by binding and sequestering editing substrates of ADAR1 and ADAR2 [15, 16].

How does RNA editing work? The hydrolytic deamination of adenosine by the catalytic activity of ADAR1 and 2 disrupts the canonical Watson and crick base pairing of adenosine and as a result the edited inosine will be interpreted by the translational machinery as a guanosine (see Fig. 3.2). Therefore, RNA A/I editing events that fall within protein-coding regions can potentially alter the codon and allow the translational machinery to introduce amino acid changes into the protein that were not encoded by the genome. This can allow for important variation of protein products produced by a single strand of RNA (e.g. serotonin receptor [17]). Editing also occurs in noncoding regions of the transcriptome where the location of the edited nucleotide can regulate splicing, retain edited mRNA in the nucleus, or prevent micro-RNA processing [7, 18–25]. Historically, the estimation of total RNA editing sites was difficult and RNA editing was studied utilizing the serendipitous discovery of A/I sites [26]. With the ever-increasing capabilities of sequencing technologies, it is now possible to analyze RNA editing sites with far greater detail [19, 26, 27]. There are conflicting reports on the total number of RNA editing events in the human genome with reports claiming over one hundred million editing sites spanning the majority of the transcriptome [7, 18, 19, 26, 27]. The majority of these RNA editing events are found within *Alu* repetitive elements. These genomic elements are approximately 300 bp in lengths and are primate-specific transposable elements that comprise approximately 10% of the human genome [28]. These repetitive elements form long dsRNA secondary structures that make them ideal targets for ADARs. ADARs edited sites and levels of RNA editing, as well as ADAR proteins themselves are thought to be evolutionary conserved and play a role in environmental adaptation [29].

A/I RNA editing has been recognized as a significant event during CNS cortical development [30]. An increasing RNA editing pattern is observed during deep cortical layer formation suggesting these events occur at a critical period in neuronal maturation. As previously stated knockout mouse models of both ADAR1 and ADAR2 have shown that mice deficient in these deaminases form severe developmental phenotypes, emphasizing the importance of A/I RNA editing during CNS

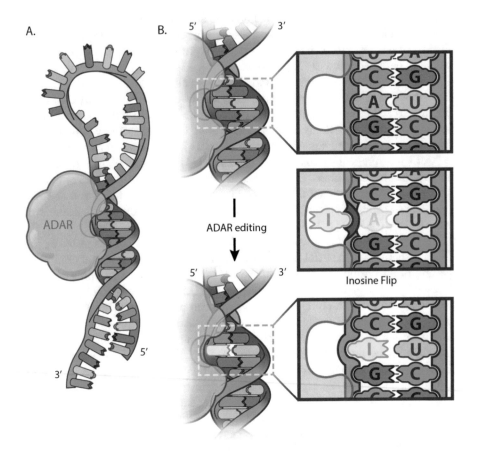

Fig. 3.2 ADAR A/I RNA editing. (**a**) ADAR enzyme (light green structure) acting on double-stranded RNA. (**b**) ADAR dsRNA-binding domains act on dsRNA editing sites and its catalytic domain converts adenine to inosine. Within the catalytic domain an amino group on the adenine base is replaced by an oxygen and converted to inosine

development [13]. At mature states, neurons show higher ADAR expression and editing activity than non-neuronal cells suggesting a limited involvement of other brain cells in RNA editing [4, 30]. Editing events may occur in response to environmental factors or to maintain normal CNS physiology. It can alter the function of target genes such as α-Amino-3-hydroxy-5-methyl-4-isoxazolepropionic acid (AMPA) receptors for fast excitatory neurotransmission, serotonin-5HT$_{2C}$ receptors, or potassium channels K$_v$1.1 for modulation of neuronal excitability [31, 32]. Due to the regulation of these ion channels by ADARs, RNA A/I editing is considered crucial for proper neuronal function.

3.1.1 Major CNS RNA Editing Targets

To illustrate the importance of RNA editing in the CNS, we decided to introduce briefly three major RNA editing targets, which have shown to play a role in disease pathogenesis of several of the CNS disorders discussed below.

3.1.1.1 AMPA Receptors

AMPA receptors are ionotropic glutamate receptors responsible for fast synaptic transmission in the CNS [33]. The functional properties of AMPA receptors are greatly dependent on its subunit composition, GluA1–4, determining its role in synapse formation, stabilization, and synaptic plasticity [34]. The GluA2 subunit has the ability to regulate the calcium (Ca^{2+})-permeability of AMPA receptors [35–37]. Most AMPA receptors become permeable to Ca^{2+} by lacking the GluA2 subunit, and these GluA2-lacking receptors are thought to contribute to normal brain function, especially synaptic plasticity [33, 37–41]. However, there are numerous reports suggesting that GluA2-containing AMPA receptors become Ca^{2+}-permeable due to a lack of editing of the GluA2 Q/R site, although in the brain, almost 100% of GluA2 mRNA is present in its edited form [42–46] (see Fig. 3.3). This unique element of GluA2 is regulated by ADAR2-mediated A/I RNA editing [31]. Mice lacking ADAR2 can be rescued by expression of a forced edited GluA2 subunit [13]. This provides evidence that this single editing event is essential for normal development and survival. It further supports the idea that unedited Ca^{2+}-permeable GluA2-containing AMPA receptors do not have a physiological role similar to GluA2-lacking AMPA receptors. In this chapter, we will discuss the role of AMPA receptor GluA2 Q/R editing in the context of the role of glutamate excitotoxicity in neurodegenerative diseases, especially Amyotrophic lateral sclerosis (ALS; see below).

3.1.1.2 Serotonin Receptors

Serotonin 5-hydroxytryptamine (5-HT) receptors are a family of chemical messengers that produce a wide variety of physiological responses including circadian rhythms, mood, memory, cognition, and possibly peristalsis in the gastrointestinal tract [47–49]. There are 15 unique receptors divided into seven subgroups (5-HT1–7), all subgroups are classified as G-protein coupled receptors with the exception being the 5-HT3 receptors that are ionotropic [50–52]. The 5-HT2C receptor subtype is expressed throughout the CNS [53, 54]. There are five ADAR-meditated RNA editing sites on the 5-HT2C mRNA, designated sites A through E [17]. These five editing sites are located within 13 base pairs and are responsible for three codons allowing for significant variation in the protein isoforms [17, 55]. With only 7% of 5-HT2C mRNA lack editing at any of the five sites, the majority of transcripts are exposed to ADAR-mediated A/I editing, the most prevalent showing editing at

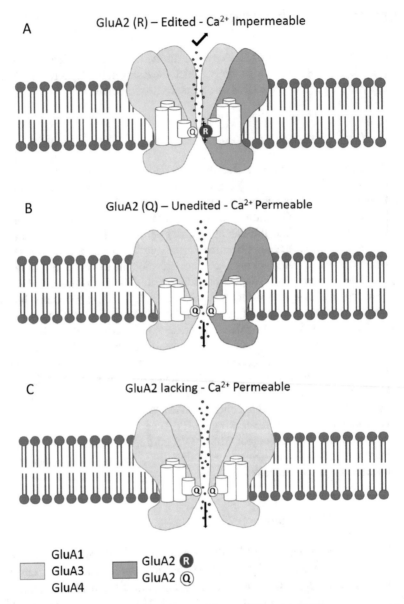

Fig. 3.3 Role of GluA2 in AMPA receiver Ca²⁺ permeability. (**a**) AMPA receptors containing fully edited GluA2 (R) are impermeable to calcium due to the positively charged arginine in the channel pore. (**b**) When GluA2 (Q) is unedited, this positive charge is removed with the presence of the glutamine, and AMPA receptors become permeable to calcium. (**c**) Calcium permeability is also present when AMPA receptors lack GluA2 (Q) altogether and are composed of other AMPA receptor subunits instead

the ABC and D sites [17]. Editing of this receptor alters binding affinity and functional potency of receptor agonists, and thereby affection receptor function during synaptic signaling. The fully edited 5-HT2C receptor isoforms have been shown to have a 40-fold decrease in serotonergic potency, decreasing inositol phosphate accumulation and calcium release [56–58]. The role of serotonin receptor editing is mostly relevant for neuropsychiatric disorders, such as schizophrenia and depression (see below).

3.1.1.3   Voltage Gated Potassium Channels

Voltage gated potassium channels (Kv channels) are the largest subgroup of potassium channels [59, 60]. Comprised of 12 subgroups (Kv1–12) these six transmembrane domain subunits form tetrameric Kv channels containing an inner pore and external voltage sensor domains allowing for the conversion of voltage across the membrane to be transferred into mechanical work [60]. The Kv1 family (Kv 1.1,2, and 4) has been shown to localize to soma, axons, synaptic terminals, and proximal dendrites [59, 61]. The Kv1.1 channel plays an important role in the regulation of neuronal excitability [62]. An ADAR2-mediated A/I editing site lies within the ion pore of the Kv1.1 subunit, mediating an isoleucine to valine substitution [63]. No differences were observed between the voltage-dependent activation of edited and unedited Kv1.1 channels [63]. In contrast A/I editing at this site has been proposed to target the process of fast inactivation [63]. Fast inactivation of Kv1 channels is mediated by the inactivating ball domain on the Kvβ1 subunit [64]. Regulation of this mechanism by RNA editing will have profound effects on regulation of neuronal excitability.

3.2   RNA Editing Deficits in Neurodegeneration

As summarized above, the post-transcriptional modification of RNA transcripts by ADARs through RNA editing generates protein diversity regulating many critical aspects of CNS function. Therefore, if the RNA editing process fails it could lead to CNS diseases, or exacerbate acute injury and chronic disorders. In the following sections, we will discuss RNA editing deficits for chronic and acute neurodegenerative disorders, neuropsychiatric diseases, and brain cancers.

3.2.1   RNA Editing in Chronic Neurodegenerative Diseases

Alzheimer's disease (AD) accounts for 60–80% of dementia cases [65]. It is the most prevalent form of dementia characterized by a progressive loss of memory and cognitive dysfunction. The neuropathological hallmarks comprise of plaques and tangles known to play a critical role in neurodegeneration [65, 66]. Areas of the hippocampus, pre-frontal, and temporal cortex play a significant role in AD patho-physiology [67]. Studies done by Akbarian et al. associated deficits in RNA editing of the α-Amino-3-hydroxy-5-methyl-4-isoxazolepropionic acid (AMPA) receptor subunit GluA2 in the pre-frontal cortex of AD patients with changes in intracellular Ca^{2+} which could lead to neuronal dysfunction and neurodegeneration due to excessive Ca^{2+} permeability [68]. The authors showed that the pre-frontal cortex of Alzheimer's patients has approximately 1.0% of all GluA2 RNA molecules unedited. In healthy states the pre-frontal cortex shows less than 0.1% of all GluA2 RNA molecules are unedited and more than 99.9% are edited. Other studies found lower RNA editing levels at the GluA2 Q/R site in the hippocampus and caudate of sporadic AD patients and Apo E4 carriers, independent of clinical diagnosis. Interestingly, ADAR levels were decreased only in the caudate region of the patient's brains [69]. The E4 allele of the apolipoprotein ApoE gene has been recognized as a major genetic risk factor for AD and it has been suggested that ApoE plays a role in hippocampus AMPA receptor dynamics and glutamate regulation [70–72]. Interestingly, studies performed in the triple-transgenic AD mouse model (3×Tg-AD, PS1(M146 V); APP(Swe); tau (P301L)), a widely used transgenic mouse model for AD which exhibits both plaques and tau pathology, showed decreased levels of all AMPA receptor subunits, except for GluA2, while no editing deficiencies were detected [73]. A study aimed at analyzing the hippocampal transcriptome of normal aged mice using RNA sequencing, also examined age-related RNA editing changes as a mechanism to generate alternative transcripts [74]. In 29 months old mice, 41 out of 682 editing sites were significantly changed, which corresponded to 35 genes. One of the genes exhibiting increased editing was the serotonin receptor 2c, which has previously been found showing altered RNA editing in a mouse model of impaired memory function [75]. A comprehensive study on RNA editing in postmortem AD patient tissue revealed significant loss of RNA editing in the hippocampus, and to a lesser extent in the temporal and frontal lobes [76]. Most of the editing changes showed hypo-editing, including the serotonin receptor 2c, which in contrast to what was found in the aging mice discussed above showed less RNA editing in the hippocampus, temporal and frontal lobes. Surprisingly, the authors were unable to find a true correlation between the editing deficits and the expression levels of neither ADAR1 nor ADAR2, suggesting that ADAR dysfunction could be caused by mechanisms other than decreased transcription.

Amyotrophic lateral sclerosis (ALS) is a fatal neurodegenerative disorder where the progressive death of both the upper and lower motor neurons leads to atrophy of skeletal muscles and ultimately death due to respiratory failure [77]. The known genetic contribution to ALS is relatively little, only 10% of the ALS cases are believed to be familial. The remaining 90% of ALS is designated as sporadic ALS in which there is no familial history of the disease [78]. While the etiology remains largely unknown there have been great strides in understanding the pathology of the disease attributed to the advances in genomic sequencing capabilities [79].

Early studies identified the dysregulation of astrocytic glutamate transporters in ALS as the leading cause for increased levels of glutamate at the synapse [80]. Pyramidal tract projection into the spinal cord uses glutamate as the excitatory neurotransmitter and motor neurons expressing abundant glutamate receptors are most vulnerable to exaggerated glutamate stimulation, supporting excitotoxicity as a major mechanism for motor neuron loss in ALS [81]. A likely contributor to the mechanism behind neuronal excitotoxicity is the dysfunction of the AMPA receptor leading to exaggerated calcium influx and slow neuronal death [82, 83]. As mentioned previously, elevated calcium influx through the AMPA receptors can occur through the absence of GluA2 from the receptor complex or through RNA editing of the GluA2 Q/R site. Initial studies addressing the role of AMPA receptors in motor neuron cell death supported both of these mechanisms [84–88]. Over the years, Kwak and colleagues provided accumulating evidence that spinal motor neurons from sporadic ALS patients showed reduced GluA2 Q/R editing efficiencies, leading to increased Ca^{2+} permeability of AMPA receptors and subsequent excitotoxic motor neuron cell death [84, 89, 90]. The group further showed that these editing deficits are accompanied by a downregulation of ADAR2 [91], and transgenic mice with specific motor neuron knockdown of ADAR2 exhibited inefficient GluA2 Q/R editing and decreased motor function, which was rescued when the mice were crossed with transgenic mice overexpressing a fully edited version of GluA2 [92]. Interestingly, the oculomotor neurons, which are generally not affected in ALS patients, of these mice were not degenerated despite a loss of ADAR2 and a decrease in GluA2 Q/R editing. Also, the motor neurons of the ADAR2 conditional knockout mice exhibited classical TDP-43 pathology, and similar co-pathologies were found in sporadic ALS patient spinal cord motor neurons [93]. The authors propose that Ca^{2+} influx via unedited GluA2 containing AMPA receptors leads to activation of calpain, which in turn triggers TDP-43 pathology and nucleocytoplasmic transport deficits, in addition to excitotoxicity [94, 95].

The loss of GluA2 Q/R editing efficiency has not been demonstrated in other subgroups of ALS, while decreased ADAR2 levels were reported in spinal motor neurons of a single patient carrying a FUS mutation [96]. A recent transcriptome study using deep RNA sequencing technology reported that while spinal cord tissue shows decreased GluA2 Q/R editing efficiencies compared to other brain regions, there was no detectable difference of GluA2 Q/R editing deficits between control spinal cord patient tissue and sporadic ALS patient tissue [97]. One explanation for this discrepancy could be the use of spinal cord tissue lysate versus laser-captured motor neuron analysis, or, the use of RNA sequencing versus a restriction digest-based RNA editing technique. Future studies are required to address these conflicting results. Finally, Donnelly et al. described sequestration of ADAR3 to C9orf72 repeat RNAs in postmortem C9orf72 ALS patient tissue and patient-derived human-induced pluripotent stem cells differentiated into motor neurons (hiPSC-MNs) [98]. Additionally, the hiPSC-MNs showed increased susceptibility to glutamate toxicity, which was mimicked by siRNA knockdown of ADAR3. Ongoing studies in our laboratory are aimed at understanding how ADAR3 dysfunction could regulate ADAR2 function and subsequent excitotoxicity in C9orf72 ALS/Frontotemporal Dementia (FTD), and whether ADAR2 function itself is altered in C9orf72 ALS/FTD patients.

Huntington's disease (HD) and Parkinson's disease (PD) have not been investigated much in regards to RNA editing deficits. HD is an autosomal dominant mutation caused by an abnormal trinucleotide CAG repeat expansion in the huntingtin gene (HTT). Carriers of this mutation produce an unusual polyglutamine sequence that causes disease by a toxic gain of function of the protein huntingtin. Even though HD impacts the entire brain, the most affected regions are the basal ganglia and striatum composed of the caudate nucleus and putamen. To a lesser extent areas of the cerebellum, substantia nigra, hippocampus, and layer III, V and IV of the cerebral cortex are affected [99]. Very early research in HD, often referred to as Huntington's chorea, suggested that aberrant glutamate homeostasis might be involved in HD disease pathogenesis [100]. As an example, researchers used intrastriatial injections of glutamate or kainic acid to mimic biochemical changes observed in HD [101, 102]. With the cloning and discovery of glutamate receptors, the role of glutamate and excitotoxicity becomes a major disease mechanism for HD [103] and the first study examining RNA editing of glutamate receptors subunits GluA2, 5 and 6 noted no difference in the RNA editing efficiency between healthy control and HD patient brain tissue samples [104]. A later study provided the first evidence to support little, yet significant changes in GluA2 editing in the striatum on HD patient tissue [68]. Nearly 5% of GluA2 Q/R was unedited, which still leaves a large percentage of edited GluA2, but could nevertheless contribute to increased Ca2+ permeability and neuronal death. Interestingly, a more recent study decreased immunostaining for GluA2 in the striatum of HD patient tissue when compared to control tissue, suggesting that an overall lack of GluA2 might further contribute to glutamate excitotoxicity in HD [105].

PD is the second most common neurodegenerative disorder affecting nearly 1% of the population [106]. Patients with PD exhibit crippling motor deficits or bradykinesia (or slowness of movement), rigidity, resting tremor, and postural instability also known as the four cardinal manifestations of PD [107]. These symptoms arise due to the degeneration of dopaminergic neurons in the substantia nigra [106, 108]. Similar to HD, among the many proposed cellular dysfunctions [108] excitotoxicity has been suggested to play a role in the degeneration of the dopaminergic neurons [109]. However, despite the proposed role of excitotoxicity there has been little evidence that suggests any known RNA editing deficits [110]. With the increase in RNA sequencing capabilities the ability to study RNA editing events by whole transcriptome sequencing is allowing for more complex analysis of A/I editing sites in disease. One whole transcriptome study associated Parkinson's disease with changes in Alu insertions the largest target of the ADAR family of proteins [111]. Due to ADARs RNA editing of micro RNAs and Long noncoding RNA and alterations in these RNAs in PD, RNA editing is hypothesized to play a role in disease pathogenesis [110], but only future studies will prove whether this hypothesis is correct. An intriguing new concept has just been proposed in regards to utilizing endogenous ADAR2 editing activity to repair a PD disease causing mutation in PINK1 [112]. A G-to-A mutation in PINK1 introduces a premature stop codon and shortens the protein's C-terminus including its kinase domain. The authors designed guideRNAs to enable endogenous ADAR2 to edit and recode the user-defined mRNA target.

This was successfully achieved in mammalian cell lines and showed a functional rescue of PINK1/Parkin-mediated mitophagy [112].

3.2.2 RNA Editing in Acute Neurodegeneration

Epilepsy is a neurological disorder characterized by abnormal neuronal hyperexcitability of a subpopulation of cells resulting in unprovoked recurrent seizures [113, 114]. The mechanisms responsible for this neuronal hyperexcitability are multifaceted and include genetic predispositions, acute brain injuries, as well as epigenetic changes alterations. Overstimulated cells have a prolonged increase in intracellular Ca^{2+} concentrations, which has been suggested to contribute to the mechanisms of hyperexcitability seen in epilepsy. AMPA receptors are involved in fast excitatory neurotransmission and are therefore thought to play a key role in the generation of seizures. Various studies present evidence that connects deficits in AMPA receptor editing with seizure vulnerability. Transgenic mice expressing a fully unedited GluA2 Q/R site die around 3 weeks of age and develop severe seizures [115]. Interestingly, GluA2 knockout mice, while similarly showing premature death, do not show signs of seizures, but instead show increased susceptibility to absence seizures [116]. ADAR2 knockout mice behave very similar to the GluA2 Q/R unedited mice and develop seizures before prematurely dying at 21 days of age [13]. These mice are rescued by crossing the ADAR2 KO mice with transgenic mice overexpressing a fully edited GluA2 Q/R site [13]. RNA editing analyses of epileptic brain tissue resulted in contradictory results, with studies showing no altered RNA editing at the GluA2 Q/R site (while there were RNA editing changes in GluA5 and GluA6) [117]. Only one study examined ADAR2 expression from needle biopsy samples obtained from hypothalamic hamartoma tissue and found loss of nuclear immunostaining of ADAR2 concomitant with lower RNA editing efficiency at the GluA2 Q/R site [118]. A recent genome-wide analysis of epileptic and healthy mouse hippocampus revealed a correlation between seizure frequency and differential RNA editing [119]. Functional enrichment analysis revealed that pathways relevant for epilepsy showed the highest degree of differential RNA editing, e.g., neuron projection, synapse, seizures. More work needs to be done to fully understand whether RNA editing plays a significant role in this disorder.

Stroke patients suffer from a spontaneously disrupted blood supply to the brain resulting in a loss of oxygen and nutrients to affected regions. Accounting for 85% of all strokes an ischemic stroke occurs when blood flow to part of the brain is obstructed. After an ischemic attack and loss of blood supply, cells are immediately unable to sustain normal homeostasis leading to massive irreversible cell death [120]. Because of the rapid neuronal loss in stroke victims immediate and effective treatment is crucial to minimize damage [121, 122]. Post-ischemic excitotoxicity results from consumption of ATP, failure of ATP synthesis, and dysregulation of the ionic concentration across the plasma membrane leading to rapid rise in intracellular calcium concentrations and death of the cell [123]. Historically, the increase in calcium permeability of neurons affected by ischemia was thought to be due the downregulation of GluA2 following ischemia commonly referred to as "The GluA2 hypothesis" [36]. However, in 2006 unedited GluA2 was found in the CA1 pyramidal neurons of rats following ischemia [124]. The calcium permeability of AMPA receptors in the CA1 pyramidal neurons is 18-fold higher following ischemia when compared to control groups [125]. In addition, loss of ADAR2 expression increases neuronal sensitivity to ischemia and can be rescued by expression of a fully edited GluA2(R) [124, 125]. These studies suggest that loss of RNA editing contributes to the disruption in neuronal homeostasis following ischemic stroke and immediate prevention of these deficits may protect against neuronal damage.

Spinal Cord Injury (SCI) is defined as damage to the spinal cord causing reduced or complete loss of motor function [126, 127]. It generally affects glutamatergic tracts descending from varying brain regions and serotonergic tracts descending from the brainstem. Serotonin signaling is critical in the spinal cord by providing neuromodulation to motor neuron and recent studies showed reduced A → I RNA editing of the $5HT_{2c}R$ serotonin receptor after SCI, which was suggested to contribute to loss of motor neuron function [126–129]. These studies demonstrated that RNA editing deficiency for $5HT_{2c}R$ was due to a decrease in the ADAR2 expression suggested to be caused by a continuous inflammatory response during injury. In addition to $5HT_{2c}R$, the authors also found reduced RNA editing of potassium channel Kv1.1, an additional ADAR2 target. Additional studies strongly support the fact that microglial cells and immune infiltrating cells are involved in the dysfunction of A → I RNA editing in SCI [4, 128, 129], suggesting that at least during spinal cord injury, RNA deficits of neuronal targets are triggered by non-cell autonomous mechanisms. Future studies are needed to test the hypothesis that these non-cell autonomous mechanisms also occur in other neurodegenerative diseases characterized by RNA editing deficits.

3.3 A → I RNA Editing Dysfunction in Psychiatric Diseases

Depression and Schizophrenia. Depression is a long term mood disorder that affects a person's thoughts and feelings as well as daily activities such as working, eating and sleeping [130]. This disorder is caused by a combination of genetic, biological and environmental factors. Serotonin or 5-hydroxytryptamine (5HT), a monoamine neurotransmitter has been implicated in this psychiatric disease [131, 132]. Patients suffering from depression have lower levels of serotonin or an increase in the number of serotonin receptors. Selective serotonin reuptake inhibitors are frequently used to treat depression to maintain serotonin for longer periods at the synapse. Schizophrenia is classified as a chronic mental disorder where the patients lose contact with reality and present psychotic behaviors (positive symptoms), disruption of normal behaviors (negative symptoms), poor executive function and poor working memory (cognitive symptoms) [133]. Similar to depression, schizophrenia is caused by genetic aberrations and environmental factors.

5HT-serotonergic receptors are relevant to mental disorders such as depression, anxiety, and schizophrenia. The $5HT_{2c}R$, a G-protein couple receptor, is known to undergo RNA editing post-transcriptional modification [32, 134–136]. Altered editing of $5HT_{2c}R$ pre-mRNA occurs in the pre-frontal cortex of depressive and schizophrenic patients. A/I RNA editing of the $5HT_{2c}R$ occurs at five sites (A-to-E) causing protein and functional diversity. Previous studies have shown that depressive and schizophrenic patients have reduced expression of ADAR2 with a decrease or increase in RNA editing in some of the five $5HT_{2c}R$ sites [137, 138] making it difficult to elucidate how RNA editing is associated with these psychiatric disorders. These studies suggest that RNA editing is not only disease-specific, but it may also be determined by the severity of the psychiatric diseases.

Cocaine addiction. An estimated 18.3 million people between the ages of 16–64 used cocaine in 2014 making it one of the most common illicit drugs in the world (National Institute on Drug Abuse 2016; [139]). Numerous health risks are associated with cocaine use such as cognitive impairment, respiratory disease, cardiovascular disease, congenital malformations, and premature mortality [140]. Approximately 20% of recreational users will develop a dependence for cocaine within 5 years [141]. Drug-seeking behavior is thought to be influenced by limbic cortical-ventral striatal circuitry which afferents to the basolateral amygdala and nucleus accumbens providing the circuitry for stimulus-reward pathway that reinforces drug seeking behavior [142]. Increased calcium permeable AMPA receptors in the nucleus accumbens have been associated with drug-seeking behavior [143]. These alterations may be due to increased GluA1 in the nucleus accumbens [144]. However, downregulation of ADAR2 and GluA2 Q/R editing deficits have been identified in the nucleus accumbens shell in rats following cocaine self-administration [145]. Both upregulation of GluA1 and misediting of the GluA2 Q/R site could explain alterations in the nucleus accumbens that leads to the reinforcement of drug-seeking behavior.

Considered a multi-factorial disorder *autism spectrum disorder* (ASD) is a range of neurological abnormalities affecting one in 68 children in the United States [146]. Children affected by ASD exhibit reduced eye contact, facial expression, and body gestures [147]. Due to the heterogeneity of the classification of the disease the etiology is still widely unknown. Genetic causes have only been identified in 10–20% of individuals. Deep whole transcriptome sequencing of 30 patients with ASD identified RNA A/I editing alterations in 20 of 25 sites analyzed [148]. In contrast to other neurodegenerative disorders discussed in this chapter, RNA editing levels in ASD were found to be significantly higher than control groups [148]. Interestingly, the editing at the GluA2 Q/R site is not altered in ASD [148, 149]. Alterations in RNA A/I editing in ASD have been explained by alterations in ADAR2 self-regulation and loss of fmr1 [148–150].

3.4 Brain Cancer

Glioblastoma multiforme (GBM) is a tumor generated from astroglial cells generally localized in the cerebral hemispheres, and to a lesser extent in other regions of the brain or spinal cord. A transcriptome study using RNA sequencing for global A-to-I editing events in human revealed that genes with predicted editing events were significantly enriched for cancer-related genes, suggesting that RNA editing plays a significant role in the development of cancer [151]. This was later confirmed by Hwang and colleagues, who showed via gene ontology analyses that there was a selective change in the pattern of RNA editing in gliobastomas [30] (also recently reviewed in [152]). Indeed, early studies found a significant reduction in the GluA2 Q/R and the serotonin receptor 5-HT(2C) editing efficiency in malignant human brain tumors, which correlated with decreased ADAR2 self-editing activity [153]. These studies were confirmed when significantly reduced editing in Alu sequences was found in brain tissues [154]. All three ADAR genes showed lower RNA levels and the reduced ADAR3 levels correlated with the grade of malignancy of glioblastoma multiforme. Along those lines, high grade astrocytomas equally show lack of ADAR2 editing activity when grown in vitro, as well as in vivo via a flank tumor growth model in nude mice [155, 156].

As previously discussed, A-to-I editing also affects miRNAs, ~22 nucleotide long noncoding RNAs known to silence gene expression by binding to the 3′untranslated region (3′UTR) of mRNAs. miRNAs can undergo A-to-I RNA editing at premature states when the miRNA has a double-stranded structure. Analyses of high grade gliomas revealed reduced editing of miRNA-376 [157]. The authors found a strong correlation between the extent of unedited miRNA-376 and tumor spread, which was measured using magnetic resonance imaging of the patient's brains. The authors further confirmed these results in xenograft mouse models, showing that unedited miRNA-376 promoted glioma growth and spread, while edited miRNA-376 was protective. Similar results were recently reported on miRNA-589-3p [158]. A more recent study showed that A-to-I miRNA editing is enhanced at the seed region of the miRNA, an area critical to bind its target mRNA [159]. The authors further confirmed by RNA sequencing of GBM patient tissue that a significant reduction of miRNA editing occurs in GBM tissue and is correlated with the reduction of ADAR2 expression [159].

Interestingly, one study found elevated levels of ADAR3 in GBMs when compared to control brain tissue [16]. The authors suggested ADAR3 as a potential regulator of the Q/R editing site by binding to GluA2 subunit pre-mRNA and thereby inhibiting editing by ADAR2 in GBM. They hypothesized that an elevated expression of ADAR3 and reduced GluA2 editing will induce calcium permeability through the glutamate receptor, which in turn accelerates cell migration and tumor invasion into surrounding peri-tumoral tissue.

3.5 Conclusions

RNA editing, with now an estimate of over a million editing sites in primates and humans, has gained increasing interest as an important mechanism of RNA processing, not only during development, but also in disease. Given its ability to contribute to the molecular complexity in the human body, including the brain, it is of importance that we learn more about the regulation of RNA editing and how it can contribute to disease pathogenesis. It will be important to fully understand temporal and spatial regulation, of specific brain regions and likely also cell types, of the individual ADAR editing enzymes. This knowledge will be especially critical if we consider targeting ADAR enzymes for therapeutic purposes in any of the discussed diseases, as well as any non-CNS disorders.

References

1. Rueter SM, Dawson TR, Emeson RB. Regulation of alternative splicing by RNA editing. Nature. 1999;399(6731):75–80.
2. Gerber AP, Keller W. RNA editing by base deamination: more enzymes, more targets, new mysteries. Trends Biochem Sci. 2001;26(6):376–84.
3. Paul MS, Bass BL. Inosine exists in mRNA at tissue-specific levels and is most abundant in brain mRNA. EMBO J. 1998;17(4):1120–7.
4. Gal-Mark N, et al. Abnormalities in A-to-I RNA editing patterns in CNS injuries correlate with dynamic changes in cell type composition. Sci Rep. 2017;7:43421.
5. Lee SY, et al. RCARE: RNA sequence comparison and annotation for RNA editing. BMC Med Genomics. 2015;8(Suppl 2):S8.
6. Gu T, et al. Canonical A-to-I and C-to-U RNA editing is enriched at 3′UTRs and microRNA target sites in multiple mouse tissues. PLoS One. 2012;7(3):e33720.
7. Kim DD, et al. Widespread RNA editing of embedded alu elements in the human transcriptome. Genome Res. 2004;14(9):1719–25.
8. Yang JH, et al. Intracellular localization of differentially regulated RNA-specific adenosine deaminase isoforms in inflammation. J Biol Chem. 2003;278(46):45833–42.
9. Patterson JB, Samuel CE. Expression and regulation by interferon of a double-stranded-RNA-specific adenosine deaminase from human cells: evidence for two forms of the deaminase. Mol Cell Biol. 1995;15(10):5376–88.
10. Wang Q, et al. Stress-induced apoptosis associated with null mutation of ADAR1 RNA editing deaminase gene. J Biol Chem. 2004;279(6):4952–61.
11. Hartner JC, et al. Liver disintegration in the mouse embryo caused by deficiency in the RNA-editing enzyme ADAR1. J Biol Chem. 2004;279(6):4894–902.
12. Yao L, et al. Large-scale prediction of ADAR-mediated effective human A-to-I RNA editing. Brief Bioinform. 2017;PMID:28968662.
13. Higuchi M, et al. Point mutation in an AMPA receptor gene rescues lethality in mice deficient in the RNA-editing enzyme ADAR2. Nature. 2000;406(6791):78–81.
14. Chen CX, Cho DS, Wang Q, Lai F, Carter KC, Nishikura K. A third member of the RNA-specific adenosine deaminase gene family, ADAR3, contains both single- and double-stranded RNA binding domains. RNA. 2000;6:755–67.
15. Galipon J, et al. Differential binding of three major human ADAR isoforms to coding and long non-coding transcripts. Genes (Basel). 2017;8(2):pii:E68.

16. Oakes E, et al. Adenosine deaminase that acts on RNA 3 (ADAR3) binding to glutamate receptor subunit B pre-mRNA inhibits RNA editing in glioblastoma. J Biol Chem. 2017;292(10):4326–35.
17. Fitzgerald LW, Iyer G, Conklin DS, Krause CM, Marshall A, Patterson JP, Tran DP, Jonak GJ, Hartig PR. Messenger RNA editing of the human serotonin 5-HT 2C receptor. Europsychopharmacology. 1999;21(2S):82S–90S.
18. Bazak L, et al. A-to-I RNA editing occurs at over a hundred million genomic sites, located in a majority of human genes. Genome Res. 2014;24(3):365–76.
19. Ramaswami G, et al. Identifying RNA editing sites using RNA sequencing data alone. Nat Methods. 2013;10(2):128–32.
20. Ramaswami G, et al. Accurate identification of human Alu and non-Alu RNA editing sites. Nat Methods. 2012;9(6):579–81.
21. Neeman Y, et al. RNA editing level in the mouse is determined by the genomic repeat repertoire. RNA. 2006;12(10):1802–9.
22. Rodriguez J, Menet JS, Rosbash M. Nascent-seq indicates widespread cotranscriptional RNA editing in drosophila. Mol Cell. 2012;47(1):27–37.
23. Levanon EY, et al. Systematic identification of abundant A-to-I editing sites in the human transcriptome. Nat Biotechnol. 2004;22(8):1001–5.
24. Blow M, et al. A survey of RNA editing in human brain. Genome Res. 2004;14(12):2379–87.
25. Athanasiadis A, Rich A, Maas S. Widespread A-to-I RNA editing of Alu-containing mRNAs in the human transcriptome. PLoS Biol. 2004;2(12):e391.
26. Tan MH, et al. Dynamic landscape and regulation of RNA editing in mammals. Nature. 2017;550(7675):249–54.
27. Bahn JH, et al. Genomic analysis of ADAR1 binding and its involvement in multiple RNA processing pathways. Nat Commun. 2015;6:6355.
28. Korenberg JR, Rykowski MC. Human genome Orginazation: alu, lines, and the molecular structure of metaphase chromosome bands. Cell. 1988;53:391–400.
29. Yablonovitch AL, et al. The evolution and adaptation of A-to-I RNA editing. PLoS Genet. 2017;13(11):e1007064.
30. Hwang T, et al. Dynamic regulation of RNA editing in human brain development and disease. Nat Neurosci. 2016;19(8):1093–9.
31. Higuchi M, et al. RNA editing of AMPA receptor subunit GluR-B: a base-paired intron-exon structure determines position and efficiency. Cell. 1993;75(7):1361–70.
32. Burns CM, et al. Regulation of serotonin-2C receptor G-protein coupling by RNA editing. Nature. 1997;387(6630):303–8.
33. Henley JM, Wilkinson KA. Synaptic AMPA receptor composition in development, plasticity and disease. Nat Rev Neurosci. 2016;17(6):337–50.
34. Huganir RL, Nicoll RA. AMPARs and synaptic plasticity: the last 25 years. Neuron. 2013;80(3):704–17.
35. Hollmann M, Hartley M, Heinemann SF. Calcium permeability of KA-AMPA-gated glutamate receptor channels depnds on subunit composition. Science. 1991;252:851–3.
36. Bennett MV, et al. The GluR2 hypothesis: ca(++)-permeable AMPA receptors in delayed neurodegeneration. Cold Spring Harb Symp Quant Biol. 1996;61:373–84.
37. Wright A, Vissel B. The essential role of AMPA receptor GluR2 subunit RNA editing in the normal and diseased brain. Front Mol Neurosci. 2012;5:34.
38. Wenthold RJ, et al. Evidence for multiple AMPA receptor complexes in hippocampal CA1/CA2 neurons. J Neurosci. 1996;16(6):1982–9.
39. Isaac JT, Ashby MC, McBain CJ. The role of the GluR2 subunit in AMPA receptor function and synaptic plasticity. Neuron. 2007;54(6):859–71.
40. Cull-Candy S, Kelly L, Farrant M. Regulation of Ca2+-permeable AMPA receptors: synaptic plasticity and beyond. Curr Opin Neurobiol. 2006;16(3):288–97.
41. Sanderson JL, Gorski JA, Dell'Acqua ML. NMDA receptor-dependent LTD requires transient synaptic incorporation of ca(2+)-permeable AMPARs mediated by AKAP150-anchored PKA and calcineurin. Neuron. 2016;89(5):1000–15.

42. Nishikura K. Functions and regulation of RNA editing by ADAR deaminases. Annu Rev Biochem. 2010;79:321–49.
43. Washburn MC, et al. The dsRBP and inactive editor ADR-1 utilizes dsRNA binding to regulate A-to-I RNA editing across the C. elegans transcriptome. Cell Rep. 2014;6(4):599–607.
44. Melcher T, et al. RED2, a brain-specific member of the RNA-specific adenosine deaminase family. J Biol Chem. 1996;271(50):31795–8.
45. Melcher T, et al. Editing of alpha-amino-3-hydroxy-5-methylisoxazole-4-propionic acid receptor GluR-B pre-mRNA in vitro reveals site-selective adenosine to inosine conversion. J Biol Chem. 1995;270(15):8566–70.
46. Sommer B, et al. RNA editing in brain controls a determinant of ion flow in glutamate-gated channels. Cell. 1991;67(1):11–9.
47. Ray RS, et al. Impaired respiratory and body temperature control upon acute serotonergic neuron inhibition. Science. 2011;333(6042):637–42.
48. Gershon MD, et al. 5-HT receptor subtypes outside the central nervous system. Roles in the physiology of the gut. Neuropsychopharmacology. 1990;3(5–6):385–95.
49. Spencer NJ, Keating DJ. Is there a role for endogenous 5-HT in gastrointestinal motility? How recent studies have changed our understanding. Adv Exp Med Biol. 2016;891:113–22.
50. Hood JL, Emeson RB. Editing of neurotransmitter receptor and ion channel RNAs in the nervous system. Curr Top Microbiol Immunol. 2012;353:61–90.
51. Bockaert J, et al. Neuronal 5-HT metabotropic receptors: fine-tuning of their structure, signaling, and roles in synaptic modulation. Cell Tissue Res. 2006;326(2):553–72.
52. Hoyer D, Clarke DE, Fozard JR, Hartig PR, Martin GR, Mylecharane EJ, Saxena PR, Humphrey PP. International Union of Pharmacology classification of receptors for 5-hydroxytryptamine (serotonin). Pharmacol Rev. 1994;46(2):157–203.
53. McCorvy JD, Roth BL. Structure and function of serotonin G protein-coupled receptors. Pharmacol Ther. 2015;150:129–42.
54. Helton L, Thor KB, Baez M. 5-Hydroxytryptamine2A, 5-hydroxytryptamine2B, and 5-hydroxytryptamine2C receptor mRNA expression in the spinal cord of rat, cat, monkey and human. Mol Neurosci. 1994;5:2617–20.
55. Iwamoto K, et al. Measuring RNA editing of serotonin 2C receptor. Biochemistry (Mosc). 2011;76(8):912–4.
56. Price RD, et al. RNA editing of the human serotonin 5-HT2C receptor alters receptor-mediated activation of G13 protein. J Biol Chem. 2001;276(48):44663–8.
57. Niswender CM, Copeland SC, Herrick-Davis K, Emeson RB, Sanders-Bush E. RNA editing of the human serotonin 5-hydroxytryptamine 2C receptor silences constitutive activity. J Biol Chem. 1999;274(14):9472–8.
58. Sanders-Bush E, Price RD. RNA editing of the human serotonin 5-HT2C receptor delays agonist-stimulated calcium release. Mol Pharmacol. 2000;58(4):859–62.
59. Coetzee WA, Amarillo Y, Chiu J, Chow A, Lau D, McCormack T, Moreno H, Nadal MS, Ozaita A, Pountney D, Saganich M, Vega-Saenz de Miera E, Rudy B. Molecular diversity of K+ channels. Ann N Y Acad Sci. 1999;868:233–85.
60. Tian C, et al. Potassium channels: structures, diseases, and modulators. Chem Biol Drug Des. 2014;83(1):1–26.
61. Wang H, Kunkel DD, Schwartzkroin PA, Tempel BL. Localization of Kv1.1 and Kv1.2, two K channel proteins, to synaptic terminals, somata, and dendrites in the mouse brain. J Neurosci. 1994;14(8):4588–99.
62. Robbins CA, Tempel BL. Kv1.1 and Kv1.2: similar channels, different seizure models. Epilepsia. 2012;53(Suppl 1):134–41.
63. Bhalla T, et al. Control of human potassium channel inactivation by editing of a small mRNA hairpin. Nat Struct Mol Biol. 2004;11(10):950–6.
64. Rettig J, Heinemann SH, Wunder F, Lorra C, Parcej DN, Dolly JO, Pongs O. Inactivation properties of voltage-gated K+ channels altered by presence of beta-subunit. Nature. 1994;369(6478):289–94.

65. Kumar A, Singh A, Ekavali. A review on Alzheimer's disease pathophysiology and its management: an update. Pharmacol Rep. 2015;67(2):195–203.
66. Area-Gomez E, Schon EA. Alzheimer disease. Adv Exp Med Biol. 2017;997:149–56.
67. Calderon-Garciduenas AL, Duyckaerts C. Alzheimer disease. Handb Clin Neurol. 2017;145:325–37.
68. Akbarian S, Smith MA, Jones EG. Editing for an AMPA receptor subunit RNA in prefrontal cortex and striatum in Alzheimer's disease, Huntington's disease and schizophrenia. Brain Res. 1995;699(2):297–304.
69. Gaisler-Salomon I, et al. Hippocampus-specific deficiency in RNA editing of GluA2 in Alzheimer's disease. Neurobiol Aging. 2014;35(8):1785–91.
70. Payami H, et al. Apolipoprotein E genotype and Alzheimer's disease. Lancet. 1993;342(8873):738.
71. Saunders AM, et al. Association of apolipoprotein E allele epsilon 4 with late-onset familial and sporadic Alzheimer's disease. Neurology. 1993;43(8):1467–72.
72. Valastro B, et al. AMPA receptor regulation and LTP in the hippocampus of young and aged apolipoprotein E-deficient mice. Neurobiol Aging. 2001;22(1):9–15.
73. Cantanelli P, et al. Age-dependent modifications of AMPA receptor subunit expression levels and related cognitive effects in 3xTg-AD mice. Front Aging Neurosci. 2014;6:200.
74. Stilling RM, et al. De-regulation of gene expression and alternative splicing affects distinct cellular pathways in the aging hippocampus. Front Cell Neurosci. 2014;8:373.
75. Stilling RM, et al. K-lysine acetyltransferase 2a regulates a hippocampal gene expression network linked to memory formation. EMBO J. 2014;33(17):1912–27.
76. Khermesh K, et al. Reduced levels of protein recoding by A-to-I RNA editing in Alzheimer's disease. RNA. 2016;22(2):290–302.
77. Vucic S, Rothstein JD, Kiernan MC. Advances in treating amyotrophic lateral sclerosis: insights from pathophysiological studies. Trends Neurosci. 2014;37(8):433–42.
78. Taylor JP, Brown RH Jr, Cleveland DW. Decoding ALS: from genes to mechanism. Nature. 2016;539(7628):197–206.
79. Bettencourt C, Houlden H. Exome sequencing uncovers hidden pathways in familial and sporadic ALS. Nat Neurosci. 2015;18(5):611–3.
80. Chien-Liang Glenn Lin LAB, Lin J, Margaret Dykes-Hoberg TC, Lora-Clawson JDR. 1-s2.0-S0896627300809976-main.Pdf. Neuron. 1998;20:589–602.
81. Rothstein JD, et al. Abnormal excitatory amino acid metabolism in amyotrophic lateral sclerosis. Ann Neurol. 1990;28(1):18–25.
82. Lu YM, Yin HZ, Chiang J, Weiss JH. Ca2+-permeable AMPA/Kainate and NMDA channels: high rate of Ca2+ influx underlies potent induction of injury. J Neurosci. 1996;16(17):5457–65.
83. Carriedo SG, Yin HZ, Weiss JH. Motor neurons are selectively vulnerable to AMPA/kainate receptor-mediated injury in vitro. J Neurosci. 1996;16(13):4069–79.
84. Takuma H, et al. Reduction of GluR2 RNA editing, a molecular change that increases calcium influx through AMPA receptors, selective in the spinal ventral gray of patients with amyotrophic lateral sclerosis. Ann Neurol. 1999;46(6):806–15.
85. Kawahara Y, et al. Human spinal motoneurons express low relative abundance of GluR2 mRNA: an implication for excitotoxicity in ALS. J Neurochem. 2003;85(3):680–9.
86. Williams TL, et al. Calcium-permeable alpha-amino-3-hydroxy-5-methyl-4-isoxazole propionic acid receptors: a molecular determinant of selective vulnerability in amyotrophic lateral sclerosis. Ann Neurol. 1997;42(2):200–7.
87. Van Den Bosch L, et al. Ca(2+)-permeable AMPA receptors and selective vulnerability of motor neurons. J Neurol Sci. 2000;180(1–2):29–34.
88. Morrison BM, et al. Light and electron microscopic distribution of the AMPA receptor subunit, GluR2, in the spinal cord of control and G86R mutant superoxide dismutase transgenic mice. J Comp Neurol. 1998;395(4):523–34.
89. Kawahara Y, et al. Glutamate receptors: RNA editing and death of motor neurons. Nature. 2004;427(6977):801.

90. Kwak S, Kawahara Y. Deficient RNA editing of GluR2 and neuronal death in amyotropic lateral sclerosis. J Mol Med (Berl). 2005;83(2):110–20.
91. Hideyama T, et al. Profound downregulation of the RNA editing enzyme ADAR2 in ALS spinal motor neurons. Neurobiol Dis. 2012;45(3):1121–8.
92. Hideyama T, et al. Induced loss of ADAR2 engenders slow death of motor neurons from Q/R site-unedited GluR2. J Neurosci. 2010;30(36):11917–25.
93. Aizawa H, et al. TDP-43 pathology in sporadic ALS occurs in motor neurons lacking the RNA editing enzyme ADAR2. Acta Neuropathol. 2010;120(1):75–84.
94. Yamashita T, et al. Calpain-dependent disruption of nucleo-cytoplasmic transport in ALS motor neurons. Sci Rep. 2017;7:39994.
95. Yamashita T, Akamatsu M, Kwak S. Altered intracellular milieu of ADAR2-deficient motor neurons in amyotrophic lateral sclerosis. Genes (Basel). 2017;8(2):pii: E60.
96. Aizawa H, et al. Deficient RNA-editing enzyme ADAR2 in an amyotrophic lateral sclerosis patient with a FUSP525L mutation. J Clin Neurosci. 2016;32:128–9.
97. D'Erchia AM, et al. Massive transcriptome sequencing of human spinal cord tissues provides new insights into motor neuron degeneration in ALS. Sci Rep. 2017;7(1):10046.
98. Donnelly CJ, et al. RNA toxicity from the ALS/FTD C9ORF72 expansion is mitigated by antisense intervention. Neuron. 2013;80(2):415–28.
99. Bates GP, et al. Huntington disease. Nat Rev Dis Primers. 2015;1:15005.
100. Olney JW, de Gubareff T. Glutamate neurotoxicity and Huntington's chorea. Nature. 1978;271(5645):557–9.
101. McGeer EG, McGeer PL. Duplication of biochemical changes of Huntington's chorea by intrastriatal injections of glutamic and kainic acids. Nature. 1976;263(5577):517–9.
102. Coyle JT, Schwarcz R. Lesion of striatal neurones with kainic acid provides a model for Huntington's chorea. Nature. 1976;263(5574):244–6.
103. Beal MF. Huntington's disease, energy, and excitotoxicity. Neurobiol Aging. 1994;15(2):275–6.
104. Paschen W, Hedreen JC, Ross CA. RNA editing of the glutamate receptor subunits GluR2 and GluR6 in human brain tissue. J Neurochem. 1994;63(5):1596–602.
105. Fourie C, et al. Differential changes in postsynaptic density proteins in postmortem Huntington's disease and Parkinson's disease human brains. J Neurodegener Dis. 2014;2014:938530.
106. Garrett E, Alexander M. Biology of Parkinson's disease: pathogenesis and pathophysiology of a multisystem neurodegenerative disorder. Dialogues Clin Neurosci. 2004;6(3):259–80.
107. Jankovic J. Parkinson's disease: clinical features and diagnosis. J Neurol Neurosurg Psychiatry. 2008;79(4):368–76.
108. Maiti P, Manna J, Dunbar GL. Current understanding of the molecular mechanisms in Parkinson's disease: targets for potential treatments. Transl Neurodegener. 2017;6:28.
109. Dong XX, Wang Y, Qin ZH. Molecular mechanisms of excitotoxicity and their relevance to pathogenesis of neurodegenerative diseases. Acta Pharmacol Sin. 2009;30(4):379–87.
110. Labbe C, Lorenzo-Betancor O, Ross OA. Epigenetic regulation in Parkinson's disease. Acta Neuropathol. 2016;132(4):515–30.
111. Paz-Yaacov N, et al. Adenosine-to-inosine RNA editing shapes transcriptome diversity in primates. Proc Natl Acad Sci U S A. 2010;107(27):12174–9.
112. Wettengel J, et al. Harnessing human ADAR2 for RNA repair – recoding a PINK1 mutation rescues mitophagy. Nucleic Acids Res. 2017;45(5):2797–808.
113. Chen T, et al. Genetic and epigenetic mechanisms of epilepsy: a review. Neuropsychiatr Dis Treat. 2017;13:1841–59.
114. Wang J, et al. Epilepsy-associated genes. Seizure. 2017;44:11–20.
115. Brusa R, et al. Early-onset epilepsy and postnatal lethality associated with an editing-deficient GluR-B allele in mice. Science. 1995;270(5242):1677–80.
116. Hu RQ, et al. Gamma-hydroxybutyric acid-induced absence seizures in GluR2 null mutant mice. Brain Res. 2001;897(1–2):27–35.

117. Kortenbruck G, et al. RNA editing at the Q/R site for the glutamate receptor subunits GLUR2, GLUR5, and GLUR6 in hippocampus and temporal cortex from epileptic patients. Neurobiol Dis. 2001;8(3):459–68.
118. Kitaura H, et al. Ca(2+)-permeable AMPA receptors associated with epileptogenesis of hypothalamic hamartoma. Epilepsia. 2017;58(4):e59–63.
119. Srivastava PK, et al. Genome-wide analysis of differential RNA editing in epilepsy. Genome Res. 2017;27(3):440–50.
120. RAB JN, Allen SG, Fujikawa DG, Wasterlain CG. Hypoxic neuronal necrosis: protein synthesisindependent activation of a cell death program. PNAS. 2003;100(5):2825–30.
121. Saver JL. Time is brain—quantified. Stroke. 2006;37(1):263–6.
122. Vilela P, Rowley HA. Brain ischemia: CT and MRI techniques in acute ischemic stroke. Eur J Radiol. 2017;96:162–72.
123. Khoshnam SE, et al. Pathogenic mechanisms following ischemic stroke. Neurol Sci. 2017;38(7):1167–86.
124. Peng PL, et al. ADAR2-dependent RNA editing of AMPA receptor subunit GluR2 determines vulnerability of neurons in forebrain ischemia. Neuron. 2006;49(5):719–33.
125. Liu S, et al. Expression of ca(2+)-permeable AMPA receptor channels primes cell death in transient forebrain ischemia. Neuron. 2004;43(1):43–55.
126. Ahuja CS, et al. Traumatic spinal cord injury. Nat Rev Dis Primers. 2017;3:17018.
127. Ahuja CS, et al. Traumatic spinal cord injury-repair and regeneration. Neurosurgery. 2017;80(3S):S9–S22.
128. Di Narzo AF, et al. Decrease of mRNA editing after spinal cord injury is caused by downregulation of ADAR2 that is triggered by inflammatory response. Sci Rep. 2015;5:12615.
129. Murray KC, et al. Recovery of motoneuron and locomotor function after spinal cord injury depends on constitutive activity in 5-HT2C receptors. Nat Med. 2010;16(6):694–700.
130. Huang YJ, Lane HY, Lin CH. New treatment strategies of depression: based on mechanisms related to neuroplasticity. Neural Plast. 2017;2017:4605971.
131. Palacios JM, Pazos A, Hoyer D. A short history of the 5-HT2C receptor: from the choroid plexus to depression, obesity and addiction treatment. Psychopharmacology. 2017;234(9–10):1395–418.
132. Lin SH, Lee LT, Yang YK. Serotonin and mental disorders: a concise review on molecular neuroimaging evidence. Clin Psychopharmacol Neurosci. 2014;12(3):196–202.
133. Owen MJ, Sawa A, Mortensen PB. Schizophrenia. Lancet. 2016;388(10039):86–97.
134. Baxter G, et al. 5-HT2 receptor subtypes: a family re-united? Trends Pharmacol Sci. 1995;16(3):105–10.
135. Sergeeva OA, Amberger BT, Haas HL. Editing of AMPA and serotonin 2C receptors in individual central neurons, controlling wakefulness. Cell Mol Neurobiol. 2007;27(5):669–80.
136. Barnes NM, Sharp T. A review of central 5-HT receptors and their function. Neuropharmacology. 1999;38(8):1083–152.
137. Lyddon R, et al. Serotonin 2c receptor RNA editing in major depression and suicide. World J Biol Psychiatry. 2013;14(8):590–601.
138. Kubota-Sakashita M, et al. A role of ADAR2 and RNA editing of glutamate receptors in mood disorders and schizophrenia. Mol Brain. 2014;7:5.
139. Park TM, Haning WF III. Stimulant use disorders. Child Adolesc Psychiatr Clin N Am. 2016;25(3):461–71.
140. Mathuru AS. A little rein on addiction. Semin Cell Dev Biol. 2017. https://doi.org/10.1016/j.semcdb.2017.09.030.
141. Lopez-Quintero C, et al. Probability and predictors of transition from first use to dependence on nicotine, alcohol, cannabis, and cocaine: results of the National Epidemiologic Survey on alcohol and related conditions (NESARC). Drug Alcohol Depend. 2011;115(1–2):120–30.
142. Everitt BJ. Neural and psychological mechanisms underlying compulsive drug seeking habits and drug memories—indications for novel treatments of addiction. Eur J Neurosci. 2014;40(1):2163–82.

143. Carr KD, et al. AMPA receptor subunit GluR1 downstream of D-1 dopamine receptor stimulation in nucleus accumbens shell mediates increased drug reward magnitude in food-restricted rats. Neuroscience. 2010;165(4):1074–86.

144. Zheng D, et al. Nucleus accumbens AMPA receptor involvement in cocaine-conditioned place preference under different dietary conditions in rats. Psychopharmacology. 2015;232(13):2313–22.

145. Schmidt HD, et al. ADAR2-dependent GluA2 editing regulates cocaine seeking. Mol Psychiatry. 2015;20(11):1460–6.

146. Deborah L, Christensen P, et al. Prevalence and characteristics of autism spectrum disorder among children aged 8 years – autism and developmental disabilities monitoring network, 11 sites, United States, 2012. MMWR. 2016;65(3):1–23.

147. Park HR, et al. A short review on the current understanding of autism spectrum disorders. Exp Neurobiol. 2016;25(1):1–13.

148. Eran A, et al. Comparative RNA editing in autistic and neurotypical cerebella. Mol Psychiatry. 2013;18(9):1041–8.

149. Shamay-Ramot A, et al. Fmrp interacts with adar and regulates RNA editing, synaptic density and locomotor activity in zebrafish. PLoS Genet. 2015;11(12):e1005702.

150. Feng Y, et al. Altered RNA editing in mice lacking ADAR2 autoregulation. Mol Cell Biol. 2006;26(2):480–8.

151. Bahn JH, et al. Accurate identification of A-to-I RNA editing in human by transcriptome sequencing. Genome Res. 2012;22(1):142–50.

152. Wang C, et al. Mechanisms and implications of ADAR-mediated RNA editing in cancer. Cancer Lett. 2017;411:27–34.

153. Maas S, et al. Underediting of glutamate receptor GluR-B mRNA in malignant gliomas. Proc Natl Acad Sci U S A. 2001;98(25):14687–92.

154. Paz N, et al. Altered adenosine-to-inosine RNA editing in human cancer. Genome Res. 2007;17(11):1586–95.

155. Galeano F, et al. ADAR2-editing activity inhibits glioblastoma growth through the modulation of the CDC14B/Skp2/p21/p27 axis. Oncogene. 2013;32(8):998–1009.

156. Cenci C, et al. Down-regulation of RNA editing in pediatric astrocytomas: ADAR2 editing activity inhibits cell migration and proliferation. J Biol Chem. 2008;283(11):7251–60.

157. Choudhury Y, et al. Attenuated adenosine-to-inosine editing of microRNA-376a* promotes invasiveness of glioblastoma cells. J Clin Invest. 2012;122(11):4059–76.

158. Cesarini V, et al. ADAR2/miR-589-3p axis controls glioblastoma cell migration/invasion. Nucleic Acids Res. 2017;46(4):2045–59.

159. Paul D, et al. A-to-I editing in human miRNAs is enriched in seed sequence, influenced by sequence contexts and significantly hypoedited in glioblastoma multiforme. Sci Rep. 2017;7(1):2466.

Chapter 4
RNA Nucleocytoplasmic Transport Defects in Neurodegenerative Diseases

Ashley Boehringer and Robert Bowser

Abstract In eukaryotic cells, transcription and translation are compartmentalized by the nuclear membrane, requiring an active transport of RNA from the nucleus into the cytoplasm. This is accomplished by a variety of transport complexes that contain either a member of the exportin family of proteins and translocation fueled by GTP hydrolysis or in the case of mRNA by complexes containing the export protein NXF1. Recent evidence indicates that RNA transport is altered in a number of different neurodegenerative diseases including Huntington's disease, Alzheimer's disease, frontotemporal dementia, and amyotrophic lateral sclerosis. Alterations in RNA transport predominately fall into three categories: Alterations in the nuclear membrane and mislocalization and aggregation of the nucleoporins that make up the nuclear pore; alterations in the Ran gradient and the proteins that control it which impacts exportin based nuclear export; and alterations of proteins that are required for the export of mRNA leading nuclear accumulation of mRNA.

Keywords RNA · TREX · Exportin · Nuclear pore complex · Amyotrophic lateral sclerosis · Alzheimer's disease · Huntington's disease · Frontotemporal dementia

4.1 Introduction

In eukaryotic cells, transcription and translation are compartmentalized by the nuclear membrane, or nuclear envelope. The nuclear membrane separates the nucleus, where transcription takes place, from the cytoplasm, where translation

A. Boehringer
Department of Neurobiology, Barrow Neurological Institute, Phoenix, AZ, USA

School of Life Sciences, Arizona State University, Phoenix, AZ, USA

R. Bowser (✉)
Department of Neurobiology, Barrow Neurological Institute, Phoenix, AZ, USA
e-mail: robert.bowser@dignityhealth.org

© Springer International Publishing AG, part of Springer Nature 2018
R. Sattler, C. J. Donnelly (eds.), *RNA Metabolism in Neurodegenerative Diseases*, Advances in Neurobiology 20,
https://doi.org/10.1007/978-3-319-89689-2_4

occurs. Transport between the two compartments is tightly regulated via trafficking through nuclear pores that are contained within the nuclear membrane. While molecules, including both proteins and nucleic acids, with a molecular mass below 40 kDa may diffuse freely through the pores, most larger molecules are actively transported using numerous carrier proteins [1–3]. RNA transport is predominantly mediated by either NXF1 (Nuclear RNA Export Factor 1), also known as TAP, or members of the exportin family of proteins. The export adaptor used is largely dependent on the type of RNA, with mRNA predominantly relying on NXF1 [4, 5]. All other types of RNA require a member of the exportin family as an adaptor, along with a gradient of the GTPase Ran. rRNA [6–8], snRNA, and some mRNAs utilize CRM1 (exportin-1, XPO1) [9, 10], and tRNA and miRNA require exportin-t and exportin-5 respectively [11–14]. In each case, the transport carrier protein is required to move the RNA, in the form of a ribonucleoprotein particle (RNP), through the nuclear pore and release it on the cytoplasmic side. In this chapter, we will review the canonical pathways for transport of RNA from the nucleus to the cytoplasm under normal conditions, as well as explore the alterations in RNA transport that have been identified in neurodegenerative diseases. These alterations predominantly fall into three categories; alterations in the nuclear envelope as well as mislocalization of the proteins making up the nuclear pore, alterations in the Ran gradient, and deficits in the export of mRNA, identified in both models of neurodegenerative disease and tissue from patients who suffered from these diseases.

4.2 Export of RNA from the Nucleus

4.2.1 Nuclear Pores Regulate Transport Between the Nucleus and the Cytoplasm

Transport between the nucleoplasm and cytoplasm is controlled by a protein structure called the nuclear pore complex (NPC). The NPC is approximately 125 MDa in size in vertebrates and comprised of a group of proteins known as nucleoporins [15]. The geometric structure of the nuclear pore consists of eight spokes connecting radially to form concentric rings and exhibits an eightfold symmetry, formed from over 500 copies of up to 30 different nucleoporins [15, 16]. The NPC can be broken into three regions; the central channel, nuclear basket, and cytoplasmic filaments. The central channel that is embedded within the nuclear envelope allows cargoes to move in and out of the nucleus. The nuclear basket is found on the nuclear side of the pore, and functions to bind transport competent mRNPs (messenger ribonucleoprotein particles) and direct them to the pore. Cytoplasmic filaments guide both proteins into the nuclear pore, and RNA cargoes which are exiting the pore, toward the translational machinery. The pore forms a central channel approximately 50–100 kDa/40 nm in size and is lined with nucleoporins-containing phenylalanine-glycine (FG) repeat domains. These FG repeats both fill the channel of the pore as well as comprise both the cytoplasmic filaments and nuclear basket. An estimated 6 MDa of FG repeats are found in a single pore and these domains provide both a barrier to diffusion, as well as docking sites for transport factors as they are trafficked through the pore [17].

4.2.2   Exportin-Mediated Export of RNA

Different types of molecules (proteins, mRNA, rRNA, tRNA, miRNA) rely on a host of different transport factors to transverse through the nuclear pore. Some mRNA as well as most other types of RNA, including rRNA, tRNA, and miRNA, require an exportin protein to facilitate their export [18]. Exportins are a family of seven proteins including: CRM1 (XPO1), CSE1L (XPO2), XPOt (XPO3), XPO4, XPO5, XPO6, and XPO7 which function in export from the nucleus (Fig. 4.1). Much like the nuclear import transporters importins, exportins require a small GTPase called RanGTPase (Ran) to function. Export via exportins requires a gradient of Ran to exist in which GTP bound Ran (RanGTP) is concentrated in the nucleus, and both GDP bound Ran (RanGDP) and its GTPase activator RanGAP1 are concentrated in the cytoplasm [19]. Of the seven known exportins, CRM1 is required for the export of some mRNAs as well as rRNA, in addition to being a primary transporter of proteins [18]. CRM1 does not directly bind RNA but instead relies on a series of RNA-binding adaptor proteins which bind RNA and then CRM1 for RNA export (Fig. 4.1) [20–22]. These adaptor proteins require a Nuclear Export Sequence (NES), for CRM1 is HX2–3HX2–3HXH, where H is a hydrophobic

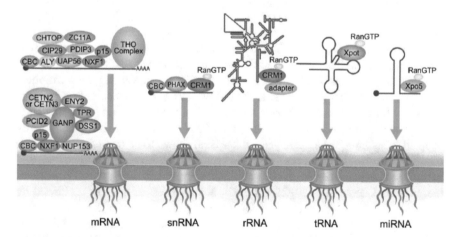

Fig. 4.1 Canonical RNA export pathways. Export of mRNA predominantly requires the TREX and TREX-2 pathways. snRNA and rRNA export requires exportin CRM1 bound to RanGTP along with the adaptor PHAX for snRNA and specific adaptors for different subunits of rRNA. Export of tRNA and miRNA require the RanGTP bound exportins, XPOt and XPO5 respectively

amino acid (i.e., isoleucine, leucine, methionine, phenylalanine, or valine) X is any amino acid [23, 24]. The binding of CRM1 to a NES-containing protein is cooperative with its binding to RanGTP [25]. After transport through the nuclear pore, GTP hydrolysis occurs which helps to dissociate its cargoes. In addition to a small subset of mRNAs, CRM1 is necessary for the export of rRNAs. Both the pre-60S subunit and the pre-40S subunit can be exported via CRM1 and an adaptor (Nmd3 or Lvt1 respectively). The pre-60S can also be exported by exportin-5 while the pre-40S subunit seems to rely solely on CRM1 [6, 7]. Other types of RNAs are also exported in a similar Ran-dependent process using other exportins, with export of tRNA requiring exportin-t (XPOt) and export of miRNA requiring exportin-5 (XPO5) [11–14, 26] (Fig. 4.1). Binding between pre-miRNA and XPO5 is mediated by the pre-miRNA structure rather than sequence with the recognition of a two-nucleotide 3′ end overhang structure and the double-stranded stem found in pre-miRNA [27]. In both cases the RNA is bound by a GTP-bound exportin that allows for its trafficking through the pore.

4.2.3   NXF1 Is the Primary Transporter of mRNA

Nucleocytoplasmic trafficking of mRNA through the nuclear pore mainly occurs via the transport factor NXF1. NXF1 is loaded onto mRNA via a series of handoffs involving the TREX (TRanscription and EXport) complex. Transport of mRNA is intricately linked with transcription and all stages of pre-mRNA processing including splicing. The TREX complex is made up of the THO complex-containing Thoc1 (Hpr1), Thoc2, Thoc3 (hTEX1), Thoc5, Thoc6, and Thoc7 as well as UAP56 (ddx39b), and Aly (AlyRef) [28] (Fig. 4.1). Unlike exportin-mediated export, TREX does not rely on a Ran gradient, but rather ATP hydrolysis.

The specificity of mRNA to TREX is mediated by its link to RNA polymerase II transcription, as well as a length requirement mediated by hnRNPC. hnRNPC interacts with the 5′ end of RNA if it is longer than 300 bp, preventing the recruitment of export factors other than TREX to the mRNP [29]. During transcription, proteins necessary for capping of the 5′ end, splicing, 3′ end cleavage, and polyadenylation bind to the nascent RNA. In metazoans, TREX has been shown to be predominantly coupled to splicing, whereas in yeast it has been shown to be more associated with transcription [30]. In human cells, TREX proteins have been shown to be recruited to the 5′ end of pre-mRNA near the cap-binding complex (CBC) which consists of the proteins CBP80 and CBP20 [31]. Aly binds closest to the CBC followed by UAP56 which binds downstream of Aly but upstream of the exon junction complex (EJC) (Fig. 4.1). This interaction is thought to be mediated by protein-protein interactions between Aly and CBP80 [31]. Interestingly, binding of mRNA to Aly and TREX complex member Thoc2 has been shown to require capped and spliced mRNA, suggesting that the recruitment of Aly to mRNA requires more than just binding to CBP80 [31].

Binding of Aly and RNA to UAP56 has been shown to stimulate the intrinsic ATPase activity of UAP56, which aids in its dissociation from the complex. The dissociation of UAP56 from the mRNP constitutes the handover of the mRNP to Aly. Aly along with a co-activator, Thoc5 or Chtop, are required for the binding of NXF1 to RNA [32]. NXF1 functions as a heterodimer with p15 (NXT1), and has very little RNA-binding activity in its native state. Upon binding with Aly and a co-activator, NXF1 is remodeled to expose its RNA-binding domains [32]. At this stage, the mRNP is turned over to NXF1 for trafficking though the nuclear pore.

Another export complex, TREX-2, also has a role in the export of mRNA via the NXF1 transporter. TREX-2 is built upon a scaffold protein GANP (Germinal-center-associated nuclear protein), which subsequently binds ENY2, PCID2, and DSS1 [33] (Fig. 4.1). The exact role of TREX-2 is unclear, though in yeast it has been shown to be involved in localizing a subset of actively transcribing genes to the pore [34]. In metazoans however, it has been shown to be involved in chaperoning mature mRNPs from processing centers to the pore for export [33]. It is unclear whether TREX and TREX-2 work cooperatively on the same mRNPs or transport different subsets of mRNPs, though some cooperation between the two complexes is thought to occur in mammalian cells [33]. One proposed model suggests that TREX-2 attaches to the mRNP after it is transferred from Aly to NXF1 and mediates its transport to and interaction with the nuclear pore [33].

4.3   Alterations in RNA Export and in Proteins Required for RNA Export Have Been Identified in Neurodegenerative Diseases

Many groups, including ours, have recently emphasized the role that alterations in nucleocytoplasmic trafficking play in a number of neurodegenerative diseases [35–40]. While initial studies focused on defects in protein trafficking, likely due to the common pathology of protein aggregation in the cytoplasm observed in many of these diseases, evidence for defects in RNA trafficking has recently come to light [39, 41]. These RNA trafficking alterations in disease states predominantly fall into three categories of defects; alterations in the localization of nucleoporins and abnormal nuclear envelope architecture, defects in the Ran gradient and alterations in the proteins that are responsible for maintaining it, and alterations in TREX proteins as well as mRNA retention within the nucleus (Fig. 4.2). It is important to note that alterations in protein trafficking are intricately linked to alterations in RNA trafficking due to the use of common regulatory proteins in nuclear export of proteins and RNA. Alterations in nucleoporins and the nuclear envelope as well as loss of the Ran gradient are likely to influence all forms of transport in and out of the nucleus. While export of mRNA via the TREX/NXF1 pathway is Ran independent, it requires members of the export process to be imported back into the nucleus to function, which is a Ran-dependent process.

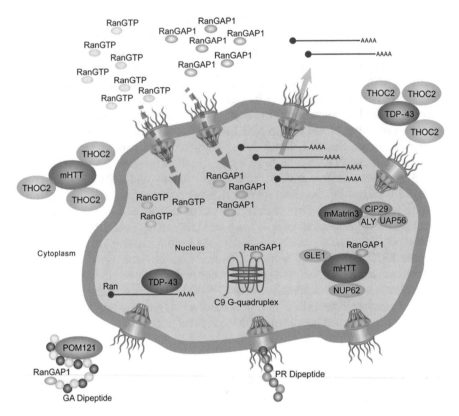

Fig. 4.2 RNA export defects identified in neurodegenerative diseases. Altered nuclear membrane morphology, interactions between DPRs and the nuclear pore, mislocalization and sequestration of export proteins into pathological aggregates, mutant proteins or pathological RNA species, interactions of TDP-43 with Ran mRNA as well as modifications to the Ran and RanGAP1 gradients, and decreased levels of mRNA export are all seen in a host of different neurodegenerative diseases

4.3.1 Nuclear Pore Alterations in Neurodegenerative Diseases

The earliest evidence for RNA transport alterations is the mislocalization of nucleoporins away from the nuclear envelope where they function, as well as abnormal nuclear envelope morphology which is often highlighted by nucleoporin immunostaining. These phenotypes have been identified in both animal models and patient tissue from several different neurodegenerative diseases, including Alzheimer's disease, Huntington's disease, Amyotrophic Lateral Sclerosis (ALS), and Frontotemporal Dementia (FTD) (Fig. 4.2) [35–37, 39, 40].

In Alzheimer's disease tissue, nuclear envelope abnormalities were noted in the hippocampus by immunostaining for Nup62, an FG-containing nucleoporin normally localized to the central channel of the nuclear pore [35]. In control tissue, Nup62 immunoreactivity forms a smooth circle in the nuclear envelope whereas in Alzheimer's patients it forms a tortuous and uneven nuclear envelope. It is important to note that these alterations in the nuclear envelope were not accompanied by positive staining for caspase-3 or TUNEL suggesting that this is not a consequence of apoptotic cell death [35].

Nuclear envelope defects were detected in two mouse models of Huntington's disease. The first, expressing physiological levels of ~175 CAG trinucleotide repeat expansion within one or both huntingtin (Htt) alleles which survive on average approximately 90 weeks and exhibit a tremor by 33 weeks, exhibited an age-dependent increase in the number of cells with abnormal nuclear envelopes, which was more severe in mice expressing two copies of Q175 compared to those expressing a single expanded copy, as observed using staining against Lamin B1 in the cortex and striatum. In this model, 89% of cells in the cortex and 62% in striatum have abnormal nuclear envelopes at 24 months of age [36]. This phenotype was also present in the cortex of mice expressing a 23 kDa human exon 1 fragment of Htt with a 120–125 repeat polyglutamine expansion (R6/2 mice) which have an age of onset of 9–11 weeks and a lifespan of approximately 10–13 weeks. In R6/2 mice, 89% of cortical cells had altered nuclear envelopes by 3 months of age [36]. This same mouse model of Htt was shown by others to exhibit intranuclear inclusions of Nup62 that colocalized with mHtt aggregates in the striatum and cortex [37]. In the zQ175 mouse model of Huntington's disease which contains the human Htt exon 1 sequence with a 193 CAG repeat which replaces the mouse Htt exon 1 within the mouse Htt gene, the nucleoporin Nup88 was identified in intracellular inclusions that colocalized with mHtt aggregates [37]. Abnormal nuclear envelopes were also seen in iPSC (induced pluripotent stem cell) derived neural progenitors from Huntington's patients, and in the motor cortex of patient tissue [36]. Components of the nuclear pore complex including Dbp5, a protein necessary at the terminal step of mRNA export to remove proteins from mRNAs after they have been transported through the pore, and RanBP3, a Ran-binding protein that acts a cofactor for CRM1-mediated export, were also identified in polyglutamine aggregates isolated from a cell culture model of Huntington's disease [42].

In ALS mutant SOD1 mouse models of ALS, alterations of NPC components include increased immunoreactivity of the nucleoporins GP210 and Nup205 [38]. This staining was reminiscent of staining patterns in sporadic ALS patients which showed increased staining for GP210 in the nuclear envelope and cytoplasm [38]. Others have also identified nuclear envelope irregularities as denoted by Nup62, Nup88, and Nup153 immunoreactivity in SOD1 mice that worsened with age as well as in both sporadic ALS (sALS) and familial ALS (fALS) patient tissue [43].

In a genetic screen performed in a Drosophila model of C9orf72, loss of function of Nup50 enhanced the phenotype of the C9 repeat expansion, as did a dominant negative form of Ran, whereas loss of function of Nup107 and Nup160 suppressed the phenotype [39]. When a Drosophila model expressing codon optimized PR DPRs was used for a genetic screen, knockdown of fly orthologs of the nucleoporins TPR, SEH1, NUP62, and NUP93 enhanced the phenotype while NUP50, NUP197, and NUP155 suppressed the phenotype [44]. Interestingly, NUP50 appears to modify the phenotype of the repeat expansion differently than the phenotype of the PR DPRs. These results suggest that altered subcellular distribution of nucleoporins may have a functional role in disease pathogenesis rather than being a consequence of the disease pathology, and that these alterations could have both loss of function and toxic gains of function phenotypes. This phenotype was accompanied by nuclear envelope irregularities as well as puncta of Nup107 in the salivary glands of flies [39]. As Nup107 is both found in aggregates and puncta, and its loss of function suppresses the disease phenotype in flies, it is possible that these aggregates and puncta of NPC can be toxic to cells. The mechanism by which these alterations in the nuclear envelope and mislocalization of NPC proteins induce disease is unknown, but a number of hypotheses have been proposed. PR dipeptides, formed from RAN (repeat-associated non-ATG) translation of the C9orf72 repeat expansion (DPRs), were found to bind to the FG repeat of the central channel of the nuclear pore complex and keep them in a polymerized state, possibly physically blocking movement through the nuclear pore (Fig. 4.2) [45]. Nuclear transport proteins including nuclear pore complex components and transport proteins such as CRM1 were found to interact with the DPRs PR and GR, produced from the C9orf72 repeat expansion, and CRM1 was also found to be an enhancer of a GR viability phenotype in *Drosophila* [46]. Another group suggests that cytoplasmic protein aggregates lead to the mislocalization of NPC proteins [47]. This hypothesis was tested using an artificial, aggregation prone β-sheet protein which led to the accumulation of NPC proteins in the cytoplasm and defects in both protein import and export [47].

4.3.2   Alterations in the RanGTPase Gradient Have Been Identified in Neurodegenerative Diseases

Another theme common among neurodegenerative diseases is alterations in the Ran gradient or its binding partners and regulators. As noted above, a high nuclear to cytoplasmic ratio of RanGTP is required for nuclear export where RanGTP is needed to bind to the exportin family of proteins within the nucleus.

In mice expressing mutant *Htt,* Gle1, part of the terminal step of mRNA export, as well as RanGAP1 are found co-aggregated with *Htt* [36]. RanGAP1 (Ran GTPase Activating Protein) is necessary for activating the GTPase function of Ran leading to its conversion to a GDP bound state Both RanGAP1 and Nup62 were found in inclusions in Htt R6/2 mice and RanGAP1 and Nup88 were found in mHtt inclusions in zQ175 Htt mice [37]. RanGAP1 was also mislocalized and concentrated in perinuclear puncta, and Nup62 was mislocalized in the frontal cortex and striatum of Huntington's patients [37]. Higher levels of RanGTP are required in the nucleus compared to the cytoplasm to fuel active transport via exportins. In iPSC derived neurons from Huntington's patients, the nuclear to cytoplasmic ratio of Ran is decreased [37]. Interestingly, expression of either RanGAP1 or Ran ameliorated cell death in cells expressing mutant Huntingtin, suggesting that at least part of the mechanism of action may be a loss of function of these proteins [37].

In Alzheimer's disease, cytoplasmic aggregates of NTF2, part of the import pathway required for importing Ran into the nucleus, were found in patient tissue [35]. Nuclear levels of Ran were also found to be decreased both in a mouse model of FTD based on knockout of the gene-encoding progranulin, GRN [48]. Mice lacking GRN exhibit increased levels of ubiquitin immunoreactivity in the form of amorphous granular cytoplasmic staining in neurons of the posterior thalamus, CA2-4 regions of the hippocampus, midbrain and brainstem. Mice also exhibited increased lipofusin granules and vacuolation in the habenular nucleus and CA2-3 regions of the hippocampus. Both microgliosis and astrogliosis were found in mice lacking GRN which was most evident in the brainstem and thalamus, and focal neuronal loss was found in the CA2-3 region of the hippocampus at 23 months of age [48]. These defects in the nuclear levels of Ran were also present in FTD patients carrying GRN mutations [49, 50].

In a model of Parkinson's disease based on administration of the drug 1-methyl-4-phenyl-1,2,3,6-tetrahydropyridine (MPTP), mice that lacked one copy of the Ran-binding protein, Ranbp2, had a more severe disease course and slower recovery [51]. Interestingly, in mice lacking any other genetic modifications, knockdown of Ranbp2 in Thy1 positive motor neurons led to motor deficits, respiratory distress, and premature death [52].

Many models of ALS also exhibit similar defects in either the Ran gradient or Ran-binding proteins. TDP-43 is a protein mutated in rare forms of fALS as well as present in pathological aggregates in most ALS, FTD and subsets of patients in a number of other neurodegenerative diseases, and has been shown to bind the 3' UTR of Ran mRNA and regulate its levels (Fig. 4.2) [50]. Loss of nuclear TDP-43 correlated with loss of Ran in the frontal gyrus of patients with FTD caused by mutations in progranulin (GRN) and led to overall decreased levels of Ran in the cortex [49, 50]. In addition, knockdown of TDP-43 in SH-SY5Y cells, which models the loss of nuclear TDP-43 commonly seen in ALS patients, leads to decreased levels of RanBP1 [53], and TDP-43 knockdown in Neuro2a cells led to decreased levels of Ran mRNA and protein [50]. In mice expressing mutant SOD1 an upregulation and nucleoplasmic mislocalization of RanGAP1 were observed [38]. A similar increase in RanGAP1 staining was seen in tissue from sALS patients [38].

The RanGAP1 protein has also been shown to bind to the G-quadruplex structure formed by the RNA of the C9orf72 repeat expansion, and there is a reduced nuclear to cytoplasmic ratio of Ran in iPSC motor neurons derived from C9-ALS patients and immortalized cell lines (S2 cells) expressing the 30 G_4C_2 repeats (Fig. 4.2) [40]. Both iPSC derived motor neurons and motor cortex tissue from ALS patients carrying the C9orf72 expansion exhibited discontinuous nuclear envelope staining for RanGAP1 as well as mislocalization and puncta that occasionally colocalized with Nup107 and Nup205 [40]. In a mouse model of C9orf72 expressing the GA DPR, both RanGAP and Pom121, a transmembrane nucleoporin involved in anchoring the NPC to the membrane, were found in nuclear and cytoplasmic puncta that often colocalized with the poly(GA) aggregates (Fig. 4.2) [54]. Interestingly, in two *Drosophila* models of C9orf72, genetic screens identified RanGAP1 as a modifier of the disease phenotype. Both screens were performed by expressing constructs in the eye and then co-expressing targets and looking for modification of the eye phenotype. Codon optimized, ATG-mediated expression of the DPR PR was coupled with expression of RNAi lines leading to the discovery that knockdown of RanGAP1 enhanced the toxicity caused by PR [44]. When 30 G_4C_2 repeats were similarly expressed in the *Drosophila* eye, RanGAP1 overexpression suppressed the toxicity accompanied by the repeat, whereas RanGEF enhanced the toxicity [40]. Importantly, in this model system the phenotype of the altered Ran gradient (which likely inhibits the export of both proteins and RNA) could be partially rescued by a variety of treatments. The Ran gradient phenotype was rescued with antisense oligonucleotides against the C9orf72 repeat, by destabilizing the G quadruplex structure the repeat forms, or by inhibiting CRM1, suggesting both that these defects may be induced by the repeat, and that drug strategies currently being employed for the repeat might modulate these defects [40]. Conversely, knockdown of the splicing factor SRSF1, which also has a role in TREX-mediated mRNA export, has been proposed as a possible therapeutic strategy in C9orf72 ALS as cytoplasmic localization of the repeat transcript is necessary for the production of DPRs and both knockdown of SRSF1 or blockage of the interaction between SFSF1 and NXF1 was protective in multiple models of C9orf72 [55].

4.3.3   Nuclear mRNA Retention in Neurodegenerative Diseases

While mislocalization of nucleoporins and defects in the Ran gradient and Ranbinding proteins are likely to cause alterations in the nuclear export of RNA, recent studies have identified deficits in the export of mRNA in models of neurodegenerative disease.

In Huntington R6/2 mice, the TREX complex component Thoc2 is mislocalized and found in inclusions, and mRNA was found to be retained within the nucleus of these cells (Fig. 4.2) [36, 47]. The same phenotypes of Thoc2 aggregation and nuclear mRNA retention were found in cells expressing Htt86Q as well as C-terminal fragments of TDP-43 or even an artificial aggregation prone β-sheet construct [47]. In mice expressing a ~175 CAG trinucleotide repeat of *Htt* (Htt^{Q165}), mRNA accumulated within nuclei by RNA-FISH (fluorescence in situ hybridization) using an oligo dT probe, in a dose-dependent manner [36]. In addition to phenotypes in models of neurodegenerative disease, this phenotype of mRNA nuclear accumulation has been identified in the cortex in tissue from Huntington's patients [36].

Some rare forms of fALS are caused by mutations in Gle1, a protein that is an integral component of the release of mRNA from transport machinery in the cytoplasm. While the mechanism by which these mutations cause disease is not completely understood, it has been suggested that haploinsufficiency of Gle1 is to blame, suggesting a role for mRNA transport defects in this disease [56].

Expression of an ALS causing variant of SOD1 (G93A) in NSC-34 cells causes retention of RNA within the nucleus, as measured by an increased nuclear to cytoplasmic ratio of RNA transcripts identified using RNA-seq [57]. This retention was not accompanied by an increase in transcripts-containing introns suggesting that the nuclear retention was not linked to defects in splicing, but rather likely due to defects in nuclear trafficking [57].

Recently, multiple groups have shown interactions between the C9orf72 repeat or its products with proteins involved in mRNA nuclear export. Multiple nucleoporins, as well as CRM1 and SRSF7 have been identified as protein interactors of the dipeptide repeats PR and GR [46]. In a genetic screen in *Drosophila,* in which 8, 28, or 58 copies of the G_4C_2 repeat are expressed in the eye using the GMR-GAL4 driver, aimed at discovering modifiers of the C9orf72 phenotype, proteins involved in mRNA export were identified. The strongest suppressor was found to be Aly, with partial loss of function of NXF1, CHTOP, NCBP2, ARS2, Gle1, and CRM1 enhancing the phenotype. Importantly, expression of the repeat in cells led to an accumulation of poly(A) + mRNA within the nucleus, which can be decreased with Aly knockdown [39]. Others have also shown the accumulation of poly(A) + mRNA within the nucleus of cells transfected with the C9orf72 repeat accompanied by the nuclear accumulation of PABPC1 with binds to the C9orf72 RNA. PABPC1 accumulation is a phenomenon reminiscent of viral infection where nuclear PABPC1 nuclear accumulation is sufficient to cause nuclear mRNA retention [58].

Recently, we have shown that Matrin 3, a nuclear matrix protein mutated in rare forms of ALS, binds to many TREX components and proteins involved in nuclear RNA export including, Aly, UAP56 and Sarnp in cell culture as well as nuclear spinal cord lysates [41]. The expression of ALS linked mutations in Matrin 3 in cell lines also causes the accumulation of poly(A) + mRNA within the nucleus. These mutations also caused nuclear accumulation of mRNAs of ALS-relevant proteins TDP-43 and FUS linking mRNA nuclear retention to disease pathology (Fig. 4.2) [41].

4.4   Conclusions

Alterations in nucleocytoplasmic transport have been identified by numerous groups in a wide range of neurodegenerative disorders including Alzheimer's disease, Huntington's disease, FTD, and ALS (Table 4.1). The identification of these alterations in such a wide span of neurodegenerative diseases suggests neuronal survival depends upon proper regulation of trafficking to and from the nucleus. While altered protein nucleocytoplasmic transport has been well documented in many neurodegenerative diseases, the only direct evidence for defective RNA transport has been the accumulation of poly(A) + mRNA within the nucleus in patient-derived tissue and various disease models. However, the alterations in both the localization and levels of nucleoporins and the loss of the Ran gradient and mislocalization of Ran-binding proteins strongly suggests defects occur in the transport of all RNA subtypes. Further studies are necessary to explore how other RNA subtypes are mislocalized in neurodegenerative diseases. While it is unclear why defects in nucleocytoplasmic trafficking preferentially affect neurons, there is evidence to

Table 4.1 Summary of defects identified in RNA export in neurodegenerative diseases

Phenotype	Disease/model supporting data	Citation
Nuclear membrane structural abnormalities	AD: Identified by Nup62 immunostaining in patient hippocampal tissue	[35]
	HD: Identified with Lamin B1 immunostaining in cortex and striatum of 175 CAG repeat expansion mice, in cortex of R6/2 mice, in iPSC-derived neuronal progenitors, and in the motor cortex of HD patients	[36]
	ALS: Nuclear envelope abnormalities denoted by Nup62, Nup88, and Nup153 immunoreactivity in spinal cord tissue from mutant SOD1 G93A mice and ALS patients	[43]
	ALS: Identified with Lamin C immunostaining in salivary gland cells of *Drosophila* expressing the C9orf72 repeat expansion	[39]
Puncta and aggregates of nucleoporins and RNA export proteins	HD: R6/2 mice exhibited Nup62 positive inclusions that co-localize with mHtt aggregates in striatum and cortex	[37]
	HD: zQ175 mice exhibited Nup88 and RanGAP1 positive intracellular inclusions that co-localized with mHtt aggregates	[37]
	ALS: Intranuclear Nup107 positive puncta identified in fly salivary gland cells expressing the C9orf72 repeat expansion	[39]
	ALS + HD: Expression of an aggregation prone C-terminal fragment of TDP-43 or the 96 CAG repeat Huntington results in mislocalization of THOC2 to cytoplasmic puncta	[47]
	HD: Dbp5 and RanBP3 identified in polyglutamine aggregates isolated from cells expressing a 96 CAG repeat expansion	[42]
	HD: Gle1 and RanGAP1 identified in mHtt aggregates in cortex of 175 CAG repeat expansion mice	[36]
	HD: RanGAP1 was mislocalized and concentrated in perinuclear puncta and Nup62 was mislocalized in frontal cortex and striatum of HD patients	[36]
	ALS: Discontinuous nuclear immunostaining as well as mislocalization and puncta that occasionally co-localized with Nup107 and Nup205 in iPSC derived motor neurons and motor cortex tissue from ALS patients carrying the C9orf72 repeat expansion	[40]
	ALS: In mice expressing the GA DPR (C9orf72), RanGAP1 and Pom121 were identified in nuclear and cytoplasmic puncta which often co-localized with GA aggregates	[54]

(continued)

Table 4.1 (continued)

Phenotype	Disease/model supporting data	Citation
Increased NPC protein immunoreactivity	*ALS*: SOD1 G93A mice showed increased immunoreactivity for GP210 and Nup205 in spinal cord tissue and patient tissue showed increased GP210 immunoreactivity in the nuclear envelope and cytoplasm	[38]
NPC and RNA export proteins identified as disease modifiers in *Drosophila* genetic screens	*ALS*: Loss of function of Nup50 and dominant negative Ran enhanced an eye phenotype in flies expressing the C9orf72 repeat expansion, loss of function of Nup107, and Nup160 suppressed phenotype	[39]
	ALS: Knockdown of fly orthologs of TPR, SEH1, NUP62, and NUP93 enhanced an eye phenotype in flies expressing the DPR PR (which is created by the C9orf72 repeat expansion), Nup50, Nup197, and Nup155 suppressed the phenotype	[44]
	ALS: CRM1 knockdown enhanced viability phenotype of flies expressing the DPR GR (from C9orf72 repeat expansion) and Nup205 knockdown suppressed phenotype	[46]
	ALS: RanGAP1 was identified as a modifier of an eye phenotype in two Drosophila models of C9orf72: 1) Flies expressing the PR DPR in which RanGAP1 knockdown enhanced toxicity; 2) Flies expressing 30 G_4C_2 repeats in which RanGAP1 overexpression suppressed toxicity and RanGEF enhanced toxicity	[40, 44]
	ALS: Loss of function of Aly, NXF1, CHTOP, NCBP2, ARS2, Gle1, and CRM1 suppressed the eye phenotype of the C9orf72 repeat	[39]
Polymerization of FG nucleoporins	*ALS*: PR DPR formed from C9orf72 repeat expansion bound to FG repeat of the central channel of the nuclear pore keeping them in a polymerized state	[45]
RNA export proteins interact with disease linked proteins or dipeptide repeats	*ALS*: CRM1 and Aly identified as interactors of the DPRs PR and GR (from C9orf72 repeat expansion) expressed in HEK293 cells	[46]
	ALS: Aly, UAP56, and Sarnp identified as interactors of mutant Matrin 3 in NSC-34 cells and spinal cord extracts	[41]

(continued)

Table 4.1 (continued)

Phenotype	Disease/model supporting data	Citation
Altered levels or localization of Ran and Ran binding proteins	*HD*: iPSC derived neurons from Huntington's patients exhibited decreased nuclear to cytoplasmic Ran ratios	[37]
	FTD: FTD patients with progranulin mutations as well as a mouse model of FTD based on progranulin knockout exhibited decreased nuclear Ran	[49, 50]
	ALS: Knockdown of TDP-43 in SH-SY5Y cells led to decreased levels of Ranbp1 and TDP-43 knockdown in Neuro2a cells lied to decreased levels of Ran mRNA and protein	[50, 53]
	PD: Mice lacking one copy of Ranbp2 had a more severe disease course and slower recovery in MPTP model of Parkinson's	[51]
	ALS: RanGAP1 immunostaining is increased in sALS patient tissue and is upregulated and mislocalized to the nucleoplasm in mice expressing mutant SOD1	[38]
	ALS: Reduced nuclear to cytoplasmic ratio of Ran was identified in iPSC motor neurons derived from ALS patients carrying the C9orf72 repeat expansion as well as immortalized cells expressing 30 G_4C_2 repeats	[40]
Nuclear retention of RNA and mRNA	*HD*: mRNA is retained within the nucleus of cells in both the R6/2 and 175 CAG repeat mouse models of Huntington's disease, as well as in cells expressing 86 polyglutamine repeats and in the cortex of tissue from Huntington's patients	[36, 47]
	ALS: mRNA was retained within the nucleus of cells expressing a C-terminal fragment of TDP-43 or an artificial aggregation prone β-sheet construct	[47]
	ALS: An increased nuclear to cytoplasmic ratio of RNA was identified by RNA-seq when G93A mutant SOD1 was expressed in NSC-34 cells	[57]
	ALS: Increased nuclear retention of RNA was identified in *Drosophila* salivary gland cells from flies expressing the C9orf72 repeat expansion. This phenotype was partially rescued by Aly knockdown	[39]
	ALS: Accumulation of mRNA within the nucleus of cells was identified in cells expressing the C9orf72 repeat expansion	[58]
	ALS: Expression of ALS linked mutations in Matrin 3 leads to increased nuclear to cytoplasmic ratios of both total mRNA and TDP-43 and FUS mRNA	[41]
Mutations of Gle1 in ALS patients	*ALS*: mutations in Gle1 which likely lead to haploinsufficiency are a rare genetic cause of ALS	[56]

AD Alzheimer's disease, *HD* Huntington's disease, *FTD* frontotemporal dementia, *ALS* amyotrophic lateral sclerosis, *PD* Parkinson's disease

suggest that post-mitotic cells including neurons may be more susceptible to age-related defects in nucleocytoplasmic transport. The proteins of the NPC are normally replaced during cell division where they are disassembled and reassembled with newly synthesized proteins during mitosis [59]. In post-mitotic cells such as neurons, the NPC is not completely disassembled and proteins such as Nup107 and Nup160 do not appear to turn over, suggesting that they are some of the longest-lived proteins in the body [60, 61]. The longevity of the NPC makes it vulnerable to the buildup of damage over time and unsurprisingly is subject to age-related dysfunction [60]. The susceptibility of neurons as post-mitotic cells to defects in the NPC, as well as the age-related nature of neurodegenerative diseases, could explain the contribution of nucleocytoplasmic trafficking defects in these diseases. While there is clear evidence that these defects are present in neurodegenerative diseases such as Alzheimer's disease, Huntington's disease, FTD, and ALS, the mechanism by which these defects occur as well as the role that these defects play in disease onset and pathogenesis remains unknown and merits continued study.

References

1. Paine PL. Nucleocytoplasmic movement of fluorescent tracers microinjected into living salivary gland cells. J Cell Biol. 1975;66(3):652–7.
2. De Robertis EM, Longthorne RF, Gurdon JB. Intracellular migration of nuclear proteins in Xenopus oocytes. Nature. 1978;272(5650):254–6.
3. Dingwall C, Sharnick SV, Laskey RA. A polypeptide domain that specifies migration of nucleoplasmin into the nucleus. Cell. 1982;30(2):449–58.
4. Segref A, et al. Mex67p, a novel factor for nuclear mRNA export, binds to both poly(A)+ RNA and nuclear pores. EMBO J. 1997;16(11):3256–71.
5. Herold A, Klymenko T, Izaurralde E. NXF1/p15 heterodimers are essential for mRNA nuclear export in Drosophila. RNA. 2001;7(12):1768–80.
6. Thomas F, Kutay U. Biogenesis and nuclear export of ribosomal subunits in higher eukaryotes depend on the CRM1 export pathway. J Cell Sci. 2003;116(Pt 12):2409–19.
7. Wild T, et al. A protein inventory of human ribosome biogenesis reveals an essential function of exportin 5 in 60S subunit export. PLoS Biol. 2010;8(10):e1000522.
8. Rouquette J, Choesmel V, Gleizes PE. Nuclear export and cytoplasmic processing of precursors to the 40S ribosomal subunits in mammalian cells. EMBO J. 2005;24(16):2862–72.
9. Fornerod M, et al. CRM1 is an export receptor for leucine-rich nuclear export signals. Cell. 1997;90(6):1051–60.
10. Watanabe M, et al. Involvement of CRM1, a nuclear export receptor, in mRNA export in mammalian cells and fission yeast. Genes Cells. 1999;4(5):291–7.
11. Kutay U, et al. Identification of a tRNA-specific nuclear export receptor. Mol Cell. 1998;1(3):359–69.
12. Arts GJ, Fornerod M, Mattaj IW. Identification of a nuclear export receptor for tRNA. Curr Biol. 1998;8(6):305–14.
13. Lund E, et al. Nuclear export of microRNA precursors. Science. 2004;303(5654):95–8.
14. Bohnsack MT, Czaplinski K, Gorlich D. Exportin 5 is a RanGTP-dependent dsRNA-binding protein that mediates nuclear export of pre-miRNAs. RNA. 2004;10(2):185–91.
15. Reichelt R, et al. Correlation between structure and mass distribution of the nuclear pore complex and of distinct pore complex components. J Cell Biol. 1990;110(4):883–94.

16. Cronshaw JM, et al. Proteomic analysis of the mammalian nuclear pore complex. J Cell Biol. 2002;158(5):915–27.
17. Frey S, Gorlich D. A saturated FG-repeat hydrogel can reproduce the permeability properties of nuclear pore complexes. Cell. 2007;130(3):512–23.
18. Fukuda M, et al. CRM1 is responsible for intracellular transport mediated by the nuclear export signal. Nature. 1997;390(6657):308–11.
19. Bischoff FR, Ponstingl H. Catalysis of guanine nucleotide exchange on Ran by the mitotic regulator RCC1. Nature. 1991;354(6348):80–2.
20. Brennan CM, Gallouzi IE, Steitz JA. Protein ligands to HuR modulate its interaction with target mRNAs in vivo. J Cell Biol. 2000;151(1):1–14.
21. Topisirovic I, et al. Molecular dissection of the eukaryotic initiation factor 4E (eIF4E) export-competent RNP. EMBO J. 2009;28(8):1087–98.
22. Yang J, et al. Two closely related human nuclear export factors utilize entirely distinct export pathways. Mol Cell. 2001;8(2):397–406.
23. Henderson BR, Eleftheriou A. A comparison of the activity, sequence specificity, and CRM1-dependence of different nuclear export signals. Exp Cell Res. 2000;256(1):213–24.
24. Kalderon D, et al. A short amino acid sequence able to specify nuclear location. Cell. 1984;39(3 Pt 2):499–509.
25. Petosa C, et al. Architecture of CRM1/Exportin1 suggests how cooperativity is achieved during formation of a nuclear export complex. Mol Cell. 2004;16(5):761–75.
26. Yi R, et al. Exportin-5 mediates the nuclear export of pre-microRNAs and short hairpin RNAs. Genes Dev. 2003;17(24):3011–6.
27. Okada C, et al. A high-resolution structure of the pre-microRNA nuclear export machinery. Science. 2009;326(5957):1275–9.
28. Strasser K, et al. TREX is a conserved complex coupling transcription with messenger RNA export. Nature. 2002;417(6886):304–8.
29. McCloskey A, et al. hnRNP C tetramer measures RNA length to classify RNA polymerase II transcripts for export. Science. 2012;335(6076):1643–6.
30. Reed R, Cheng H. TREX, SR proteins and export of mRNA. Curr Opin Cell Biol. 2005;17(3):269–73.
31. Cheng H, et al. Human mRNA export machinery recruited to the 5′ end of mRNA. Cell. 2006;127(7):1389–400.
32. Viphakone N, et al. TREX exposes the RNA-binding domain of Nxf1 to enable mRNA export. Nat Commun. 2012;3:1006.
33. Wickramasinghe VO, Stewart M, Laskey RA. GANP enhances the efficiency of mRNA nuclear export in mammalian cells. Nucleus. 2010;1(5):393–6.
34. Kohler A, et al. Yeast Ataxin-7 links histone deubiquitination with gene gating and mRNA export. Nat Cell Biol. 2008;10(6):707–15.
35. Sheffield LG, et al. Nuclear pore complex proteins in Alzheimer disease. J Neuropathol Exp Neurol. 2006;65(1):45–54.
36. Gasset-Rosa F, et al. Polyglutamine-expanded huntingtin exacerbates age-related disruption of nuclear integrity and nucleocytoplasmic transport. Neuron. 2017;94(1):48–57e4.
37. Grima JC, et al. Mutant Huntingtin disrupts the nuclear pore complex. Neuron. 2017;94(1):93–107e6.
38. Shang J, et al. Aberrant distributions of nuclear pore complex proteins in ALS mice and ALS patients. Neuroscience. 2017;350:158–68.
39. Freibaum BD, et al. GGGGCC repeat expansion in C9orf72 compromises nucleocytoplasmic transport. Nature. 2015;525(7567):129–33.
40. Zhang K, et al. The C9orf72 repeat expansion disrupts nucleocytoplasmic transport. Nature. 2015;525(7567):56–61.
41. Boehringer A, et al. ALS associated mutations in matrin 3 alter protein-protein interactions and impede mRNA nuclear export. Sci Rep. 2017;7(1):14529.

42. Suhr ST, et al. Identities of sequestered proteins in aggregates from cells with induced polyglutamine expression. J Cell Biol. 2001;153(2):283–94.
43. Kinoshita Y, et al. Nuclear contour irregularity and abnormal transporter protein distribution in anterior horn cells in amyotrophic lateral sclerosis. J Neuropathol Exp Neurol. 2009;68(11):1184–92.
44. Boeynaems S, et al. Drosophila screen connects nuclear transport genes to DPR pathology in c9ALS/FTD. Sci Rep. 2016;6:20877.
45. Shi KY, et al. Toxic PRn poly-dipeptides encoded by the C9orf72 repeat expansion block nuclear import and export. Proc Natl Acad Sci U S A. 2017;114(7):E1111–7.
46. Lee KH, et al. C9orf72 dipeptide repeats impair the assembly, dynamics, and function of membrane-less organelles. Cell. 2016;167(3):774–788e17.
47. Woerner AC, et al. Cytoplasmic protein aggregates interfere with nucleocytoplasmic transport of protein and RNA. Science. 2016;351(6269):173–6.
48. Ahmed Z, et al. Accelerated lipofuscinosis and ubiquitination in granulin knockout mice suggest a role for progranulin in successful aging. Am J Pathol. 2010;177(1):311–24.
49. Chen-Plotkin AS, et al. Variations in the progranulin gene affect global gene expression in frontotemporal lobar degeneration. Hum Mol Genet. 2008;17(10):1349–62.
50. Ward ME, et al. Early retinal neurodegeneration and impaired Ran-mediated nuclear import of TDP-43 in progranulin-deficient FTLD. J Exp Med. 2014;211(10):1937–45.
51. Cho KI, et al. Ranbp2 haploinsufficiency mediates distinct cellular and biochemical phenotypes in brain and retinal dopaminergic and glia cells elicited by the Parkinsonian neurotoxin, 1-methyl-4-phenyl-1,2,3,6-tetrahydropyridine (MPTP). Cell Mol Life Sci. 2012;69(20):3511–27.
52. Cho KI, et al. Loss of Ranbp2 in motoneurons causes disruption of nucleocytoplasmic and chemokine signaling, proteostasis of hnRNPH3 and Mmp28, and development of amyotrophic lateral sclerosis-like syndromes. Dis Model Mech. 2017;10(5):559–79.
53. Stalekar M, et al. Proteomic analyses reveal that loss of TDP-43 affects RNA processing and intracellular transport. Neuroscience. 2015;293:157–70.
54. Zhang YJ, et al. C9ORF72 poly(GA) aggregates sequester and impair HR23 and nucleocytoplasmic transport proteins. Nat Neurosci. 2016;19(5):668–77.
55. Hautbergue GM, et al. SRSF1-dependent nuclear export inhibition of C9ORF72 repeat transcripts prevents neurodegeneration and associated motor deficits. Nat Commun. 2017;8:16063.
56. Kaneb HM, et al. Deleterious mutations in the essential mRNA metabolism factor, hGle1, in amyotrophic lateral sclerosis. Hum Mol Genet. 2015;24(5):1363–73.
57. Kim JE, et al. Altered nucleocytoplasmic proteome and transcriptome distributions in an in vitro model of amyotrophic lateral sclerosis. PLoS One. 2017;12(4):e0176462.
58. Rossi S, et al. Nuclear accumulation of mRNAs underlies G4C2-repeat-induced translational repression in a cellular model of C9orf72 ALS. J Cell Sci. 2015;128(9):1787–99.
59. Rabut G, Lenart P, Ellenberg J. Dynamics of nuclear pore complex organization through the cell cycle. Curr Opin Cell Biol. 2004;16(3):314–21.
60. D'Angelo MA, et al. Age-dependent deterioration of nuclear pore complexes causes a loss of nuclear integrity in postmitotic cells. Cell. 2009;136(2):284–95.
61. Savas JN, et al. Extremely long-lived nuclear pore proteins in the rat brain. Science. 2012;335(6071):942.

Chapter 5
RNA Degradation in Neurodegenerative Disease

Kaitlin Weskamp and Sami J. Barmada

Abstract Ribonucleic acid (RNA) homeostasis is dynamically modulated in response to changing physiological conditions. Tight regulation of RNA abundance through both transcription and degradation determines the amount, timing, and location of protein translation. This balance is of particular importance in neurons, which are among the most metabolically active and morphologically complex cells in the body. As a result, any disruptions in RNA degradation can have dramatic consequences for neuronal health. In this chapter, we will first discuss mechanisms of RNA stabilization and decay. We will then explore how the disruption of these pathways can lead to neurodegenerative disease.

Keywords RNA · Decay · Alternative splicing · Transport · Stress granule · Exosome · Disease · Neurodegeneration

5.1   Mechanisms to Maintain RNA Stability

Following transcription, the newly formed transcript can be stabilized in several ways (Fig. 5.1). Most RNA that codes for protein, also referred to as coding or messenger RNA (mRNA), undergoes several processing steps that prevent degradation, assist in export from the nucleus, and aid in translation. Additionally, both coding and noncoding RNA (ncRNA) are stabilized by the adoption of unique secondary structures or sequestration in cytoplasmic ribonucleoprotein particles when the cell is under stress.

K. Weskamp · S. J. Barmada (✉)
Neuroscience Graduate Program and Department of Neurology, University of Michigan School of Medicine, Ann Arbor, MI, USA
e-mail: sbarmada@umich.edu

© Springer International Publishing AG, part of Springer Nature 2018
R. Sattler, C. J. Donnelly (eds.), *RNA Metabolism in Neurodegenerative Diseases*, Advances in Neurobiology 20,
https://doi.org/10.1007/978-3-319-89689-2_5

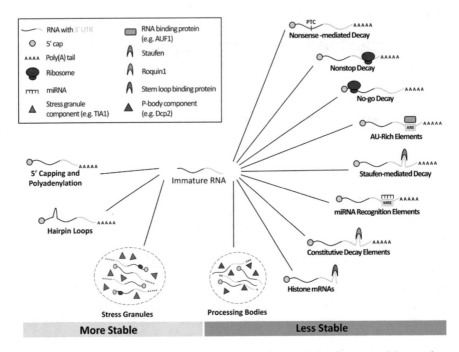

Fig. 5.1 Pathways responsible for RNA homeostasis. RNA stability is promoted by two key mechanisms (left). Following transcription, nascent RNA is stabilized by the addition of a 5′ cap and poly(A) tail, as well as the formation of secondary structures. Transcripts are also sequestered and stabilized in stress granules upon exposure to cellular stress. In contrast, RNA degradation pathways target faulty transcripts for removal (right). Transcripts that contain premature stop codons are targeted by nonsense-mediated decay. When translation fails to stop or start, the associated transcripts are degraded by nonstop decay and no-go decay, respectively. RNA decay mechanisms also regulate transcript abundance through several elements located within the 3′ UTR, including AU-rich elements, Staufen binding sites, miRNA recognition elements, and constitutive decay elements. Lastly, P-bodies sequester and destabilize RNA transcripts

5.1.1 Polyadenylation

Polyadenylation refers to the addition of a series of adenosine monophosphates to the 3′ end of mRNA transcripts [1]. This poly(A) tail protects nascent mRNA from enzymatic degradation [2, 3], facilitates nuclear export [4], and assists in translation [3]. Polyadenylation begins when a complex of several proteins recognizes a binding site on the mRNA transcript. An enzyme in this complex, cleavage/poyadenylation specificity factor (CPSF), cleaves the 3′ end of the transcript, and a second component, polyadenylate polymerase, adds sequential adenosine monophosphate units to create the poly(A) tail [5]. As the poly(A) tail grows longer, polyadenylate-binding protein 2 (PAB2) is recruited, which further increases the affinity of polyadenylate polymerase to the RNA [6]. Additional poly(A)-binding proteins then associate with the tail and facilitate nuclear export, stabilization of the RNA, and translation [7].

Many transcripts harbor more than one polyadenylation site. The site that is ultimately utilized primarily affects the length of the 3′ untranslated region (UTR), with little direct influence on protein translation or function [8]. However, the 3′ UTR may also encode microRNA recognition elements [9], DNA methylation sites [10], or motifs recognized by regulatory RNA-binding proteins [11, 12]. Thus, where a poly(A) tail starts can significantly influence the likelihood of transcript degradation. Moreover, in some cases alternative poly(A)-binding sites occur within the coding region, and their usage results in truncation of the translated protein [13]. Poly(A) tails are gradually eroded over time, and transcripts with shorter tails are both less likely to be transcribed and more likely to be degraded [14]. This process can be accelerated by the binding of microRNA to the 3′ UTR or through the removal or degradation of poly(A)-binding proteins [15].

5.1.2 Methylguanine Cap

The majority of coding RNAs undergo a second processing step that involves the addition of a methylguanine cap to the 5′ end of the transcript. This cap stabilizes the transcript by preventing exonuclease-mediated degradation [16–18], and is also required for the translation of most mRNAs [19, 20]. Additionally, the 5′ cap assists in splicing [21–25], nuclear export [24, 25], and possibly polyadenylation [26].

The capping process is initiated before transcription is complete, and begins when RNA triphosphatase removes one of the 5′ terminal phosphate groups [27]. mRNA guanylyltransferase then catalyzes the addition of guanosine triphosphate to the remaining terminal biphosphate to create an unusual 5′–5′ triphosphate linkage. This guanosine is then methylated by a methyltransferase [27]. The cap-binding complex (CBC) binds to the methylated 5′ cap, which is in turn recognized by the nuclear pore complex and exported into the cytoplasm [28, 29]. Once there, the CBC is replaced by the translation factors eIF4E and eIF4F, which are recognized by other translation initiation machinery components, including the ribosome [30, 31].

Binding of the CBC and translation factors also stabilize transcripts by blocking the binding of decapping enzymes [32–34]. When these decapping enzymes outcompete the translation factors, they hydrolyze the 5′ cap and expose the 5′ monophosphate. The resulting decapped transcripts are subject to rapid degradation by 5′ exonucleases [35].

5.1.3 Secondary Structure

DNA primarily forms double helices, but the single-stranded nature of RNA and its propensity to form hydrogen bonds allows it to form more complex structures that can directly affect transcript stability. The most common RNA secondary structure is the hairpin loop, created when two complementary regions of the same strand base-pair to form a double helix that ends in an unpaired loop [36]. These loops are found in pre-microRNA, transfer RNA (tRNA), and mRNA, and their stability depends on several factors, including length, degree of complementarity in the stem, and guanine to cytosine base pair content. Hairpin loops stabilize mRNA [37–40] and in many cases increase translation efficiency [39, 40]. This may occur by blocking exonuclease activity, but the precise mechanism remains unclear. Hairpin loops may also act as binding sites for proteins that direct mRNA transport and localization [41–43].

The combination of several hairpin loops forms a multiloop; the most abundant example of this structure is found in the cloverleaf-shaped tRNAs that assist in protein translation. The relative stabilities of multiloops vary based on size, number of loops, and complementarity [44]. Hairpin loops can also form pseudoknots, in which at least two hairpin loops are linked by single stranded loops. Pseudoknots are relatively stable, they form the catalytic core of some ribozymes [45, 46] and telomerases [47], and may also be involved in translation, though little is known about their functional significance [48]. Other structures, such as G-quadruplexes and R-loops, are more often associated with disease and will be discussed below.

5.1.4 Stress Granules

Cells undergo a wide range of molecular changes in response to environmental stressors, including the inhibition of conventional translation [49, 50] and the formation of stress granules (SGs). SGs are cytoplasmic ribonucleoprotein particles rich in mRNA, RNA-binding proteins, and stalled translation initiation complexes [51–53]. SG coalescence effectively sequesters the attached mRNAs and the 40S ribosome subunit [54, 55], preventing further translation and stabilizing the bound mRNAs. Proteins unrelated to the original translation initiation complex are also recruited, and their composition helps determine SG dynamics and longevity [56]. Which proteins participate is often dependent on their posttranslational modifications and the specific stressor involved [57–61], providing a rapid and reversible way for the cell to modulate SG formation and composition. Many RNA-binding proteins found in SGs contain low-complexity domains that are inherently flexible; the ability of these domains to form reversible homo- and heterotypic interactions with one another via their low-complexity domains may be responsible for the dynamics of SG formation and dissociation [62, 63]. Additionally, SGs often contain a number of proteins that promote RNA stability and regulate translation [64]. Moreover, deadenylation is largely inhibited in stress granules [65–67]. When the stressor has passed, several RNA-binding proteins catalyze SG disassembly [68–70], and the transcript is either degraded or released to resume translation. These observations suggest that SGs serve two basic functions: preventing the translation of unnecessary transcripts during stress, and protecting these transcripts from degradation until the stress has subsided.

5.2 Mechanisms of RNA Decay

The typical life of an mRNA transcript includes a complex sequence of events including transcription, capping, adenylation, splicing, and export. When mistakes occur during this process, quality control mechanisms exist to recognize and eliminate defective transcripts that may give rise to dysfunctional or toxic proteins (Fig. 5.1). However, these pathways do more than ensure the fidelity of RNA transcripts. They also serve important regulatory roles, enabling rapid modulation of steady-state RNA levels—and therefore protein production—in response to changes in the intracellular or extracellular environment.

5.2.1 RNA Degradation Machinery

There are three major classes of intracellular RNA-degrading enzymes: endonucleases that cut RNA internally, 5′–3′ exonucleases that degrade RNA from the 5′ end, and 3′–5′ exonucleases that hydrolyze RNA from the 3′ end. These enzymes may work independently or within a complex such as the exosome, a versatile structure for the degradation of immature or abnormal RNA. The core of the eukaryotic exosome complex is formed by nine proteins, six of which are members of the RNase PH-like family [71]. These form a ring that is capped by three additional proteins with RNA-binding domains [72]; this structure bears remarkable similarity to the 26S proteasome [73], which consists of a central proteolytic barrel (the 20S core) capped on either end by 19S regulatory subunits. The exosome is primarily composed of 3′–5′ exoribonucleases, and RNAs are degraded by removing terminal nucleotides from the 3′ end of the transcript. This occurs through the cleavage of phosphodiester bonds, either through RNase PH-like protein-mediated phosphorolytic cleavage or hydrolytic cleavage by proteins associated with the exosome [74]. Several other proteins bind to the exosome to regulate its activity and specificity [75–77]. The exosome also processes small nuclear RNAs, small nucleolar RNAs, and ribosomal RNAs [78], though how these molecules are targeted to and released from the exosome remains unclear.

5.2.2 Nonsense-Mediated Decay

Occasionally, errors introduced during transcription, insertions, deletions, or nonsense mutations uncover premature stop codons (PTCs) within the coding sequence of an mRNA. If translated, PTC-containing transcripts would encode truncated proteins that may have toxic gain-of-function or dominant-negative activities. Nonsense-mediated decay (NMD) is a surveillance mechanism that eliminates transcripts containing PTCs, thereby preventing the synthesis of proteins that could be detrimental to the cell.

mRNA transcripts undergo splicing following transcription, during which introns are removed and exons are spliced together. The resulting exon-exon junctions (EEJs) are occupied by a complex of proteins (the exon junction complex, or EJC) that assist in splicing until they are displaced by the ribosome during the first, or pioneer, round of translation. If the stop codon is downstream or within about 50 nucleotides of the final EJC, the transcript is translated normally. According to the EJC model of NMD, a stop codon that occurs upstream of an EJC is recognized as a PTC, triggering transcript degradation [79, 80]. When the ribosome stalls at a PTC, the protein UPF1, along with the eukaryotic release factors eRF1 and eRF3, forms the surveillance complex (SURF) and binds adjacent to the PTC. SURF then interacts with two components of the nearby EJC, UPF2, and UPF3B [81–83]. This triggers UPF1 phosphorylation, which causes the complex to move along the mRNA, resolving secondary structure and removing adherent proteins that may inhibit degradation [84, 85]. Phosphorylated UPF1 also binds to SMG6, an endonuclease that directly cleaves the mRNA [86, 87], as well as SMG5 and SMG7, which trigger deadenylation [88], decapping, and further degradation [89]. Additionally, UPF1 may be recruited to transcripts independent of a PTC or adjacent EJC, particularly within long 3′ UTRs [90]. A working theory is that UPF1 preferentially binds long 3′ UTRs and is phosphorylated via an unknown mechanism, triggering transcript decay. However, more work is required to identify the pathway resulting in destabilization of transcripts bearing long 3′ UTRs.

5.2.2.1 Alternative Exon Inclusion and Exclusion

Though NMD is an important quality control mechanism, it also helps regulate the expression of functional mRNA [91], predominantly through alternative mRNA splicing. This phenomenon is remarkably widespread: NMD-related regulation of transcript abundance is involved in cell proliferation [92, 93], immunity [94], stress [95], viral response [96], and neuronal activity [97, 98]. The differential inclusion or exclusion of exons (alternative splicing) enables a single gene to encode multiple transcript and protein isoforms, and in many cases alternatively spliced transcripts are subject to NMD. Because changes in the splicing environment determine which isoforms are produced [99, 100], alternative splicing can regulate gene expression by creating transcripts that are more or less stable. An estimated 33% of alternative transcripts contain PTCs [101], and between 12% and 45% of alternatively spliced transcripts are estimated to be NMD targets [101]. Regulated unproductive splicing (RUST) of this type regulates RNA abundance in relation to neuronal activity levels [102], developmental stage, and cell type [103]. Moreover, there is growing evidence that RUST is utilized by several RNA-binding proteins to regulate their own expression (autoregulation), particularly components of the splicing machinery [104–108].

5.2.2.2 Upstream Open Reading Frames

Upstream open reading frames (uORFs) are mRNA elements that include a start codon in the 5′ UTR that is out-of-frame with the main coding sequence. Because ribosomes bind to the 5′ cap of the mRNA and scan for start codons, uORFs can disrupt or interfere with translation of the downstream coding sequence [109, 110]. Moreover, a stop codon at the 3′ uORF end may be viewed as a PTC within the context of the whole transcript. As predicted by the EJC model of NMD, the presence of uORFs correlates with lower expression levels of the downstream ORF [111, 112], and uORF-bearing transcripts are particularly susceptible to degradation by NMD [113–115].

5.2.3 Nonstop Decay

Nonstop decay (NSD) is a surveillance mechanism involved in the detection and degradation of mRNA transcripts that lack stop codons [77, 116] due to premature polyadenylation or point mutations that disrupt existing terminal codons. Without a recognizable stop codon, the ribosome translates into the poly(A) tail and then stalls, unable to release the mRNA transcript [117].

NSD is activated when Ski7, a component of the exosome complex, binds the empty aminoacyl (A) site of the stalled ribosome via its C-terminal domain [76, 77]. This is supported by the fact that C-terminal deletions of Ski7 result in impaired NSD but do not affect general exosome function [116]. Additionally, the Ski7 C-terminal domain strongly resembles other proteins that bind the ribosome during normal translation, elongation, and termination such as EF1a and eRF3 [118]. After binding, Ski7 releases the stalled ribosome and recruits the exosome to rapidly deadenylate the transcript [77, 116, 119, 120].

5.2.4 No-Go Decay

No-go decay (NGD) is a mechanism that recognizes mRNA transcripts stalled during translation [121–123] due to damaged RNA, stress [124], or strong secondary structure that blocks the progress of translation machinery [121]. NGD is the most recently discovered RNA surveillance pathway, and as such little is known about its mechanism. However, evidence suggests that NGD may degrade mRNA in a manner that resembles translation termination. Two proteins that promote NGD, Hbs1 and Dom34, strongly resemble eRF1 and eRF3, two factors that catalyze the end of translation [121, 125].

Analogous to Ski7 in NSD, Hbs1 possesses the same C-terminal domain that allows EF1a, eRF3, and Ski7 to bind the empty A site on the stalled ribosome [126, 127]. Dom34 is homologous to eRF1 and binds directly to Hbs1 [126, 128]. Upon binding, the Dom34/Hbs1 complex triggers the release of the nascent peptide and the ribosome is released or degraded. Likewise, the mRNA transcript is targeted for endonucleolytic cleavage and the fragments are subsequently degraded via the exosome or exonucleases [121, 125]. It is not currently known how the Dom34/Hbs1 complex releases the mRNA from the ribosome, but the close relation between Hbs1 and Ski7 suggests that ribosome release may occur in the same manner as NSD. Moreover, NGD can occur independently of the Dom34/Hbs1 complex; further work is needed to identify the other factors involved.

Additionally, it remains unclear why some transcripts are targeted by NGD and not others. Pausing during translation is a normal occurrence [129] and may even serve biological functions [130–132], but only a fraction of transcripts are NGD substrates. Potentially important factors include the degree of ribosome stalling and whether or not the A site is empty to allow Dom34/Hbs1 complex binding. Further studies are needed to clarify this mechanism.

5.2.5   Adenylate-Uridylate-Rich Elements

While some mRNA decay pathways target faulty transcripts, others allow the cell to rapidly modulate gene expression in response to intracellular and extracellular stimuli. Several of these pathways regulate transcript levels via binding sites within the 3′ UTR, including adenylate-uridylate-rich elements (AREs), Staufen-mediated decay, microRNAs, and constitutive decay elements.

AREs are 50–150 nucleotide regions with frequent adenine and uridine bases that generally target the mRNA for rapid degradation [133, 134]. The mechanism underlying this pathway is not well understood, but several RNA-binding proteins interact with these sites and modulate transcript stability. For example, overexpression of hnRNP D, also known as ARE RNA-binding protein 1 (AUF1), destabilizes mRNA-containing AREs [135, 136]. Conversely, AUF1 depletion increases both ARE-containing mRNA stability and abundance of the corresponding proteins [137, 138]. Similarly, ablation of tristetraprolin (TTP), an RNA-binding protein that also recognizes AREs, increases mRNA and protein levels in a variety of cell types [139–141] and transcripts [142–147].

Though the exact mechanism is unclear, the association of ARE-binding proteins to AREs is followed by deadenylation [148–151], decapping, and 3′–5′ degradation via the exosome [152]. Certain subunits of the exosome bind to AREs directly, and several ARE-binding proteins including TTP associate with the exosome in vitro [75, 153], ensuring rapid and preferential elimination of ARE-continuing transcripts. Many ARE-binding proteins are also associated with SGs and P-bodies (discussed later in this chapter), suggesting that 5′–3′ exonuclease-mediated degradation may contribute to the turnover of ARE-containing transcripts as well [154, 155]. However, not all ARE-binding proteins trigger mRNA decay. For example, the Hu family of proteins stabilize bound ARE-containing transcripts [156–159], suggesting that the effect of AREs on RNA stability depends on a combination of factors, including the ARE-binding protein, transcript, and environment.

5.2.6 Staufen-Mediated Decay

Staufen-mediated decay (SMD) also regulates transcript levels via the 3′ UTR. SMD is triggered when Staufen-1 (Stau1) recognizes double-stranded RNA structures that form sufficiently downstream of the termination codon [160, 161]. Staufen-binding sites (SBS) are created by <u>intra</u>molecular hairpin loop formation within the 3′ UTR [161], or <u>inter</u>molecular base-pairing of the 3′ UTR with partially complementary long noncoding RNA [162]. Upon binding to the SBS, Stau1 recruits UPF1, which in turn stimulates mRNA decay [160], likely in much the same way as in NMD. Moreover, given that UPF1 is critical for both SMD and NMD, there may be competition between the two pathways based on the availability of UPF1 [163].

5.2.7 microRNAs

microRNAs (miRNAs) are small, noncoding RNAs that base-pair with complementary sequences within RNA transcripts to trigger their decay and/or translational repression. These 20–25 nt RNAs are produced from an RNA precursor (pri-miRNA) that forms a hairpin loop shortly after transcription [164, 165]. This structure is recognized by the nuclear protein DGCR8, which recruits the enzyme Drosha to cleave the hairpin from the rest of the transcript [166, 167]. The resulting molecule (pre-miRNA) is then exported to the cytoplasm [168] where the enzyme DICER cuts away the looped end [169], leaving a duplex of two short, complementary RNA strands behind. Though either strand can function as a mature miRNA, one is usually degraded [170, 171]. The remaining miRNA associates with the RNA-induced silencing complex (RISC), which assists in orienting the miRNA to its mRNA target, repressing translation of the target transcript and triggering its degradation.

The bound miRNA guides RISC to its binding site (miRNA recognition element or MRE) on the target transcript, most often within the 3′ UTR, though binding can occur within coding regions as well [172, 173]. The degree of miRNA-mRNA complementarity is a major predictor of transcript fate [174]. High degrees of sequence complementarity allow the Argonaute family of proteins—components of RISC [175]—to catalyze RNA decay through an unknown mechanism that may involve deadenylation, decapping, or exonucleolytic degradation [176, 177]. In contrast, miRNAs that bind weakly or with less complementarity induce translational repression [174] through a mechanism that remains unclear.

5.2.8 Constitutive Decay Elements

In addition to AREs, SBSs, and MREs, structured RNA degradation motifs also directly lead to transcript turnover. Constitutive decay elements (CDEs) are stem loop structures located within the 3′ UTR that trigger mRNA decay [178, 179] through recruitment of the RNA-binding protein Roquin1 [179, 180]. Roquin1 binds to the CDE stem loop structure via two binding sites in its ROQ domain [180], triggering degradation by recruiting the Ccr4-Caf1-Not deadenylation complex [179]. A transcriptome-wide search of 3′ UTRs in mice revealed several unique CDEs that are frequent and highly conserved across vertebrate species. Many, but not all, of these CDEs are Roquin1-associated [179], indicative of potential novel and unexplored pathways responsible for RNA decay.

5.2.9 Histone mRNAs

Much like CDE-containing transcripts, histone mRNAs encode highly conserved stem loop structures within their 3′ UTRs. These hairpins are essential for the rapid synthesis and degradation of histone mRNA during the S phase of the cell cycle, during which the cell undergoes DNA replication and chromosome remodeling [181]. At the end of S phase, histone hairpin loops are recognized by stem loop-binding protein (SLBP), which recruits the proteins necessary to add a short, oligo-nucleotide tail to histone mRNAs [182]. The oligonucleotide tail forms a binding site for LSM1–7, which triggers degradation via the exosome and endonucleases [182]. Interestingly, histone mRNA decay also requires UPF1 and its interaction with SLBP [183], though the exact role of UPF1 in histone mRNA metabolism remains unclear.

5.2.10 Processing Bodies

Processing bodies (P-bodies) are dynamic cytoplasmic foci comprised of mRNA and RNA-binding proteins. While SGs primarily sequester and protect mRNA until it can resume translation, P-bodies target associated transcripts for translational repression, decapping, and decay. Although P-body assembly is not required for RNA decay [184], it may directly compete with translation initiation; only transcripts that are not engaged in translation can be recruited to P-bodies [185–187], and upon translational inhibition P-bodies increase in number [185, 188]. Conversely, a decrease in P-body components leads to an increase in mRNAs associated with actively-translating polysomes [189]. P-bodies lack translation initiation machinery [185, 187], and are instead primarily composed of proteins associated with translational repression and mRNA decay, including decapping enzymes, exonucleases, and NMD components [190]. This suggests that functional transcripts undergo active translation before they are recruited to P-bodies. Once transferred, the mRNA is no longer translated [189, 191] and is instead degraded by decapping enzymes [192, 193] or other nucleases. However, mRNAs may also escape P-bodies and resume translation [187, 194], and regulated expression of proteins such as NoBody and MLN51 can drive P-body disassembly [195, 196]. Together, these observations indicate that P-bodies are part of a highly dynamic process characterized by constant flux between pools of mRNA transcripts that are being actively translated, those that are stalled or sequestered in SGs, and those that are being degraded within P-bodies.

5.3 RNA Turnover in Neurodegenerative Disease

The regulation of RNA is critical to cell health, and increasing evidence indicates that disruption of RNA stability may underlie neurodegenerative disease. Alterations in RNA turnover have been identified in several pathways, including RNA sequestration in stress granules or foci, RNA transport, the exosome, alternative splicing, and retrotransposons (Fig. 5.2).

5.3.1 RNA Sequestration

During times of stress, the cell diverts its energy and resources toward survival and recovery. A powerful mechanism to conserve resources is the sequestration of mRNAs in SGs to limit the translation of nonessential proteins. Typically, when the stressor passes, SGs dissolve and stalled mRNAs are released for translation. However, during prolonged periods of stress or disease, SGs sometimes fail to disassemble. This extended sequestration of mRNAs could effectively disrupt the delicate balance between SGs, polysomes, and P-bodies, effectively interrupting mRNA homeostasis, interfering with protein synthesis, and potentially contributing to downstream toxicity in neurodegenerative diseases.

5.3.1.1 Disruption of Stress Granule Dynamics

Of the ~125 proteins identified as components of human SGs, 60% are RNA-binding proteins [197]. This group of proteins is also highly enriched for the low complexity domains that facilitate the reversible aggregation of proteins into membraneless organelles such as SGs. The mutation or mislocalization of several RNA-binding proteins stabilizes SGs, sometimes driving them to form irreversible aggregates that sequester mRNA and RNA-binding proteins indefinitely and disrupt SG homeostasis. Conversely, though the machinery that drives SG disassembly remains unclear, any errors within this pathway may likewise lead to RNA dyshomeostasis and subsequent disease.

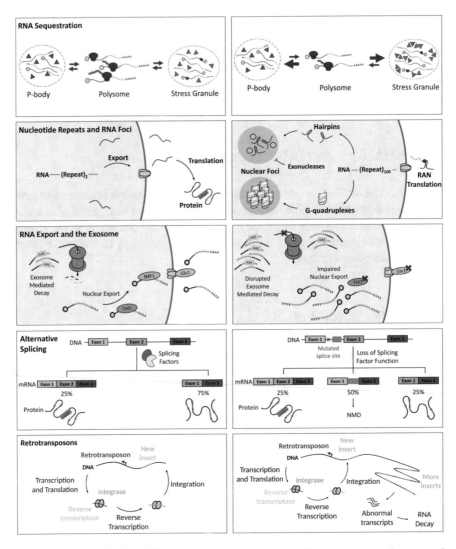

Fig. 5.2 Abnormal RNA stability in neurodegenerative disease. Here, we compare how normal pathways (left column) are disrupted in disease (right column). *RNA Sequestration*: There is constant flux between pools of RNA transcripts that are actively being translated (the polysome), those sequestered in stress granules, and those associated with P-bodies. In disease states, increased stress granule formation or reduced stress granule dissociation disrupts the equilibrium, resulting in fewer transcripts undergoing translation. *Repeat Expansions and RNA Foci:* Transcripts containing repeat expansions form secondary structures such as hairpin loops and G-quadruplexes that are often stabilized in nuclear foci, which also sequester RNA-binding proteins (green circles). These transcripts also generate proteins via RAN translation that can disrupt membraneless organelles involved in RNA splicing and processing. *RNA Transport and the Exosome*: Mutations in *THO*, *Gle1*, and other components of the RNA export pathway result in nuclear RNA retention and degradation via the exosome complex. Mutations in exosome components can inhibit RNA turnover and further disrupt RNA homeostasis. *Alternative Splicing*: Mutations that disrupt splice sites, or splicing regulators such as TDP43, result in the inclusion of unannotated or "cryptic" exons (pink). These transcripts are often targeted for nonsense-mediated decay. *Retrotransposons:* These transposable elements insert themselves into the genome, often disrupting open reading frames or splice sites. The transcripts that are transcribed from these regions are often faulty, and are targeted for RNA decay

RNA-Binding Proteins in Stress Granule Dynamics

TDP43 and FUS are two stress granule components that are integrally involved in neurodegenerative disease, particularly amyotrophic lateral sclerosis (ALS) and frontotemporal dementia (FTD). Both TDP43 and FUS are primarily nuclear proteins, but their cytoplasmic mislocalization [198–200] and nuclear exclusion [201–203] are characteristic features of ALS and FTD. These proteins are capable of nucleocytoplasmic shuttling in response to various stressors they associate with cytoplasmic SGs, but when the stress has passed they return to the nucleus [204]. ALS-linked mutations in the genes encoding TDP43 and FUS promote increased association with SGs [202, 205], abnormal SG formation [206], and reduced SG dissociation [207, 208]. TDP43 and FUS play important roles in alternative splicing and the stress response, and their sequestration impacts the processing of several transcripts that are critical for neuronal viability [209, 210]. Likewise, excess cytoplasmic TDP43 and FUS may sequester related RNA-binding proteins within SGs, further disrupting RNA homeostasis [64]. Importantly, TDP43- and FUS-related toxicity relies upon the ability of these proteins to bind RNA. Deletion of the RNA recognition motifs in either protein greatly reduces toxicity without affecting localization [211, 212], suggesting that RNA binding, not localization, imparts toxicity. Furthermore, these observations indicate that the sequestration of mRNAs themselves, not just RNA-binding proteins, is particularly damaging to neurons.

ALS-linked mutations are also found in other RNA-binding proteins such as Matrin3 [213], hnRNPA1, hnRNPA2/B1 [214], and TIA1 [215], all of which associate with SGs. These mutations are often centralized within the proteins' low complexity domains, and evidence indicates that they likewise alter SG dynamics, suggesting a link between SG association/dissociation and pathogenicity.

Stress Granule Disassembly

Though relatively little is known about SG disassembly, evidence suggests that valosin-containing protein (VCP) is crucial for this phenomenon. VCP regulates several cellular processes including autophagy [216], chromatin remodeling [217], and membrane trafficking [216], as well as SG clearance [218]. VCP accumulates in SGs, and its knockdown results in the persistence of SGs even after the stressor has passed [218]. Moreover, mutations in the gene-encoding VCP cause a multisystem proteinopathy that includes ALS and FTD [219], and the overexpression of mutant VCP results in impaired SG disassembly [218]. Thus, pathogenic mutations in the genes encoding VCP, TDP43, and FUS all stabilize SGs, thereby effectively sequestering essential mRNA and RNA-binding proteins within these organelles. As such, altered SG dynamics and abnormal RNA stability may represent a conserved pathway underlying ALS, FTD, and related neurodegenerative diseases.

5.3.2 Nucleotide Repeats and RNA Foci

Microsatellites are repeated tracts of nucleic acids that compose approximately 50% of the human genome [220]. These regions are a source of genomic instability, and expansion mutations that increase the number of repeats above a certain threshold can lead to neurodegenerative diseases such as Huntington's disease (HD), myotonic dystrophy (DM), spinocerebellar ataxias, Freidrich's ataxia, fragile X syndrome, fragile X-associated tremor ataxia syndrome (FXTAS), ALS, and FTD [221, 222]. In most cases, the length of the expanded region is inversely correlated with prognosis—higher repeat number results in earlier onset and more severe symptoms. Repeat expansions have unique pathological implications—they form unique secondary structures that may disrupt translation, sequester RNAs and other proteins into nuclear foci, and serve as a substrate for noncanonical translation.

5.3.2.1 Repeat Expansion Secondary Structure

The majority of expansion mutations associated with disease are trinucleotide CNG repeats, where N is any nucleotide. Due to the high degree of complementarity, CCG, CAG, CUG, and CGG repeats readily form mismatched hairpin loops [223] whose stability increases proportionally with the number of repeats [224]. Tetra-, penta-, and hexa-nucleotide repeats also form hairpins [225], though they appear to be less stable.

Repeat expansions with a high percentage of guanine nucleotides can also form G-quadruplexes. In these structures, four guanine bases associate through Hoogsteen hydrogen bonding to form a square guanine tetrad, and two or more tetrads stack to form a G-quadruplex [226]. Whether or not G-quadruplexes exhibit a physiological function remains unknown, but some evidence indicates that they participate in transcriptional regulation and/or telomere maintenance [227]. They are also observed in association with cancer, copy number variants, and age-related disease, specifically ALS and FTD. The most common mutation responsible for inherited ALS and FTD consists of a GGGGCC (G_4C_2) repeat expansion in the first intron of *C9orf72* [228, 229]. Unaffected individuals have 2–8 (G_4C_2) repeats [230], but tracts of >32 (G_4C_2) repeats lead to ALS, FTD, or both with nearly 100% penetrance by age 80 [231]. These repeats form stable G-quadruplexes [232], which are further stabilized in longer repeat expansions [233].

(G_4C_2) repeat expansions also form structures known as R-loops at the site of transcription, composed of nascently-synthesized RNA hybridized to the complementary DNA strand [234, 235]. The unbound DNA strand may also form hairpins or G-quadruplexes, further stabilizing the loop [236]. In addition to *C9orf72*-related ALS/FTD, R-loops are also observed in fragile X syndrome and Freidrich's ataxia [237] characterized by CGG and GAA trinucleotide repeats, respectively. The abundance of R-loops in these disorders depends on the size of the repeat expansion, with higher repeat number correlating with more frequent R-loops. These structures may contribute to the pathology of expansion diseases in several ways: by blocking translation [238], disrupting chromatin remodeling [239], or promoting genomic instability at the repeat expansion site [235]. In support of the pathogenic effects of R-loops, mutations in the gene encoding senataxin (*SETX*), a helicase that helps resolve R-loops [240], cause juvenile ALS (ALS4), while SETX overexpression prevents neurodegeneration in ALS models [241].

5.3.2.2   RNA Foci

In addition to their effects on RNA stability and translation, the propensity of repeat expansions to form stable secondary structures contributes to the formation of RNA foci [242, 243]. These nuclear inclusions may drive pathogenesis through the sequestration and nuclear retention of specific RNA-binding proteins. For example, CUG repeat expansions in *DMPK* cause myotonic dystrophy type 1 (DM1), a neuromuscular disease characterized by progressive muscle loss and weakness. This repeat expansion sequesters and disrupts the splicing activity of muscleblind (MBNL) [244, 245], a protein responsible for the processing of several key downstream transcripts [246]. MBNL binds to hairpins that result from repeat expansion mutations in *DMPK* with high affinity [245, 247], and preventing MBNL sequestration via small molecules that recognize CUG hairpin loops restores its splicing activity and helps maintain RNA homeostasis in DM1 models [248]. Additionally, the RNA foci observed in DM1 [249] and myotonic dystrophy type 2 (DM2) [250] sequester several other RNA-binding proteins, suggesting that global disruption of alternative splicing may contribute to DM pathogenesis [251]. RNA foci are also observed in *C9orf72*-linked ALS/FTD [252], where the G_4C_2 repeat transcripts sequester several splicing factors including hnRNPA1, hnRNPH, and SC35, as well as the RNA-binding protein hnRNPA3 and the mRNA export receptor ALYREF [253]. The sequestration of proteins essential to multiple cellular processes by repeat expansion transcripts suggests that these diseases occur, at least in part, through an RNA gain-of-function mechanism.

5.3.2.3   Repeat-Associated Non-AUG (RAN) Translation

Nucleotide repeats can be translated into polypeptides even if they are not located within a traditional open reading frame, via a noncanonical pathway termed repeat-associated non-AUG (RAN) translation. RAN translation may be triggered by hairpin loops formed by repeat-containing stretches of DNA, which effectively stall ribosome scanning and facilitate translational initiation at near-AUG codons [254–256]. This process occurs in multiple reading frames in both the sense and antisense directions, producing several dipeptide repeat-containing proteins (DPRs) [254]. RAN translation products are detected in spinocerebellar ataxia type 8, HD [257], DM1 [254], FXTAS [256], and *C9orf72*-associated ALS/FTD [258], suggesting that RAN translation is a common phenomenon in repeat expansion diseases. In some cases, there appears to be an inverse relationship between RAN translation and RNA foci formed by repeat expansions. This observation suggests that the repeat-expanded RNA may be sequestered in nuclear foci, precluding nuclear export and subsequent translation [259]. This may serve as a coping response to prevent the translation of DPRs; failure of this coping response over time may result in increased RAN translation and subsequent neurodegeneration [260, 261]. In support of this hypothesis, RNA foci in *C9orf72* mutant mice are abundant yet rarely associated with neurodegeneration [261]. RAN peptides may also affect RNA stability by disrupting membraneless organelles such as the nucleoli [262] and Cajal body [263], which are responsible for ribosomal RNA [264] and spliceosome maturation [265], respectively. Lastly, an increase in SGs and a decrease in P-bodies is observed in neurons expressing RAN peptides [266]; in this case, RAN peptides may act similarly to small proteins such as NoBody [195] that dissolve P-bodies, releasing unstable RNAs to be sequestered by SGs. Additional studies are required to determine the effect of RAN peptides on RNA stability, P-body dynamics, and global RNA homeostasis.

5.3.3 RNA Transport

The diverse functions of RNA are determined, in part, by its subcellular localization. As a result, RNA transport mechanisms are crucial for RNA function, particularly in highly compartmentalized and morphologically complex cells such as neurons. Among the most important of these mechanisms is nucleocytoplasmic transport, in which RNA transcripts are shuttled from the nucleus to the cytoplasm. Several neurodegenerative diseases exhibit deficits in nucleocytoplasmic RNA transport, leading to RNA sequestration in the nucleus and widespread dysregulation of gene expression. Thus, interruption of nuclear export machinery can have severe consequences on neuronal health.

5.3.3.1 Impaired Nuclear Export

Nuclear mRNA export is triggered by deposition of the highly conserved translation export (TREX) complex at the 5′ end of the nascent transcript [267]. The core of this complex, THO, recruits ALYREF and several other nuclear export factors [268–271]. ALYREF then binds to nuclear export factor 1 (NXF1) [272], triggering a shift from a conformation with low RNA-binding affinity to one that readily binds the transcript [273, 274]. NXF1 directs the transcript to the nuclear pore complex (NPC), a large multimeric structure that spans the nuclear envelope and enables the transport of molecules into and out of the nucleus. NXF1 facilitates NPC docking and transcript translocation via interactions with NPC components containing low complexity domains enriched in phenylalanine and glycine residues [275].

Disruption of this pathway leads to nuclear retention of RNA, and which is then rapidly degraded by the nuclear exosome [276, 277]. Interrupting nuclear RNA export can have severe consequences for neuronal survival, and mutations in nuclear export components are linked to several neurological and neurodevelopmental disorders. Chromosomal translocation and inactivation of THOC2, a subunit of the core TREX complex, leads to cognitive impairment, cerebellar hypoplasia, and congenital ataxia in humans [278]. Additionally, missense mutations in THOC2 have been implicated in fragile X syndrome [279], and mutations in a second THO subunit, THOC6, lead to intellectual disabilities [280]. Moreover, loss-of-function mutations in *Gle1* result in ALS [281] and fetal motor neuron disease [282]. Gle1 is a nuclear export mediator located on the cytoplasmic face of the nuclear pore that facilitates both the release of the transcript from the nuclear pore and its dissociation from export adaptor proteins [283], freeing it to undergo translation. This process may be specific to mRNAs with poly(A) tails, as depletion of Gle1 results in a nuclear accumulation and subsequent degradation of polyadenylated mRNAs [284, 285].

Abnormal nucleocytoplasmic transport is also a characteristic finding in models of ALS [286–288], DM1 [289], and HD [290, 291]. Toxicity in these models can be suppressed by pharmacologic or genetic modulation of nuclear transport components, testifying to the broad significance of this pathway in disease pathogenesis. Moreover, age is a likely contributor to impaired nuclear import, as aged cells display abnormal NPCs and reduced expression of nucleocytoplasmic transport genes [292, 293]; the resulting reduced fidelity in nuclear import/export is consistent with the observed age-dependent risk of nearly every neurodegenerative disease.

5.3.3.2 Disruption of the Nuclear Pore

In addition to disruption of the recruitment of the transcript to the pore, interruption of the pore itself can alter nucleocytoplasmic transport. RAN translation of repeat expansion mutations produces several DPRs. Some of these DPRs, including arginine-rich dipeptides generated from RAN translation of the *C9orf72* G_4C_2 repeat in familial ALS/FTD, clog the nuclear pore, and inhibit the transport of RNA and other macromolecules into and out of the nucleus [294]. Again, this contributes to the nuclear retention of RNAs that are susceptible to exosome-mediated decay [276, 277]. Arginine-containing DPRs are among the most toxic of the dipeptides in ALS/FTD models [262, 295], suggesting that impaired nucleocytoplasmic transport contributes significantly to neurodegeneration in these disorders.

5.3.4 The RNA Exosome Complex

The exosome complex is an RNA degradation mechanism that contributes broadly to RNA turnover, surveillance, and processing. This complex works closely with other pathways to orchestrate the degradation of immature, abnormal, or misplaced RNA.

5.3.4.1 Exosome-Associated Mutations in Neurodegenerative Disease

Due to the importance of the exosome in regulating RNA decay, mutations in this complex can have severe implications. Mutations in *EXOSC3*, the gene encoding the core exosome component RRP40, are linked to autosomal recessive pontocerebellar hypoplasia type 1 (PCH1) [296]. This progressive neurodegenerative disease is characterized by atrophy of the pons and cerebellum and loss of spinal motor neurons, accompanied by developmental delay, muscle atrophy, and difficulty breathing [297]. Thirty-seven percent of PCH1 patients exhibit *EXOSC3* mutations, most of which are heterozygous missense mutations [297]. Disease severity correlates with genotype, as patients with homozygous missense mutation fare better and those with a combined missense and null mutation fare worse [298].

Similarly, mutations in a gene encoding a separate exosome component, *EXOSC8*, result in cerebellar hypoplasia (CH) [299]. This autosomal recessive disorder is also characterized by progressive degeneration of the cerebellum, pons, and spinal motor neurons, as well as abnormal myelination. Though the mechanism is unclear, an increase in exosome substrates, including ARE-containing mRNAs encoding myelin proteins, in CH models suggests that impaired exosome function may contribute to dysmyelination of the involved tracts and subsequent neurodegeneration [299].

5.3.5 Alternative Splicing

Between 92% and 94% of all genes in the human genome are alternatively spliced [300], and the brain expresses more alternatively spliced genes than any other organ [301, 302]. This suggests that alternative splicing is a key regulator of transcript stability and gene expression, and its misregulation can have severe effects on neuronal health [303].

5.3.5.1 Nonsense-Mediated Decay and Unannotated or "Cryptic" Exon Splicing

A primary consequence of alternative splicing is RNA destabilization [101]. As discussed above, in many cases alternative splicing may serve to regulate normal transcript levels. This is supported by the fact that over one third of RNA transcripts are spliced to include PTCs, and these transcripts are likely targeted for degradation via NMD [101]. Mutations that affect splicing and result in either the inclusion of PTC-encoding exons or a shift the reading frame that uncovers "silent" PTCs may destabilize transcripts and lead to disease via gene haploinsufficiency. For example, disease-associated missense *GRN* mutations cause ALS and FTD by altering mRNA splicing, triggering NMD of *GRN* transcripts, and consequent reductions in progranulin protein expression [304–307]. In other cases, mutations that create novel splice sites or the dysregulation of splicing factors leads to the inclusion of unannotated or "cryptic" exons and the production of faulty transcripts that are eventually targeted for decay. Several regulatory proteins suppress these unannotated exon splicing events, including TDP43. Depletion of TDP43 results in a widespread increase in cryptic exon splicing events, and the inclusion of these exons may lead to NMD [308, 309]. Many of these events are specific to neurons [310], which suggests that the disruption of TDP43-mediated cryptic exon regulation may contribute to ALS and FTD.

NMD can be manipulated through the modulation of specific pathway components: overexpression of UPF1 and UPF3B stimulates NMD, while UPF1 knockdown or the overexpression of UPF3A, an antagonistic paralog of UPF3B that sequesters UPF2, suppresses NMD [311]. Consistent with a potential link between NMD and ALS/FTD pathogenesis, overexpression of UPF1 or UPF2 prevents FUS- and TDP43-mediated neurodegeneration in model systems [312]. One possibility is that UPF1 overexpression in these models prevents cell death by boosting endogenous NMD, thereby enabling the pathway to properly metabolize an overabundance of NMD substrates. However, further investigation is required to confirm and extend these findings.

5.4   Retrotransposons

Transposable elements (TEs) are mobile genetic elements that constitute a large portion of most eukaryotic genomes. Retrotransposons, which encode a reverse transcriptase and an integrase that allow them to "copy and paste" themselves from one region to another, represent approximately 40% of the human genome [313]. Though the vast majority of retrotransposons are inactive [314], some retain the ability to mobilize. Retrotransposition occurs approximately once every 10–100 births [315], and the insertion of these elements near or within active genes is a significant source of genomic instability and cellular toxicity [316, 317]. Though transcription of these regions is downregulated [318, 319], the transcripts that are transcribed are degraded via NMD [320] and other noncanonical pathways [321]. Several mechanisms have also evolved to suppress retrotransposon expression and prevent the resultant large-scale deletions and genomic rearrangements [322], though the efficiency of these mechanisms declines with age [316, 323, 324]. Moreover, the elevated expression of retrotransposons correlates with several neurodegenerative disorders [325–327], suggesting that a reduction in retrotransposon repression may contribute to disease pathogenesis.

5.4.1   Retrotransposons in ALS

As previously discussed, TDP43 aggregation and mislocalization play a fundamental role in ALS and FTD, and TDP43 serves as a key regulator of alternative splicing for hundreds of transcripts. TDP43 also recognizes several TE-derived RNA transcripts [328], and this binding is reduced in FTD patients coincident with elevated TE expression. This suggests that TDP43 normally regulates TE expression, and the loss-of-functional TDP43 in FTD results in TE overexpression [328]. This is further supported by the finding that TEs are derepressed in ALS/FTD models involving TDP43 overexpression or knockdown [328, 329], suggesting that TE dysregulation may contribute to neurodegeneration in ALS and FTD. This may occur through activation of DNA damage-mediated programmed cell death due to the large-scale deletions and genomic rearrangements that result from de-repressed TEs [329], and there is some evidence to suggest that TDP43 pathology impairs siRNA-mediated gene silencing, an essential system that normally protects the genome from retrotransposons [329].

Human endogenous retroviruses (HERVs) represent a subclass of retrotransposons originating from ancient viral infections that resulted in the integration of viral DNA into the host genome. The most recent of the retroviruses to integrate into the human genome is HERV-K [330]. The HERV-K envelope protein is expressed in both cortical and spinal neurons of ALS patients, suggesting activation of the retrovirus in disease. Furthermore, ectopic expression of the HERV-K envelope protein triggers neurodegeneration and motor dysfunction in mice [331]. Like other retrotransposons, HERV-K is regulated by TDP43, suggesting that HERV-K derepression in TDP43-deficient cells might contribute to neurodegeneration in ALS [331].

5.4.2 Retrotransposons in Aging

Age is a major risk factor for most neurodegenerative diseases, likely due to a reduced ability to regulate protein degradation [332], oxidative stress [333], and DNA damage [334]. While retrotransposons are a significant source of genomic instability, additional evidence suggests that they are more destructive in aging brains. The expression and mobility of several TEs increase with advanced age [316, 324]; these changes, in turn, are linked to progressive, age-dependent memory impairment and shortened lifespan [324]. Thus, the derepression of retrotransposons during normal aging could contribute to the age-related increase in risk for neurodegenerative diseases.

5.5 Conclusions and Future Directions

Neurodegenerative diseases vary widely in clinical presentation, neuropathology, and genetic background. However, it is becoming increasingly clear that alterations in RNA turnover are a key contributor to disease pathogenesis. The magnitude and extent of RNA dyshomeostasis observed in neurodegenerative disease models strongly suggests a fundamental disruption of one or more of the many mechanisms that tightly regulate RNA stability. While compensatory pathways may allow cells to cope with subtle changes in SG dynamics, alternative RNA splicing, or RNA degradation, over time such pathways become less efficient and the ability of the cell to maintain RNA homeostasis slowly erodes. Mitotic cells evade toxicity by dilution and division, but for long-lived cells such as neurons, the resulting abnormalities eventually lead to cell death. Because altered RNA stability results from the disruption of several related but distinct pathways, it is unlikely that focusing on single transcripts will result in a cure. Instead, a more complete understanding of RNA degradation in both healthy and diseased conditions may highlight common mechanisms and key upstream elements that could be rationally targeted for therapeutic development.

References

1. Yang L, Duff MO, Graveley BR, Carmichael GG, Chen L-L. Genomewide characterization of non-polyadenylated RNAs. Genome Biol. 2011;12:R16. https://doi.org/10.1186/gb-2011-12-2-r16.
2. Guhaniyogi J, Brewer G. Regulation of mRNA stability in mammalian cells. Gene. 2001;265:11–23. Available: https://www.ncbi.nlm.nih.gov/pubmed/11255003
3. Gerstel B, Tuite MF, McCarthy JEG. The effects of 5′-capping, 3′-polyadenylation and leader composition upon the translation and stability of mRNA in a cell-free extract derived from the yeast Saccharomyces cerevisiae. Mol Microbiol Wiley Online Library. 1992;6:2339–48. Available: http://onlinelibrary.wiley.com/doi/10.1111/j.1365-2958.1992.tb01409.x/full
4. Huang Y, Carmichael GG. Role of polyadenylation in nucleocytoplasmic transport of mRNA. Mol Cell Biol. 1996;16:1534–42. Available: https://www.ncbi.nlm.nih.gov/pubmed/8657127
5. Bienroth S, Keller W, Wahle E. Assembly of a processive messenger RNA polyadenylation complex. EMBO J. 1993;12:585–94. Available: https://www.ncbi.nlm.nih.gov/pubmed/8440247
6. Wahle E. Poly(A) tail length control is caused by termination of processive synthesis. J Biol Chem. 1995;270:2800–8. Available: http://www.jbc.org/content/270/6/2800.abstract
7. Coller JM, Gray NK, Wickens MP. mRNA stabilization by poly(A) binding protein is independent of poly(A) and requires translation. Genes Dev. 1998;12:3226–35. Available: https://www.ncbi.nlm.nih.gov/pubmed/9784497
8. Tian B, Hu J, Zhang H, Lutz CS. A large-scale analysis of mRNA polyadenylation of human and mouse genes. Nucleic Acids Res. 2005;33:201–12. https://doi.org/10.1093/nar/gki158.
9. Shell SA, Hesse C, Morris SM Jr, Milcarek C. Elevated levels of the 64-kDa cleavage stimulatory factor (CstF-64) in lipopolysaccharide-stimulated macrophages influence gene expression and induce alternative poly(A) site selection. J Biol Chem. 2005;280:39950–61. https://doi.org/10.1074/jbc.M508848200.
10. Wood AJ, Schulz R, Woodfine K, Koltowska K, Beechey CV, Peters J, et al. Regulation of alternative polyadenylation by genomic imprinting. Genes Dev. 2008;22:1141–6. https://doi.org/10.1101/gad.473408.
11. Danckwardt S, Kaufmann I, Gentzel M, Foerstner KU, Gantzert A-S, Gehring NH, et al. Splicing factors stimulate polyadenylation via USEs at non-canonical 3′ end formation signals. EMBO J. 2007;26:2658–69. Available: http://emboj.embopress.org/content/26/11/2658?utm_source=TrendMD&utm_medium=cpc&utm_campaign=EMBO_J_TrendMD_0
12. Hall-Pogar T, Liang S, Hague LK, Lutz CS. Specific trans-acting proteins interact with auxiliary RNA polyadenylation elements in the COX-2 3′-UTR. RNA. 2007;13:1103–15. https://doi.org/10.1261/rna.577707.
13. Tian B, Pan Z, Lee JY. Widespread mRNA polyadenylation events in introns indicate dynamic interplay between polyadenylation and splicing. Genome Res. 2007;17:156–65. https://doi.org/10.1101/gr.5532707.

14. Meyer S, Temme C, Wahle E. Messenger RNA turnover in eukaryotes: pathways and enzymes. Crit Rev Biochem Mol Biol. 2004;39:197–216. https://doi.org/10.1080/10409230490513991.

15. Weidmann CA, Raynard NA, Blewett NH, Van Etten J, Goldstrohm AC. The RNA binding domain of Pumilio antagonizes poly-adenosine binding protein and accelerates deadenylation. RNA. 2014;20:1298–319. https://doi.org/10.1261/rna.046029.114.

16. Furuichi Y, LaFiandra A, Shatkin AJ. 5′-Terminal structure and mRNA stability. Nature. 1977;266:235–9. https://doi.org/10.1038/266235a0.

17. Shimotohno K, Kodama Y, Hashimoto J, Miura KI. Importance of 5′-terminal blocking structure to stabilize mRNA in eukaryotic protein synthesis. Proc Natl Acad Sci U S A. 1977;74:2734–8. Available: https://www.ncbi.nlm.nih.gov/pubmed/197518

18. Murthy KGK, Park P, Manley JL. A nuclear micrococcal-sensitive, ATP-dependent exoribonuclease degrades uncapped but not capped RNA substratesx. Nucleic Acids Res. 1991;19:2685–92. https://doi.org/10.1093/nar/19.10.2685.

19. Muthukrishnan S, Both GW, Furuichi Y, Shatkin AJ. 5′-Terminal 7-methylguanosine in eukaryotic mRNA is required for translation. Nature. 1975;255:33–7. Available: https://www.ncbi.nlm.nih.gov/pubmed/165427

20. Gillian-Daniel DL, Gray NK, Aström J, Barkoff A, Wickens M. Modifications of the 5′ cap of mRNAs during Xenopus oocyte maturation: independence from changes in poly(A) length and impact on translation. Mol Cell Biol. 1998;18:6152–63. Available: https://www.ncbi.nlm.nih.gov/pubmed/9742132

21. Konarska MM, Padgett RA, Sharp PA. Recognition of cap structure in splicing in vitro of mRNA precursors. Cell. 1984;38:731–6. Available: https://www.ncbi.nlm.nih.gov/pubmed/6567484

22. Edery I, Sonenberg N. Cap-dependent RNA splicing in a HeLa nuclear extract. Proc Natl Acad Sci U S A. 1985;82:7590–4. Available: https://www.ncbi.nlm.nih.gov/pubmed/3865180

23. Flaherty SM, Fortes P, Izaurralde E, Mattaj IW, Gilmartin GM. Participation of the nuclear cap binding complex in pre-mRNA 3′ processing. Proc Natl Acad Sci U S A. 1997;94:11893–8. Available: https://www.ncbi.nlm.nih.gov/pubmed/9342333

24. Jarmolowski A, Boelens WC, Izaurralde E, Mattaj IW. Nuclear export of different classes of RNA is mediated by specific factors. J Cell Biol. 1994;124:627–35. Available: https://www.ncbi.nlm.nih.gov/pubmed/7509815

25. Fresco LD, Buratowski S. Conditional mutants of the yeast mRNA capping enzyme show that the cap enhances, but is not required for, mRNA splicing. RNA. 1996;2:584–96. Available: https://www.ncbi.nlm.nih.gov/pubmed/8718687

26. Glover-Cutter K, Kim S, Espinosa J, Bentley DL. RNA polymerase II pauses and associates with pre-mRNA processing factors at both ends of genes. Nat Struct Mol Biol. 2008;15:71–8. https://doi.org/10.1038/nsmb1352.

27. Shatkin AJ. Capping of eucaryotic mRNAs. Cell. 1976;9:645–53. https://doi.org/10.1016/0092-8674(76)90128-8.

28. Nojima T, Hirose T, Kimura H, Hagiwara M. The interaction between cap-binding complex and RNA export factor is required for intronless mRNA export. J Biol Chem. 2007;282:15645–51. https://doi.org/10.1074/jbc.M700629200.

29. Cheng H, Dufu K, Lee C-S, Hsu JL, Dias A, Reed R. Human mRNA export machinery recruited to the 5′ end of mRNA. Cell. 2006;127:1389–400. https://doi.org/10.1016/j.cell.2006.10.044.

30. Sato H, Maquat LE. Remodeling of the pioneer translation initiation complex involves translation and the karyopherin importin beta. Genes Dev. 2009;23:2537–50. https://doi.org/10.1101/gad.1817109.

31. Dias SMG, Wilson KF, Rojas KS, Ambrosio ALB, Cerione RA. The molecular basis for the regulation of the cap-binding complex by the importins. Nat Struct Mol Biol. 2009;16:930–7. https://doi.org/10.1038/nsmb.1649.

32. Schwartz DC, Parker R. mRNA decapping in yeast requires dissociation of the cap binding protein, eukaryotic translation initiation factor 4E. Mol Cell Biol. 2000;20:7933–42. Available: https://www.ncbi.nlm.nih.gov/pubmed/11027264

33. Grudzien E, Kalek M, Jemielity J, Darzynkiewicz E, Rhoads RE. Differential inhibition of mRNA degradation pathways by novel cap analogs. J Biol Chem. 2006;281:1857–67. https://doi.org/10.1074/jbc.M509121200.

34. Jiao X, Chang JH, Kilic T, Tong L, Kiledjian M. A mammalian pre-mRNA 5′ end capping quality control mechanism and an unexpected link of capping to pre-mRNA processing. Mol Cell. 2013;50:104–15. https://doi.org/10.1016/j.molcel.2013.02.017.

35. Braun JE, Truffault V, Boland A, Huntzinger E, Chang C-T, Haas G, et al. A direct interaction between DCP1 and XRN1 couples mRNA decapping to 5′ exonucleolytic degradation. Nat Struct Mol Biol Nature Research. 2012;19:1324–31. https://doi.org/10.1038/nsmb.2413.

36. Varani G. Exceptionally stable nucleic acid hairpins. Annu Rev Biophys Biomol Struct. 1995;24:379–404. https://doi.org/10.1146/annurev.bb.24.060195.002115.

37. Emory SA, Bouvet P, Belasco JG. A 5′-terminal stem-loop structure can stabilize mRNA in Escherichia coli. Genes Dev. 1992;6:135–48. Available: https://www.ncbi.nlm.nih.gov/pubmed/1370426

38. Hambraeus G, Karhumaa K, Rutberg B. A 5′ stem–loop and ribosome binding but not translation are important for the stability of Bacillus subtilis aprE leader mRNA. Microbiol Microbiol Soc. 2002;148:1795–803. https://doi.org/10.1099/00221287-148-6-1795.

39. Higgs DC, Shapiro RS, Kindle KL, Stern DB. Small cis-acting sequences that specify secondary structures in a chloroplast mRNA are essential for RNA stability and translation. Mol Cell Biol. 1999;19:8479–91. Available: https://www.ncbi.nlm.nih.gov/pubmed/10567573

40. Zou Z, Eibl C, Koop H-U. The stem-loop region of the tobacco psbA 5′ UTR is an important determinant of mRNA stability and translation efficiency. Mol Genet Genomics Springer. 2003;269:340–9. Available: http://link.springer.com/article/10.1007/s00438-003-0842-2

41. Muslimov IA, Nimmrich V, Hernandez AI, Tcherepanov A, Sacktor TC, Tiedge H. Dendritic transport and localization of protein kinase Mζ mRNA: implications for molecular memory consolidation. J Biol Chem. 2004;279:52613–22. https://doi.org/10.1074/jbc.M409240200.

42. Chao JA, Patskovsky Y, Patel V, Levy M, Almo SC, Singer RH. ZBP1 recognition of β-actin zipcode induces RNA looping. Genes Dev. 2010;24:148–58. https://doi.org/10.1101/gad.1862910.

43. Kim HH, Lee SJ, Gardiner AS, Perrone-Bizzozero NI, Yoo S. Different motif requirements for the localization zipcode element of β-actin mRNA binding by HuD and ZBP1. Nucleic Acids Res. 2015;43:7432–46. https://doi.org/10.1093/nar/gkv699.

44. Kadrmas JL, Ravin AJ, Leontis NB. Relative stabilities of DNA three-way, four-way and five-way junctions (multi-helix junction loops): unpaired nucleotides can be stabilizing or destabilizing. Nucleic Acids Res. 1995;23:2212–22. Available: https://www.ncbi.nlm.nih.gov/pubmed/7610050

45. Ke A, Zhou K, Ding F, Cate JHD, Doudna JA. A conformational switch controls hepatitis delta virus ribozyme catalysis. Nature. 2004;429:201–5. https://doi.org/10.1038/nature02522.

46. Rastogi T, Beattie TL, Olive JE, Collins RA. A long-range pseudoknot is required for activity of the Neurospora VS ribozyme. EMBO J. 1996;15:2820–5. Available: https://www.ncbi.nlm.nih.gov/pubmed/8654379

47. Theimer CA, Blois CA, Feigon J. Structure of the human telomerase RNA pseudoknot reveals conserved tertiary interactions essential for function. Mol Cell. 2005;17:671–82. https://doi.org/10.1016/j.molcel.2005.01.017.

48. Tang CK, Draper DE. Unusual mRNA pseudoknot structure is recognized by a protein translational repressor. Cell. 1989;57:531–6. Available: https://www.ncbi.nlm.nih.gov/pubmed/2470510

49. Koritzinsky M, Rouschop KMA, van den Beucken T, Magagnin MG, Savelkouls K, Lambin P, et al. Phosphorylation of eIF2alpha is required for mRNA translation inhibition and survival during moderate hypoxia. Radiother Oncol. 2007;83:353–61. https://doi.org/10.1016/j.radonc.2007.04.031.

50. Spriggs KA, Bushell M, Willis AE. Translational regulation of gene expression during conditions of cell stress. Mol Cell. 2010;40:228–37. https://doi.org/10.1016/j.molcel.2010.09.028.

51. Kedersha NL, Gupta M, Li W, Miller I, Anderson P. RNA-binding proteins TIA-1 and TIAR link the phosphorylation of eIF-2 alpha to the assembly of mammalian stress granules. J Cell Biol. 1999;147:1431–42. Available: https://www.ncbi.nlm.nih.gov/pubmed/10613902

52. Harding HP, Novoa I, Zhang Y, Zeng H, Wek R, Schapira M, et al. Regulated translation initiation controls stress-induced gene expression in mammalian cells. Mol Cell. 2000;6:1099–108. Available: https://www.ncbi.nlm.nih.gov/pubmed/11106749

53. Mazroui R, Sukarieh R, Bordeleau M-E, Kaufman RJ, Northcote P, Tanaka J, et al. Inhibition of ribosome recruitment induces stress granule formation independently of eukaryotic initiation factor 2alpha phosphorylation. Mol Biol Cell. 2006;17:4212–9. https://doi.org/10.1091/mbc.E06-04-0318.

54. Kedersha N, Chen S, Gilks N, Li W, Miller IJ, Stahl J, et al. Evidence that ternary complex (eIF2-GTP-tRNA(i)(Met))-deficient preinitiation complexes are core constituents of mammalian stress granules. Mol Biol Cell. 2002;13:195–210. https://doi.org/10.1091/mbc.01-05-0221.

55. Kimball SR, Horetsky RL, Ron D, Jefferson LS, Harding HP. Mammalian stress granules represent sites of accumulation of stalled translation initiation complexes. Am J Physiol Cell Physiol. 2003;284:C273–84. https://doi.org/10.1152/ajpcell.00314.2002.

56. Buchan JR, Yoon J-H, Parker R. Stress-specific composition, assembly and kinetics of stress granules in Saccharomyces cerevisiae. J Cell Sci. 2011;124:228–39. https://doi.org/10.1242/jcs.078444.

57. Kwon S, Zhang Y, Matthias P. The deacetylase HDAC6 is a novel critical component of stress granules involved in the stress response. Genes Dev. 2007;21:3381–94. https://doi.org/10.1101/gad.461107.

58. Gallouzi IE, Parker F, Chebli K, Maurier F, Labourier E, Barlat I, et al. A novel phosphorylation-dependent RNase activity of GAP-SH3 binding protein: a potential link between signal transduction and RNA stability. Mol Cell Biol. 1998;18:3956–65. Available: https://www.ncbi.nlm.nih.gov/pubmed/9632780

59. Schmidlin M, Lu M, Leuenberger SA, Stoecklin G, Mallaun M, Gross B, et al. The ARE-dependent mRNA-destabilizing activity of BRF1 is regulated by protein kinase B. EMBO J. 2004;23:4760–9. https://doi.org/10.1038/sj.emboj.7600477.

60. Tourrière H, Gallouzi IE, Chebli K, Capony JP, Mouaikel J, van der Geer P, et al. RasGAP-associated endoribonuclease G3Bp: selective RNA degradation and phosphorylation-dependent localization. Mol Cell Biol. 2001;21:7747–60. https://doi.org/10.1128/MCB.21.22.7747-7760.2001.

61. Ohn T, Kedersha N, Hickman T, Tisdale S, Anderson P. A functional RNAi screen links O-GlcNAc modification of ribosomal proteins to stress granule and processing body assembly. Nat Cell Biol. 2008;10:1224–31. https://doi.org/10.1038/ncb1783.

62. Gilks N, Kedersha N, Ayodele M, Shen L, Stoecklin G, Dember LM, et al. Stress granule assembly is mediated by prion-like aggregation of TIA-1. Mol Biol Cell. 2004;15:5383–98. https://doi.org/10.1091/mbc.E04-08-0715.

63. Colombrita C, Zennaro E, Fallini C, Weber M, Sommacal A, Buratti E, et al. TDP-43 is recruited to stress granules in conditions of oxidative insult. J Neurochem Wiley Online Library. 2009;111:1051–61. Available: http://onlinelibrary.wiley.com/doi/10.1111/j.1471-4159.2009.06383.x/full

64. Buchan JR, Parker R. Eukaryotic stress granules: the ins and outs of translation. Mol Cell. 2009;36:932–41. https://doi.org/10.1016/j.molcel.2009.11.020.

65. Laroia G, Cuesta R, Brewer G, Schneider RJ. Control of mRNA decay by heat shock-ubiquitin-proteasome pathway. Science. 1999;284:499–502. Available: https://www.ncbi.nlm.nih.gov/pubmed/10205060

66. Gowrishankar G, Winzen R, Dittrich-Breiholz O, Redich N, Kracht M, Holtmann H. Inhibition of mRNA deadenylation and degradation by different types of cell stress. Biol Chem. 2006;387:323–7. https://doi.org/10.1515/BC.2006.043.

67. Hilgers V, Teixeira D, Parker R. Translation-independent inhibition of mRNA deadenylation during stress in Saccharomyces cerevisiae. RNA. 2006;12:1835–45. https://doi.org/10.1261/rna.241006.

68. Thomas MG, Martinez Tosar LJ, Desbats MA, Leishman CC, Boccaccio GL. Mammalian Staufen 1 is recruited to stress granules and impairs their assembly. J Cell Sci. 2009;122:563–73. https://doi.org/10.1242/jcs.038208.

69. Tsai N-P, Ho P-C, Wei L-N. Regulation of stress granule dynamics by Grb7 and FAK signalling pathway. EMBO J. 2008;27:715–26. https://doi.org/10.1038/emboj.2008.19.

70. Rikhvanov EG, Romanova NV, Chernoff YO. Chaperone effects on prion and nonprion aggregates. Prion. 2007;1:217–22. Available: https://www.ncbi.nlm.nih.gov/pubmed/19164915

71. Raijmakers R, Egberts WV, van Venrooij WJ, Pruijn GJM. Protein–protein interactions between human exosome components support the assembly of RNase PH-type subunits into a six-membered PNPase-like ring. J Mol Biol. 2002;323:653–63. https://doi.org/10.1016/S0022-2836(02)00947-6.

72. Schilders G, van Dijk E, Raijmakers R, Pruijn GJM. Cell and molecular biology of the exosome: how to make or break an RNA. Int Rev Cytol. 2006;251:159–208. https://doi.org/10.1016/S0074-7696(06)51005-8.

73. Budenholzer L, Cheng CL, Li Y, Hochstrasser M. Proteasome structure and assembly. J Mol Biol. 2017. https://doi.org/10.1016/j.jmb.2017.05.027.

74. Houseley J, LaCava J, Tollervey D. RNA-quality control by the exosome. Nat Rev Mol Cell Biol. 2006;7:529–39. https://doi.org/10.1038/nrm1964.

75. Chen CY, Gherzi R, Ong SE, Chan EL, Raijmakers R, Pruijn GJ, et al. AU binding proteins recruit the exosome to degrade ARE-containing mRNAs. Cell. 2001;107:451–64. Available: https://www.ncbi.nlm.nih.gov/pubmed/11719186

76. Kowalinski E, Schuller A, Green R, Conti E. Saccharomyces cerevisiae Ski7 Is a GTP-binding protein adopting the characteristic conformation of active translational GTPases. Structure. 2015;23:1336–43. https://doi.org/10.1016/j.str.2015.04.018.

77. Frischmeyer PA, van Hoof A, O'Donnell K, Guerrerio AL, Parker R, Dietz HC. An mRNA surveillance mechanism that eliminates transcripts lacking termination codons. Science. 2002;295:2258–61. https://doi.org/10.1126/science.1067338.

78. Allmang C, Kufel J, Chanfreau G, Mitchell P, Petfalski E, Tollervey D. Functions of the exosome in rRNA, snoRNA and snRNA synthesis. EMBO J. 1999;18:5399–410. https://doi.org/10.1093/emboj/18.19.5399.

79. Nagy E, Maquat LE. A rule for termination-codon position within intron-containing genes: when nonsense affects RNA abundance. Trends Biochem Sci. 1998;23:198–9. Available: https://www.ncbi.nlm.nih.gov/pubmed/9644970

80. Thermann R, Neu-Yilik G, Deters A, Frede U, Wehr K, Hagemeier C, et al. Binary specification of nonsense codons by splicing and cytoplasmic translation. EMBO J. 1998;17:3484–94. https://doi.org/10.1093/emboj/17.12.3484.

81. Kashima I, Yamashita A, Izumi N, Kataoka N, Morishita R, Hoshino S, et al. Binding of a novel SMG-1-Upf1-eRF1-eRF3 complex (SURF) to the exon junction complex triggers Upf1 phosphorylation and nonsense-mediated mRNA decay. Genes Dev. 2006;20:355–67. https://doi.org/10.1101/gad.1389006.

82. Melero R, Uchiyama A, Castaño R, Kataoka N, Kurosawa H, Ohno S, et al. Structures of SMG1-UPFs complexes: SMG1 contributes to regulate UPF2-dependent activation of UPF1 in NMD. Structure. 2014;22:1105–19. https://doi.org/10.1016/j.str.2014.05.015.

83. Deniaud A, Karuppasamy M, Bock T, Masiulis S, Huard K, Garzoni F, et al. A network of SMG-8, SMG-9 and SMG-1 C-terminal insertion domain regulates UPF1 substrate recruitment and phosphorylation. Nucleic Acids Res. 2015;43:7600–11. https://doi.org/10.1093/nar/gkv668.

84. Franks TM, Singh G, Lykke-Andersen J. Upf1 ATPase-dependent mRNP disassembly is required for completion of nonsense- mediated mRNA decay. Cell. 2010;143:938–50. https://doi.org/10.1016/j.cell.2010.11.043.

85. Fiorini F, Bagchi D, Le Hir H, Croquette V. Human Upf1 is a highly processive RNA helicase and translocase with RNP remodelling activities. Nat Commun. 2015;6:7581. https://doi.org/10.1038/ncomms8581.

86. Huntzinger E, Kashima I, Fauser M, Saulière J, Izaurralde E. SMG6 is the catalytic endonuclease that cleaves mRNAs containing nonsense codons in metazoan. RNA. 2008;14:2609–17. https://doi.org/10.1261/rna.1386208.

87. Eberle AB, Lykke-Andersen S, Mühlemann O, Jensen TH. SMG6 promotes endonucleolytic cleavage of nonsense mRNA in human cells. Nat Struct Mol Biol. 2009;16:49–55. https://doi.org/10.1038/nsmb.1530.

88. Loh B, Jonas S, Izaurralde E. The SMG5–SMG7 heterodimer directly recruits the CCR4–NOT deadenylase complex to mRNAs containing nonsense codons via interaction with POP2. Genes Dev. 2013;27:2125–38. https://doi.org/10.1101/gad.226951.113.

89. Unterholzner L, Izaurralde E. SMG7 acts as a molecular link between mRNA surveillance and mRNA decay. Mol Cell. 2004;16:587–96. https://doi.org/10.1016/j.molcel.2004.10.013.

90. Hogg JR, Goff SP. Upf1 senses 3′ UTR length to potentiate mRNA decay. Cell. 2010;143:379–89. Available: http://www.sciencedirect.com/science/article/pii/S0092867410011414

91. Schweingruber C, Rufener SC, Zünd D, Yamashita A, Mühlemann O. Nonsense-mediated mRNA decay – mechanisms of substrate mRNA recognition and degradation in mammalian cells. Biochim Biophys Acta. 1829;2013:612–23. https://doi.org/10.1016/j.bbagrm.2013.02.005.

92. Avery P, Vicente-Crespo M, Francis D, Nashchekina O, Alonso CR, Palacios IM. Drosophila Upf1 and Upf2 loss of function inhibits cell growth and causes animal death in a Upf3-independent manner. RNA. 2011;17:624–38. https://doi.org/10.1261/rna.2404211.

93. Lou CH, Shao A, Shum EY, Espinoza JL, Huang L, Karam R, et al. Posttranscriptional control of the stem cell and neurogenic programs by the nonsense-mediated RNA decay pathway. Cell Rep. 2014;6:748–64. https://doi.org/10.1016/j.celrep.2014.01.028.

94. Gloggnitzer J, Akimcheva S, Srinivasan A, Kusenda B, Riehs N, Stampfl H, et al. Nonsense-mediated mRNA decay modulates immune receptor levels to regulate plant antibacterial defense. Cell Host Microbe. 2014;16:376–90. https://doi.org/10.1016/j.chom.2014.08.010.

95. Gardner LB. Nonsense-mediated RNA decay regulation by cellular stress: implications for tumorigenesis. Mol Cancer Res. 2010;8:295–308. https://doi.org/10.1158/1541-7786.MCR-09-0502.

96. Balistreri G, Horvath P, Schweingruber C, Zünd D, McInerney G, Merits A, et al. The host nonsense-mediated mRNA decay pathway restricts mammalian RNA virus replication. Cell Host Microbe. 2014;16:403–11. https://doi.org/10.1016/j.chom.2014.08.007.

97. Giorgi C, Yeo GW, Stone ME, Katz DB, Burge C, Turrigiano G, et al. The EJC factor eIF4AIII modulates synaptic strength and neuronal protein expression. Cell. 2007;130:179–91. https://doi.org/10.1016/j.cell.2007.05.028.

98. Colak D, Ji S-J, Porse BT, Jaffrey SR. Regulation of axon guidance by compartmentalized nonsense-mediated mRNA decay. Cell. 2013;153:1252–65. https://doi.org/10.1016/j.cell.2013.04.056.

99. Weg-Remers S, Ponta H, Herrlich P, König H. Regulation of alternative pre-mRNA splicing by the ERK MAP-kinase pathway. EMBO J. 2001;20:4194–203. https://doi.org/10.1093/emboj/20.15.4194.

100. van der Houven van Oordt W, Diaz-Meco MT, Lozano J, Krainer AR, Moscat J, Cáceres JF. The MKK(3/6)-p38-signaling cascade alters the subcellular distribution of hnRNP A1 and modulates alternative splicing regulation. J Cell Biol. 2000;149:307–16. Available: https://www.ncbi.nlm.nih.gov/pubmed/10769024

101. Lewis BP, Green RE, Brenner SE. Evidence for the widespread coupling of alternative splicing and nonsense-mediated mRNA decay in humans. Proc Natl Acad Sci U S A. 2003;100:189–92. https://doi.org/10.1073/pnas.0136770100.

102. Eom T, Zhang C, Wang H, Lay K, Fak J, Noebels JL, et al. NOVA-dependent regulation of cryptic NMD exons controls synaptic protein levels after seizure. elife. 2013;2:e00178. https://doi.org/10.7554/eLife.00178.

103. Winter J, Lehmann T, Krauss S, Trockenbacher A, Kijas Z, Foerster J, et al. Regulation of the MID1 protein function is fine-tuned by a complex pattern of alternative splicing. Hum Genet. 2004;114:541–52. https://doi.org/10.1007/s00439-004-1114-x.
104. Sureau A, Gattoni R, Dooghe Y, Stévenin J, Soret J. SC35 autoregulates its expression by promoting splicing events that destabilize its mRNAs. EMBO J. 2001;20:1785–96. https://doi.org/10.1093/emboj/20.7.1785.
105. Wilson GM, Sun Y, Sellers J, Lu H, Penkar N, Dillard G, et al. Regulation of AUF1 expression via conserved alternatively spliced elements in the 3′ untranslated region. Mol Cell Biol. 1999;19:4056–64. Available: https://www.ncbi.nlm.nih.gov/pubmed/10330146
106. Lamba JK, Adachi M, Sun D, Tammur J, Schuetz EG, Allikmets R, et al. Nonsense mediated decay downregulates conserved alternatively spliced ABCC4 transcripts bearing nonsense codons. Hum Mol Genet. 2003;12:99–109. Available: https://www.ncbi.nlm.nih.gov/pubmed/12499391
107. Jones RB, Wang F, Luo Y, Yu C, Jin C, Suzuki T, et al. The nonsense-mediated decay pathway and mutually exclusive expression of alternatively spliced FGFR2IIIb and -IIIc mRNAs. J Biol Chem. 2001;276:4158–67. https://doi.org/10.1074/jbc.M006151200.
108. Lareau LF, Brooks AN, Soergel DAW, Meng Q, Brenner SE. The coupling of alternative splicing and nonsense-mediated mRNA decay. Adv Exp Med Biol. 2007;623:190–211. Available: https://www.ncbi.nlm.nih.gov/pubmed/18380348
109. Morris DR, Geballe AP. Upstream open reading frames as regulators of mRNA translation. Mol Cell Biol. 2000;20:8635–42. https://doi.org/10.1128/mcb.20.23.8635-8642.2000.
110. Kozak M. Structural features in eukaryotic mRNAs that modulate the initiation of translation. J Biol Chem. 1991;266:19867–70. Available: https://www.ncbi.nlm.nih.gov/pubmed/1939050
111. Matsui M, Yachie N, Okada Y, Saito R, Tomita M. Bioinformatic analysis of post-transcriptional regulation by uORF in human and mouse. FEBS Lett. 2007;581:4184–8. https://doi.org/10.1016/j.febslet.2007.07.057.
112. Calvo SE, Pagliarini DJ, Mootha VK. Upstream open reading frames cause widespread reduction of protein expression and are polymorphic among humans. Proc Natl Acad Sci U S A. 2009;106:7507–12. https://doi.org/10.1073/pnas.0810916106.
113. Mendell JT, Sharifi NA, Meyers JL, Martinez-Murillo F, Dietz HC. Nonsense surveillance regulates expression of diverse classes of mammalian transcripts and mutes genomic noise. Nat Genet. 2004;36:1073–8. https://doi.org/10.1038/ng1429.
114. Ramani AK, Nelson AC, Kapranov P, Bell I, Gingeras TR, Fraser AG. High resolution transcriptome maps for wild-type and nonsense-mediated decay-defective Caenorhabditis elegans. Genome Biol. 2009;10:R101. https://doi.org/10.1186/gb-2009-10-9-r101.
115. He F, Li X, Spatrick P, Casillo R, Dong S, Jacobson A. Genome-wide analysis of mRNAs regulated by the nonsense-mediated and 5′ to 3′ mRNA decay pathways in yeast. Mol Cell. 2003;12:1439–52. https://doi.org/10.1016/s1097-2765(03)00446-5.
116. van Hoof A, Frischmeyer PA, Dietz HC, Parker R. Exosome-mediated recognition and degradation of mRNAs lacking a termination codon. Science. 2002;295:2262–4. https://doi.org/10.1126/science.1067272.
117. Karzai AW, Roche ED, Sauer RT. The SsrA-SmpB system for protein tagging, directed degradation and ribosome rescue. Nat Struct Biol. 2000;7:449–55. https://doi.org/10.1038/75843.
118. Benard L, Carroll K, Valle RC, Masison DC, Wickner RB. The ski7 antiviral protein is an EF1-alpha homolog that blocks expression of non-Poly(A) mRNA in Saccharomyces cerevisiae. J Virol. 1999;73:2893–900. Available: https://www.ncbi.nlm.nih.gov/pubmed/10074137
119. Anderson JS, Parker RP. The 3′ to 5′ degradation of yeast mRNAs is a general mechanism for mRNA turnover that requires the SKI2 DEVH box protein and 3′ to 5′ exonucleases of the exosome complex. EMBO J. 1998;17:1497–506. https://doi.org/10.1093/emboj/17.5.1497.
120. Caponigro G, Parker R. Multiple functions for the poly(A)-binding protein in mRNA decapping and deadenylation in yeast. Genes Dev. 1995;9:2421–32. https://doi.org/10.1101/gad.9.19.2421.

121. Doma MK, Parker R. Endonucleolytic cleavage of eukaryotic mRNAs with stalls in translation elongation. Nature. 2006;440:561–4. https://doi.org/10.1038/nature04530.

122. Tollervey D. Molecular biology: RNA lost in translation. Nature. 2006;440:425–6. https://doi.org/10.1038/440425a.

123. Clement SL, Lykke-Andersen J. No mercy for messages that mess with the ribosome. Nat Struct Mol Biol. 2006;13:299–301. https://doi.org/10.1038/nsmb0406-299.

124. Young SK, Palam LR, Wu C, Sachs MS, Wek RC. Ribosome elongation stall directs gene-specific translation in the integrated stress response. J Biol Chem. 2016;291:6546–58. https://doi.org/10.1074/jbc.M115.705640.

125. Passos DO, Doma MK, Shoemaker CJ, Muhlrad D, Green R, Weissman J, et al. Analysis of Dom34 and its function in no-go decay. Mol Biol Cell. 2009;20:3025–32. https://doi.org/10.1091/mbc.E09-01-0028.

126. Graille M, Chaillet M, van Tilbeurgh H. Structure of yeast Dom34: a protein related to translation termination factor Erf1 and involved in No-Go decay. J Biol Chem. 2008;283:7145–54. https://doi.org/10.1074/jbc.M708224200.

127. Lee HH. Structural and functional insights into Dom34, a key component of No-Go mRNA decay, and structure of a metal Ion-Bound IS200 transposase. Mol Cell. 2007;27(6):938–50. Available: http://s-space.snu.ac.kr/handle/10371/19510

128. Carr-Schmid A, Pfund C, Craig EA, Kinzy TG. Novel G-protein complex whose requirement is linked to the translational status of the cell. Mol Cell Biol. 2002;22:2564–74. Available: https://www.ncbi.nlm.nih.gov/pubmed/11909951

129. Protzel A, Morris AJ. Gel chromatographic analysis of nascent globin chains. Evidence of nonuniform size distribution. J Biol Chem. 1974;249:4594–600. Available: https://www.ncbi.nlm.nih.gov/pubmed/4843145

130. Beelman CA, Parker R. Differential effects of translational inhibition in cis and in trans on the decay of the unstable yeast MFA2 mRNA. J Biol Chem. 1994;269:9687–92. Available: https://www.ncbi.nlm.nih.gov/pubmed/8144558

131. Nagai K, Oubridge C, Kuglstatter A, Menichelli E, Isel C, Jovine L. Structure, function and evolution of the signal recognition particle. EMBO J. 2003;22:3479–85. https://doi.org/10.1093/emboj/cdg337.

132. Wang Z, Sachs MS. Ribosome stalling is responsible for arginine-specific translational attenuation in Neurospora crassa. Mol Cell Biol. 1997;17:4904–13. Available: https://www.ncbi.nlm.nih.gov/pubmed/9271370

133. Chen CY, Shyu AB. AU-rich elements: characterization and importance in mRNA degradation. Trends Biochem Sci. 1995;20:465–70. Available: https://www.ncbi.nlm.nih.gov/pubmed/8578590

134. Shaw G, Kamen R. A conserved AU sequence from the 3′ untranslated region of GM-CSF mRNA mediates selective mRNA degradation. Cell. 1986;46:659–67. Available: https://www.ncbi.nlm.nih.gov/pubmed/3488815

135. Loflin P, Chen CY, Shyu AB. Unraveling a cytoplasmic role for hnRNP D in the in vivo mRNA destabilization directed by the AU-rich element. Genes Dev. 1999;13:1884–97. Available: https://www.ncbi.nlm.nih.gov/pubmed/10421639

136. Sarkar B, Xi Q, He C, Schneider RJ. Selective degradation of AU-rich mRNAs promoted by the p37 AUF1 protein isoform. Mol Cell Biol. 2003;23:6685–93. Available: https://www.ncbi.nlm.nih.gov/pubmed/12944492

137. Lal A, Mazan-Mamczarz K, Kawai T, Yang X, Martindale JL, Gorospe M. Concurrent versus individual binding of HuR and AUF1 to common labile target mRNAs. EMBO J. 2004;23:3092–102. https://doi.org/10.1038/sj.emboj.7600305.

138. Raineri I, Wegmueller D, Gross B, Certa U, Moroni C. Roles of AUF1 isoforms, HuR and BRF1 in ARE-dependent mRNA turnover studied by RNA interference. Nucleic Acids Res. 2004;32:1279–88. https://doi.org/10.1093/nar/gkh282.

139. Carballo E, Gilkeson GS, Blackshear PJ. Bone marrow transplantation reproduces the tristetraprolin-deficiency syndrome in recombination activating gene-2 (−/−) mice. Evidence

that monocyte/macrophage progenitors may be responsible for TNFalpha overproduction. J Clin Invest. 1997;100:986–95. https://doi.org/10.1172/JCI119649.

140. Carballo E, Lai WS, Blackshear PJ. Evidence that tristetraprolin is a physiological regulator of granulocyte-macrophage colony-stimulating factor messenger RNA deadenylation and stability. Blood. 2000;95:1891–9. Available: https://www.ncbi.nlm.nih.gov/pubmed/10706852

141. Ogilvie RL, Abelson M, Hau HH, Vlasova I, Blackshear PJ, Bohjanen PR. Tristetraprolin down-regulates IL-2 gene expression through AU-rich element-mediated mRNA decay. J Immunol. 2005;174:953–61. Available: https://www.ncbi.nlm.nih.gov/pubmed/15634918

142. Carballo E, Lai WS, Blackshear PJ. Feedback inhibition of macrophage tumor necrosis factor-alpha production by tristetraprolin. Science. 1998;281:1001–5. Available: https://www.ncbi.nlm.nih.gov/pubmed/9703499

143. Lai WS, Blackshear PJ. Interactions of CCCH zinc finger proteins with mRNA: tristetraprolin-mediated AU-rich element-dependent mRNA degradation can occur in the absence of a poly(A) tail. J Biol Chem. 2001;276:23144–54. https://doi.org/10.1074/jbc.M100680200.

144. Lai WS, Kennington EA, Blackshear PJ. Tristetraprolin and its family members can promote the cell-free deadenylation of AU-rich element-containing mRNAs by poly(A) ribonuclease. Mol Cell Biol. 2003;23:3798–812. https://www.ncbi.nlm.nih.gov/pubmed/12748283

145. Lai WS, Carballo E, Strum JR, Kennington EA, Phillips RS, Blackshear PJ. Evidence that tristetraprolin binds to AU-rich elements and promotes the deadenylation and destabilization of tumor necrosis factor alpha mRNA. Mol Cell Biol. 1999;19:4311–23. Available: https://www.ncbi.nlm.nih.gov/pubmed/10330172

146. Sawaoka H, Dixon DA, Oates JA, Boutaud O. Tristetraprolin binds to the 3'-untranslated region of cyclooxygenase-2 mRNA: a polyadenylation variant in a cancer cell line lacks the binding site. J Biol Chem. 2003;278:13928–35. https://doi.org/10.1074/jbc.M300016200.

147. Stoecklin G, Ming XF, Looser R, Moroni C. Somatic mRNA turnover mutants implicate tristetraprolin in the interleukin-3 mRNA degradation pathway. Mol Cell Biol. 2000;20:3753–63. Available: https://www.ncbi.nlm.nih.gov/pubmed/10805719

148. Wilson GM, Brewer G. The search for trans-acting factors controlling messenger RNA decay. Prog Nucleic Acid Res Mol Biol. 1999;62:257–91. Available: https://www.ncbi.nlm.nih.gov/pubmed/9932457

149. Wilson T, Treisman R. Removal of poly(A) and consequent degradation of c-fos mRNA facilitated by 3' AU-rich sequences. Nature. 1988;336:396–9. https://doi.org/10.1038/336396a0.

150. Shyu AB, Belasco JG, Greenberg ME. Two distinct destabilizing elements in the c-fos message trigger deadenylation as a first step in rapid mRNA decay. Genes Dev. 1991;5:221–31. Available: https://www.ncbi.nlm.nih.gov/pubmed/1899842

151. Brewer G, Ross J. Poly(A) shortening and degradation of the 3' A+U-rich sequences of human c-myc mRNA in a cell-free system. Mol Cell Biol. 1988;8:1697–708. https://doi.org/10.1128/mcb.8.4.1697.

152. Mukherjee D, Gao M, O'Connor JP, Raijmakers R, Pruijn G, Lutz CS, et al. The mammalian exosome mediates the efficient degradation of mRNAs that contain AU-rich elements. EMBO J. 2002;21:165–74. https://doi.org/10.1093/emboj/21.1.165.

153. Gherzi R, Lee K-Y, Briata P, Wegmüller D, Moroni C, Karin M, et al. A KH domain RNA binding protein, KSRP, promotes ARE-directed mRNA turnover by recruiting the degradation machinery. Mol Cell. 2004;14:571–83. https://doi.org/10.1016/j.molcel.2004.05.002.

154. Kedersha N, Stoecklin G, Ayodele M, Yacono P, Lykke-Andersen J, Fritzler MJ, et al. Stress granules and processing bodies are dynamically linked sites of mRNP remodeling. J Cell Biol. 2005;169:871–84. https://doi.org/10.1083/jcb.200502088.

155. Kedersha N, Anderson P. Stress granules: sites of mRNA triage that regulate mRNA stability and translatability. Biochem Soc Trans. 2002;30:963–9.

156. Levy NS, Chung S, Furneaux H, Levy AP. Hypoxic stabilization of vascular endothelial growth factor mRNA by the RNA-binding protein HuR. J Biol Chem. 1998;273:6417–23. Available: https://www.ncbi.nlm.nih.gov/pubmed/9497373

157. Peng SS, Chen CY, Xu N, Shyu AB. RNA stabilization by the AU-rich element binding protein, HuR, an ELAV protein. EMBO J. 1998;17:3461–70. https://doi.org/10.1093/emboj/17.12.3461.

158. Rodriguez-Pascual F, Hausding M, Ihrig-Biedert I, Furneaux H, Levy AP, Förstermann U, et al. Complex contribution of the 3′-untranslated region to the expressional regulation of the human inducible nitric-oxide synthase gene. Involvement of the RNA-binding protein HuR. J Biol Chem. 2000;275:26040–9. https://doi.org/10.1074/jbc.M910460199.

159. JLE D, Wait R, Mahtani KR, Sully G, Clark AR, Saklatvala J. The 3′ untranslated region of tumor necrosis factor alpha mRNA is a target of the mRNA-stabilizing factor HuR. Mol Cell Biol. 2005;25:3400. https://doi.org/10.1128/MCB.25.8.3400.2005.

160. Kim YK, Furic L, Desgroseillers L, Maquat LE. Mammalian Staufen1 recruits Upf1 to specific mRNA 3′UTRs so as to elicit mRNA decay. Cell. 2005;120:195–208. https://doi.org/10.1016/j.cell.2004.11.050.

161. Kim YK, Furic L, Parisien M, Major F, DesGroseillers L, Maquat LE. Staufen1 regulates diverse classes of mammalian transcripts. EMBO J. 2007;26:2670–81. https://doi.org/10.1038/sj.emboj.7601712.

162. Gong C, Maquat LE. lncRNAs transactivate STAU1-mediated mRNA decay by duplexing with 3′ UTRs via Alu elements. Nature. 2011;470:284–8. https://doi.org/10.1038/nature09701.

163. Gong C, Kim YK, Woeller CF, Tang Y, Maquat LE. SMD and NMD are competitive pathways that contribute to myogenesis: effects on PAX3 and myogenin mRNAs. Genes Dev. 2009;23:54–66. https://doi.org/10.1101/gad.1717309.

164. Cai X, Hagedorn CH, Cullen BR. Human microRNAs are processed from capped, polyadenylated transcripts that can also function as mRNAs. RNA. 2004;10:1957–66. https://doi.org/10.1261/rna.7135204.

165. Lee Y, Kim M, Han J, Yeom K-H, Lee S, Baek SH, et al. MicroRNA genes are transcribed by RNA polymerase II. EMBO J. 2004;23:4051–60. https://doi.org/10.1038/sj.emboj.7600385.

166. Lee Y, Ahn C, Han J, Choi H, Kim J, Yim J, et al. The nuclear RNase III Drosha initiates microRNA processing. Nature. 2003;425:415–9. https://doi.org/10.1038/nature01957.

167. Gregory RI, Chendrimada TP, Shiekhattar R. MicroRNA biogenesis: isolation and characterization of the microprocessor complex. Methods Mol Biol. 2006;342:33–47. https://doi.org/10.1385/1-59745-123-1:33.

168. Murchison EP, Hannon GJ. miRNAs on the move: miRNA biogenesis and the RNAi machinery. Curr Opin Cell Biol. 2004;16:223–9. https://doi.org/10.1016/j.ceb.2004.04.003.

169. Lund E, Dahlberg JE. Substrate selectivity of exportin 5 and Dicer in the biogenesis of microRNAs. Cold Spring Harb Symp Quant Biol. 2006;71:59–66. https://doi.org/10.1101/sqb.2006.71.050.

170. Khvorova A, Reynolds A, Jayasena SD. Functional siRNAs and miRNAs exhibit strand bias. Cell. 2003;115:209–16. Available: https://www.ncbi.nlm.nih.gov/pubmed/14567918

171. Schwarz DS, Hutvágner G, Du T, Xu Z, Aronin N, Zamore PD. Asymmetry in the assembly of the RNAi enzyme complex. Cell. 2003;115:199–208. Available: https://www.ncbi.nlm.nih.gov/pubmed/14567917

172. Hausser J, Syed AP, Bilen B, Zavolan M. Analysis of CDS-located miRNA target sites suggests that they can effectively inhibit translation. Genome Res. 2013;23:604–15. https://doi.org/10.1101/gr.139758.112.

173. Fang Z, Rajewsky N. The impact of miRNA target sites in coding sequences and in 3′UTRs. PLoS One. 2011;6:e18067. https://doi.org/10.1371/journal.pone.0018067.

174. Alemán LM, Doench J, Sharp PA. Comparison of siRNA-induced off-target RNA and protein effects. RNA. 2007;13:385–95. https://doi.org/10.1261/rna.352507.

175. Eulalio A, Huntzinger E, Izaurralde E. GW182 interaction with Argonaute is essential for miRNA-mediated translational repression and mRNA decay. Nat Struct Mol Biol. 2008;15:346–53. https://doi.org/10.1038/nsmb.1405.

176. Behm-Ansmant I, Rehwinkel J, Doerks T, Stark A, Bork P, Izaurralde E. mRNA degradation by miRNAs and GW182 requires both CCR4:NOT deadenylase and DCP1:DCP2 decapping complexes. Genes Dev. 2006;20:1885–98. https://doi.org/10.1101/gad.1424106.
177. Wu L, Fan J, Belasco JG. MicroRNAs direct rapid deadenylation of mRNA. Proc Natl Acad Sci U S A. 2006;103:4034–9. https://doi.org/10.1073/pnas.0510928103.
178. Stoecklin G, Lu M, Rattenbacher B, Moroni C. A constitutive decay element promotes tumor necrosis factor alpha mRNA degradation via an AU-rich element-independent pathway. Mol Cell Biol. 2003;23:3506–15. Available: https://www.ncbi.nlm.nih.gov/pubmed/12724409
179. Leppek K, Schott J, Reitter S, Poetz F, Hammond MC, Stoecklin G. Roquin promotes constitutive mRNA decay via a conserved class of stem-loop recognition motifs. Cell. 2013;153:869–81. https://doi.org/10.1016/j.cell.2013.04.016.
180. Tan D, Zhou M, Kiledjian M, Tong L. The ROQ domain of Roquin recognizes mRNA constitutive-decay element and double-stranded RNA. Nat Struct Mol Biol. 2014;21:679–85. https://doi.org/10.1038/nsmb.2857.
181. Marzluff WF, Wagner EJ, Duronio RJ. Metabolism and regulation of canonical histone mRNAs: life without a poly(A) tail. Nat Rev Genet. 2008;9:843–54. https://doi.org/10.1038/nrg2438.
182. Mullen TE, Marzluff WF. Degradation of histone mRNA requires oligouridylation followed by decapping and simultaneous degradation of the mRNA both 5' to 3' and 3' to 5'. Genes Dev. 2008;22:50–65. https://doi.org/10.1101/gad.1622708.
183. Kaygun H, Marzluff WF. Regulated degradation of replication-dependent histone mRNAs requires both ATR and Upf1. Nat Struct Mol Biol. 2005;12:794–800. https://doi.org/10.1038/nsmb972.
184. Eulalio A, Behm-Ansmant I, Schweizer D, Izaurralde E. P-body formation is a consequence, not the cause, of RNA-mediated gene silencing. Mol Cell Biol. 2007;27:3970–81. https://doi.org/10.1128/MCB.00128-07.
185. Teixeira D, Sheth U, Valencia-Sanchez MA, Brengues M, Parker R. Processing bodies require RNA for assembly and contain nontranslating mRNAs. RNA. 2005;11:371–82. https://doi.org/10.1261/rna.7258505.
186. Liu J, Valencia-Sanchez MA, Hannon GJ, Parker R. MicroRNA-dependent localization of targeted mRNAs to mammalian P-bodies. Nat Cell Biol. 2005;7:719–23. https://doi.org/10.1038/ncb1274.
187. Brengues M, Teixeira D, Parker R. Movement of eukaryotic mRNAs between polysomes and cytoplasmic processing bodies. Science. 2005;310:486–9. https://doi.org/10.1126/science.1115791.
188. Koritzinsky M, Magagnin MG, van den Beucken T, Seigneuric R, Savelkouls K, Dostie J, et al. Gene expression during acute and prolonged hypoxia is regulated by distinct mechanisms of translational control. EMBO J. 2006;25:1114–25. https://doi.org/10.1038/sj.emboj.7600998.
189. Coller J, Parker R. General translational repression by activators of mRNA decapping. Cell. 2005;122:875–86. https://doi.org/10.1016/j.cell.2005.07.012.
190. Parker R, Sheth U. P bodies and the control of mRNA translation and degradation. Mol Cell. 2007;25:635–46. https://doi.org/10.1016/j.molcel.2007.02.011.
191. Holmes LEA, Campbell SG, De Long SK, Sachs AB, Ashe MP. Loss of translational control in yeast compromised for the major mRNA decay pathway. Mol Cell Biol. 2004;24:2998–3010. Available: https://www.ncbi.nlm.nih.gov/pubmed/15024087
192. Sheth U, Parker R. Decapping and decay of messenger RNA occur in cytoplasmic processing bodies. Science. 2003;300:805–8. https://doi.org/10.1126/science.1082320.
193. Cougot N, Babajko S, Séraphin B. Cytoplasmic foci are sites of mRNA decay in human cells. J Cell Biol. 2004;165:31–40. https://doi.org/10.1083/jcb.200309008.
194. Bhattacharyya SN, Habermacher R, Martine U, Closs EI, Filipowicz W. Relief of microRNA-mediated translational repression in human cells subjected to stress. Cell. 2006;125:1111–24. https://doi.org/10.1016/j.cell.2006.04.031.

195. D'Lima NG, Ma J, Winkler L, Chu Q, Loh KH, Corpuz EO, et al. A human microprotein that interacts with the mRNA decapping complex. Nat Chem Biol. 2017;13:174–80. https://doi.org/10.1038/nchembio.2249.

196. Cougot N, Daguenet E, Baguet A, Cavalier A, Thomas D, Bellaud P, et al. Overexpression of MLN51 triggers P-body disassembly and formation of a new type of RNA granules. J Cell Sci. 2014;127:4692–701. https://doi.org/10.1242/jcs.154500.

197. Aulas A, Vande VC. Alterations in stress granule dynamics driven by TDP-43 and FUS: a link to pathological inclusions in ALS? Front Cell Neurosci. 2015;9:423. https://doi.org/10.3389/fncel.2015.00423.

198. Zinszner H, Sok J, Immanuel D, Yin Y, Ron D. TLS (FUS) binds RNA in vivo and engages in nucleo-cytoplasmic shuttling. J Cell Sci. 1997;110(Pt 15):1741–50. Available: https://www.ncbi.nlm.nih.gov/pubmed/9264461

199. Åman P, Panagopoulos I, Lassen C, Fioretos T, Mencinger M, Toresson H, et al. Expression patterns of the human sarcoma-associated genes FUS and EWS and the genomic structure of FUS. Genomics. 1996;37:1–8. https://doi.org/10.1006/geno.1996.0513.

200. Ling S-C, Polymenidou M, Cleveland DW. Converging mechanisms in ALS and FTD: disrupted RNA and protein homeostasis. Neuron. 2013;79:416–38. https://doi.org/10.1016/j.neuron.2013.07.033.

201. Kwiatkowski TJ Jr, Bosco DA, Leclerc AL, Tamrazian E, Vanderburg CR, Russ C, et al. Mutations in the FUS/TLS gene on chromosome 16 cause familial amyotrophic lateral sclerosis. Science. 2009;323:1205–8. https://doi.org/10.1126/science.1166066.

202. Dormann D, Rodde R, Edbauer D, Bentmann E, Fischer I, Hruscha A, et al. ALS-associated fused in sarcoma (FUS) mutations disrupt transportin-mediated nuclear import. EMBO J. 2010;29:2841–57. https://doi.org/10.1038/emboj.2010.143.

203. Van Deerlin VM, Leverenz JB, Bekris LM, Bird TD, Yuan W, Elman LB, et al. TARDBP mutations in amyotrophic lateral sclerosis with TDP-43 neuropathology: a genetic and histopathological analysis. Lancet Neurol. 2008;7:409–16. https://doi.org/10.1016/S1474-4422(08)70071-1.

204. Ayala YM, Zago P, D'Ambrogio A, Xu Y-F, Petrucelli L, Buratti E, et al. Structural determinants of the cellular localization and shuttling of TDP-43. J Cell Sci. 2008;121:3778–85. https://doi.org/10.1242/jcs.038950.

205. Zhang ZC, Chook YM. Structural and energetic basis of ALS-causing mutations in the atypical proline-tyrosine nuclear localization signal of the Fused in Sarcoma protein (FUS). Proc Natl Acad Sci U S A. 2012;109:12017–21. https://doi.org/10.1073/pnas.1207247109.

206. McDonald KK, Aulas A, Destroismaisons L, Pickles S, Beleac E, Camu W, et al. TAR DNA-binding protein 43 (TDP-43) regulates stress granule dynamics via differential regulation of G3BP and TIA-1. Hum Mol Genet. 2011;20:1400–10. https://doi.org/10.1093/hmg/ddr021.

207. Liu-Yesucevitz L, Bilgutay A, Zhang Y-J, Vanderweyde T, Vanderwyde T, Citro A, et al. Tar DNA binding protein-43 (TDP-43) associates with stress granules: analysis of cultured cells and pathological brain tissue. PLoS One. 2010;5:e13250. https://doi.org/10.1371/journal.pone.0013250.

208. Parker SJ, Meyerowitz J, James JL, Liddell JR, Crouch PJ, Kanninen KM, et al. Endogenous TDP-43 localized to stress granules can subsequently form protein aggregates. Neurochem Int. 2012;60:415–24. https://doi.org/10.1016/j.neuint.2012.01.019.

209. Polymenidou M, Lagier-Tourenne C, Hutt KR, Huelga SC, Moran J, Liang TY, et al. Long pre-mRNA depletion and RNA missplicing contribute to neuronal vulnerability from loss of TDP-43. Nat Neurosci. 2011;14:459–68. https://doi.org/10.1038/nn.2779.

210. Lagier-Tourenne C, Polymenidou M, Hutt KR, Vu AQ, Baughn M, Huelga SC, et al. Divergent roles of ALS-linked proteins FUS/TLS and TDP-43 intersect in processing long pre-mRNAs. Nat Neurosci. 2012;15:1488–97. https://doi.org/10.1038/nn.3230.

211. Voigt A, Herholz D, Fiesel FC, Kaur K, Müller D, Karsten P, et al. TDP-43-mediated neuron loss in vivo requires RNA-binding activity. PLoS One. 2010;5:e12247. https://doi.org/10.1371/journal.pone.0012247.

212. Daigle JG, Lanson NA Jr, Smith RB, Casci I, Maltare A, Monaghan J, et al. RNA-binding ability of FUS regulates neurodegeneration, cytoplasmic mislocalization and incorporation into stress granules associated with FUS carrying ALS-linked mutations. Hum Mol Genet. 2013;22:1193–205. https://doi.org/10.1093/hmg/dds526.

213. Johnson JO, Pioro EP, Boehringer A, Chia R, Feit H, Renton AE, et al. Mutations in the Matrin 3 gene cause familial amyotrophic lateral sclerosis. Nat Neurosci. 2014;17:664–6. https://doi.org/10.1038/nn.3688.

214. Kim HJ, Kim NC, Wang Y-D, Scarborough EA, Moore J, Diaz Z, et al. Mutations in prion-like domains in hnRNPA2B1 and hnRNPA1 cause multisystem proteinopathy and ALS. Nature. 2013;495:467–73. https://doi.org/10.1038/nature11922.

215. Mackenzie IR, Nicholson AM, Sarkar M, Messing J, Purice MD, Pottier C, et al. TIA1 mutations in amyotrophic lateral sclerosis and frontotemporal dementia promote phase separation and alter stress granule dynamics. Neuron. 2017;95:808–816.e9. https://doi.org/10.1016/j.neuron.2017.07.025.

216. Bug M, Meyer H. Expanding into new markets—VCP/p97 in endocytosis and autophagy. J Struct Biol. 2012;179:78–82. Available: http://www.sciencedirect.com/science/article/pii/S1047847712000810

217. Dantuma NP, Acs K, Luijsterburg MS. Should I stay or should I go: VCP/p97-mediated chromatin extraction in the DNA damage response. Exp Cell Res. 2014;329:9–17. https://doi.org/10.1016/j.yexcr.2014.08.025.

218. Buchan JR, Kolaitis R-M, Taylor JP, Parker R. Eukaryotic stress granules are cleared by autophagy and Cdc48/VCP function. Cell. 2013;153:1461–74. https://doi.org/10.1016/j.cell.2013.05.037.

219. Koppers M, van Blitterswijk MM, Vlam L, Rowicka PA, van PWJ V, EJN G, et al. VCP mutations in familial and sporadic amyotrophic lateral sclerosis. Neurobiol Aging. 2012;33:837.e7–13. https://doi.org/10.1016/j.neurobiolaging.2011.10.006.

220. Treangen TJ, Salzberg SL. Repetitive DNA and next-generation sequencing: computational challenges and solutions. Nat Rev Genet. 2011;13:36–46. https://doi.org/10.1038/nrg3117.

221. Echeverria GV, Cooper TA. RNA-binding proteins in microsatellite expansion disorders: mediators of RNA toxicity. Brain Res. 2012;1462:100–11. https://doi.org/10.1016/j.brainres.2012.02.030.

222. Mohan A, Goodwin M, Swanson MS. RNA-protein interactions in unstable microsatellite diseases. Brain Res. 2014;1584:3–14. https://doi.org/10.1016/j.brainres.2014.03.039.

223. Kiliszek A, Rypniewski W. Structural studies of CNG repeats. Nucleic Acids Res. 2014;42:8189–99. https://doi.org/10.1093/nar/gku536.

224. Napierała M, Krzyzosiak WJ. CUG repeats present in myotonin kinase RNA form metastable "slippery" hairpins. J Biol Chem. 1997;272:31079–85. Available: https://www.ncbi.nlm.nih.gov/pubmed/9388259

225. Haeusler AR, Donnelly CJ, Periz G, Simko EAJ, Shaw PG, Kim M-S, et al. C9orf72 nucleotide repeat structures initiate molecular cascades of disease. Nature. 2014;507:195–200. https://doi.org/10.1038/nature13124.

226. Burge S, Parkinson GN, Hazel P, Todd AK, Neidle S. Quadruplex DNA: sequence, topology and structure. Nucleic Acids Res. 2006;34:5402–15. https://doi.org/10.1093/nar/gkl655.

227. Paeschke K, Simonsson T, Postberg J, Rhodes D, Lipps HJ. Telomere end-binding proteins control the formation of G-quadruplex DNA structures in vivo. Nat Struct Mol Biol. 2005;12:847–54. https://doi.org/10.1038/nsmb982.

228. Renton AE, Majounie E, Waite A, Simón-Sánchez J, Rollinson S, Gibbs JR, et al. A hexanucleotide repeat expansion in C9ORF72 is the cause of chromosome 9p21-linked ALS-FTD. Neuron. 2011;72:257–68. https://doi.org/10.1016/j.neuron.2011.09.010.

229. DeJesus-Hernandez M, Mackenzie IR, Boeve BF, Boxer AL, Baker M, Rutherford NJ, et al. Expanded GGGGCC hexanucleotide repeat in noncoding region of C9ORF72 causes chromosome 9p-linked FTD and ALS. Neuron. 2011;72:245–56. https://doi.org/10.1016/j.neuron.2011.09.011.

230. Rutherford NJ, Heckman MG, Dejesus-Hernandez M, Baker MC, Soto-Ortolaza AI, Rayaprolu S, et al. Length of normal alleles of C9ORF72 GGGGCC repeat do not influence disease phenotype. Neurobiol Aging. 2012;33:2950.e5–7. https://doi.org/10.1016/j.neurobiolaging.2012.07.005.

231. van Blitterswijk M, DeJesus-Hernandez M, Niemantsverdriet E, Murray ME, Heckman MG, Diehl NN, et al. Association between repeat sizes and clinical and pathological characteristics in carriers of C9ORF72 repeat expansions (Xpansize-72): a cross-sectional cohort study. Lancet Neurol. 2013;12:978–88. https://doi.org/10.1016/S1474-4422(13)70210-2.

232. Fratta P, Mizielinska S, Nicoll AJ, Zloh M, Fisher EMC, Parkinson G, et al. C9orf72 hexanucleotide repeat associated with amyotrophic lateral sclerosis and frontotemporal dementia forms RNA G-quadruplexes. Sci Rep. 2012;2:1016. https://doi.org/10.1038/srep01016.

233. Reddy K, Zamiri B, Stanley SYR, Macgregor RB Jr, Pearson CE. The disease-associated r(GGGGCC)n repeat from the C9orf72 gene forms tract length-dependent uni- and multimolecular RNA G-quadruplex structures. J Biol Chem. 2013;288:9860–6. https://doi.org/10.1074/jbc.C113.452532.

234. Reddy K, Tam M, Bowater RP, Barber M, Tomlinson M, Nichol Edamura K, et al. Determinants of R-loop formation at convergent bidirectionally transcribed trinucleotide repeats. Nucleic Acids Res. 2011;39:1749–62. https://doi.org/10.1093/nar/gkq935.

235. Lin Y, Dent SYR, Wilson JH, Wells RD, Napierala M. R loops stimulate genetic instability of CTG.CAG repeats. Proc Natl Acad Sci U S A. 2010;107:692–7. https://doi.org/10.1073/pnas.0909740107.

236. Belotserkovskii BP, Mirkin SM, Hanawalt PC. DNA sequences that interfere with transcription: implications for genome function and stability. Chem Rev. 2013;113:8620–37. https://doi.org/10.1021/cr400078y.

237. Groh M, Lufino MMP, Wade-Martins R, Gromak N. R-loops associated with triplet repeat expansions promote gene silencing in Friedreich ataxia and fragile X syndrome. PLoS Genet. 2014;10:e1004318. https://doi.org/10.1371/journal.pgen.1004318.

238. Huertas P, Aguilera A. Cotranscriptionally formed DNA:RNA hybrids mediate transcription elongation impairment and transcription-associated recombination. Mol Cell. 2003;12:711–21. Available: https://www.ncbi.nlm.nih.gov/pubmed/14527416

239. Castellano-Pozo M, Santos-Pereira JM, Rondón AG, Barroso S, Andújar E, Pérez-Alegre M, et al. R loops are linked to histone H3 S10 phosphorylation and chromatin condensation. Mol Cell. 2013;52:583–90. https://doi.org/10.1016/j.molcel.2013.10.006.

240. Skourti-Stathaki K, Proudfoot NJ, Gromak N. Human senataxin resolves RNA/DNA hybrids formed at transcriptional pause sites to promote Xrn2-dependent termination. Mol Cell. 2011;42:794–805. https://doi.org/10.1016/j.molcel.2011.04.026.

241. Walker C, Herranz-Martin S, Karyka E, Liao C, Lewis K, Elsayed W, et al. C9orf72 expansion disrupts ATM-mediated chromosomal break repair. Nat Neurosci. 2017;20:1225–35. https://doi.org/10.1038/nn.4604.

242. de Mezer M, Wojciechowska M, Napierala M, Sobczak K, Krzyzosiak WJ. Mutant CAG repeats of Huntingtin transcript fold into hairpins, form nuclear foci and are targets for RNA interference. Nucleic Acids Res. 2011;39:3852–63. https://doi.org/10.1093/nar/gkq1323.

243. Conlon EG, Lu L, Sharma A, Yamazaki T, Tang T, Shneider NA, et al. The C9ORF72 GGGGCC expansion forms RNA G-quadruplex inclusions and sequesters hnRNP H to disrupt splicing in ALS patient brains. eLife. 2016;5:e17820. Available: https://elifesciences.org/download/aHR0cHM6Ly9jZG4uZWxpZmVzY2llbmNlcy5vcmcvYXJ0aWNsZXM-vMTc4MjAvZWxpZmUtMTc4MjAtdjEucGRm/elife-17820-v1.pdf?_hash=EoUnROvtJNg S2%2BjaKQIYxZz%2FSS%2BV8wRM%2BZhCCHCAfto%3D

244. Warf MB, Berglund JA. MBNL binds similar RNA structures in the CUG repeats of myotonic dystrophy and its pre-mRNA substrate cardiac troponin T. RNA. 2007;13:2238–51. https://doi.org/10.1261/rna.610607.

245. Kino Y, Mori D, Oma Y, Takeshita Y, Sasagawa N, Ishiura S. Muscleblind protein, MBNL1/EXP, binds specifically to CHHG repeats. Hum Mol Genet. 2004;13:495–507. https://doi.org/10.1093/hmg/ddh056.

246. Du H, Cline MS, Osborne RJ, Tuttle DL, Clark TA, Donohue JP, et al. Aberrant alternative splicing and extracellular matrix gene expression in mouse models of myotonic dystrophy. Nat Struct Mol Biol. 2010;17:187–93. https://doi.org/10.1038/nsmb.1720.
247. Miller JW, Urbinati CR, Teng-Umnuay P, Stenberg MG, Byrne BJ, Thornton CA, et al. Recruitment of human muscleblind proteins to (CUG)(n) expansions associated with myotonic dystrophy. EMBO J. 2000;19:4439–48. https://doi.org/10.1093/emboj/19.17.4439.
248. Konieczny P, Selma-Soriano E, Rapisarda AS, Fernandez-Costa JM, Perez-Alonso M, Artero R. Myotonic dystrophy: candidate small molecule therapeutics. Drug Discov Today. 2017. https://doi.org/10.1016/j.drudis.2017.07.011.
249. Taneja KL, McCurrach M, Schalling M, Housman D, Singer RH. Foci of trinucleotide repeat transcripts in nuclei of myotonic dystrophy cells and tissues. J Cell Biol. 1995;128:995–1002. Available: https://www.ncbi.nlm.nih.gov/pubmed/7896884
250. Liquori CL, Ricker K, Moseley ML, Jacobsen JF, Kress W, Naylor SL, et al. Myotonic dystrophy type 2 caused by a CCTG expansion in intron 1 of ZNF9. Science. 2001;293:864–7. https://doi.org/10.1126/science.1062125.
251. Lu X, Timchenko NA, Timchenko LT. Cardiac elav-type RNA-binding protein (ETR-3) binds to RNA CUG repeats expanded in myotonic dystrophy. Hum Mol Genet. 1999;8:53–60. Available: https://www.ncbi.nlm.nih.gov/pubmed/9887331
252. Mizielinska S, Lashley T, Norona FE, Clayton EL, Ridler CE, Fratta P, et al. C9orf72 frontotemporal lobar degeneration is characterised by frequent neuronal sense and antisense RNA foci. Acta Neuropathol. 2013;126:845–57. https://doi.org/10.1007/s00401-013-1200-z.
253. Vatovec S, Kovanda A, Rogelj B. Unconventional features of C9ORF72 expanded repeat in amyotrophic lateral sclerosis and frontotemporal lobar degeneration. Neurobiol Aging. 2014;35:2421.e1–2421.e12. https://doi.org/10.1016/j.neurobiolaging.2014.04.015.
254. Zu T, Gibbens B, Doty NS, Gomes-Pereira M, Huguet A, Stone MD, et al. Non-ATG-initiated translation directed by microsatellite expansions. Proc Natl Acad Sci U S A. 2011;108:260–5. https://doi.org/10.1073/pnas.1013343108.
255. Kearse MG, Green KM, Krans A, Rodriguez CM, Linsalata AE, Goldstrohm AC, et al. CGG repeat-associated non-AUG translation utilizes a cap-dependent scanning mechanism of initiation to produce toxic proteins. Mol Cell. 2016;62:314–22. https://doi.org/10.1016/j.molcel.2016.02.034.
256. Todd PK, Oh SY, Krans A, He F, Sellier C, Frazer M, et al. CGG repeat-associated translation mediates neurodegeneration in fragile X tremor ataxia syndrome. Neuron. 2013;78:440–55. https://doi.org/10.1016/j.neuron.2013.03.026.
257. Bañez-Coronel M, Ayhan F, Tarabochia AD, Zu T, Perez BA, Tusi SK, et al. RAN translation in huntington disease. Neuron. 2015;88:667–77. https://doi.org/10.1016/j.neuron.2015.10.038.
258. Zu T, Liu Y, Bañez-Coronel M, Reid T, Pletnikova O, Lewis J, et al. RAN proteins and RNA foci from antisense transcripts in C9ORF72 ALS and frontotemporal dementia. Proc Natl Acad Sci U S A. 2013;110:E4968–77. https://doi.org/10.1073/pnas.1315438110.
259. Zu T, Cleary JD, Liu Y, Bañez-Coronel M, Bubenik JL, Ayhan F, et al. RAN translation regulated by muscleblind proteins in myotonic dystrophy type 2. Neuron. 2017;95:1292–1305.e5. https://doi.org/10.1016/j.neuron.2017.08.039.
260. May S, Hornburg D, Schludi MH, Arzberger T, Rentzsch K, Schwenk BM, et al. C9orf72 FTLD/ALS-associated Gly-Ala dipeptide repeat proteins cause neuronal toxicity and Unc119 sequestration. Acta Neuropathol. 2014;128:485–503. https://doi.org/10.1007/s00401-014-1329-4.
261. Tran H, Almeida S, Moore J, Gendron TF, Chalasani U, Lu Y, et al. Differential toxicity of nuclear RNA foci versus dipeptide repeat proteins in a Drosophila model of C9ORF72 FTD/ALS. Neuron. 2015;87:1207–14. https://doi.org/10.1016/j.neuron.2015.09.015.
262. Kwon I, Xiang S, Kato M, Wu L, Theodoropoulos P, Wang T, et al. Poly-dipeptides encoded by the C9orf72 repeats bind nucleoli, impede RNA biogenesis, and kill cells. Science. 2014;345:1139–45. https://doi.org/10.1126/science.1254917.

263. Lee K-H, Zhang P, Kim HJ, Mitrea DM, Sarkar M, Freibaum BD, et al. C9orf72 dipeptide repeats impair the assembly, dynamics, and function of membrane-less organelles. Cell. 2016;167:774–788.e17. https://doi.org/10.1016/j.cell.2016.10.002.
264. Henras AK, Plisson-Chastang C, O'Donohue M-F, Chakraborty A, Gleizes P-E. An overview of pre-ribosomal RNA processing in eukaryotes. Wiley Interdiscip Rev RNA. 2015;6:225–42. https://doi.org/10.1002/wrna.1269.
265. Staněk D, Přidalová-Hnilicová J, Novotný I, Huranová M, Blažíková M, Wen X, et al. Spliceosomal small nuclear ribonucleoprotein particles repeatedly cycle through Cajal bodies. Mol Biol Cell. 2008;19:2534–43. https://doi.org/10.1091/mbc.E07-12-1259.
266. Wen X, Tan W, Westergard T, Krishnamurthy K, Markandaiah SS, Shi Y, et al. Antisense proline-arginine RAN dipeptides linked to C9ORF72-ALS/FTD form toxic nuclear aggregates that initiate in vitro and in vivo neuronal death. Neuron. 2014;84:1213–25. https://doi.org/10.1016/j.neuron.2014.12.010.
267. Masuda S, Das R, Cheng H, Hurt E, Dorman N, Reed R. Recruitment of the human TREX complex to mRNA during splicing. Genes Dev. 2005;19:1512–7. https://doi.org/10.1101/gad.1302205.
268. Luo ML, Zhou Z, Magni K, Christoforides C, Rappsilber J, Mann M, et al. Pre-mRNA splicing and mRNA export linked by direct interactions between UAP56 and Aly. Nature. 2001;413:644–7. https://doi.org/10.1038/35098106.
269. Hautbergue GM, Hung M-L, Walsh MJ, Snijders APL, Chang C-T, Jones R, et al. UIF, a new mRNA export adaptor that works together with REF/ALY, requires FACT for recruitment to mRNA. Curr Biol. 2009;19:1918–24. https://doi.org/10.1016/j.cub.2009.09.041.
270. Viphakone N, Cumberbatch MG, Livingstone MJ, Heath PR, Dickman MJ, Catto JW, et al. Luzp4 defines a new mRNA export pathway in cancer cells. Nucleic Acids Res. 2015;43:2353–66. https://doi.org/10.1093/nar/gkv070.
271. Chang C-T, Hautbergue GM, Walsh MJ, Viphakone N, van Dijk TB, Philipsen S, et al. Chtop is a component of the dynamic TREX mRNA export complex. EMBO J. 2013;32:473–86. https://doi.org/10.1038/emboj.2012.342.
272. Hautbergue GM, Hung M-L, Golovanov AP, Lian L-Y, Wilson SA. Mutually exclusive interactions drive handover of mRNA from export adaptors to TAP. Proc Natl Acad Sci U S A. 2008;105:5154–9. https://doi.org/10.1073/pnas.0709167105.
273. Viphakone N, Hautbergue GM, Walsh M, Chang C-T, Holland A, Folco EG, et al. TREX exposes the RNA-binding domain of Nxf1 to enable mRNA export. Nat Commun. 2012;3:1006. https://doi.org/10.1038/ncomms2005.
274. Stutz F, Bachi A, Doerks T, Braun IC, Séraphin B, Wilm M, et al. REF, an evolutionarily conserved family of hnRNP-like proteins, interacts with TAP/Mex67p and participates in mRNA nuclear export. RNA. 2000;6:638–50. Available: https://www.cambridge.org/core/journals/rna/article/ref-an-evolutionarily-conserved-family-of-hnrnp-like-proteins-interacts-with-tapmex67p-and-participates-in-mrna-nuclear-export/44856EDCAC20CFB7B5-FB1405A9B7CA6E
275. Wickramasinghe VO, McMurtrie PIA, Mills AD, Takei Y, Penrhyn-Lowe S, Amagase Y, et al. mRNA export from mammalian cell nuclei is dependent on GANP. Curr Biol. 2010;20:25–31. https://doi.org/10.1016/j.cub.2009.10.078.
276. Bergeron D, Pal G, Beaulieu YB, Chabot B, Bachand F. Regulated intron retention and nuclear pre-mRNA decay contribute to PABPN1 autoregulation. Mol Cell Biol. 2015;35:2503–17. https://doi.org/10.1128/MCB.00070-15.
277. Avendaño-Vázquez SE, Dhir A, Bembich S, Buratti E, Proudfoot N, Baralle FE. Autoregulation of TDP-43 mRNA levels involves interplay between transcription, splicing, and alternative polyA site selection. Genes Dev. 2012;26:1679–84. https://doi.org/10.1101/gad.194829.112.
278. Di Gregorio E, Bianchi FT, Schiavi A, Chiotto AMA, Rolando M, di Cantogno LV, et al. A de novo X; 8 translocation creates a PTK2-THOC2 gene fusion with THOC2 expression knockdown in a patient with psychomotor retardation and congenital cerebellar hypoplasia. J Med Genet. BMJ. 2013;50:543–51. Available: http://jmg.bmj.com/content/50/8/543.short

279. Kumar R, Corbett MA, van Bon BWM, Woenig JA, Weir L, Douglas E, et al. THOC2 mutations implicate mRNA-export pathway in X-linked intellectual disability. Am J Hum Genet. 2015;97:302–10. https://doi.org/10.1016/j.ajhg.2015.05.021.
280. Beaulieu CL, Huang L, Innes AM, Akimenko M-A, Puffenberger EG, Schwartz C, et al. Intellectual disability associated with a homozygous missense mutation in THOC6. Orphanet J Rare Dis. 2013;8:62. https://doi.org/10.1186/1750-1172-8-62.
281. Kaneb HM, Folkmann AW, Belzil VV, Jao L-E, Leblond CS, Girard SL, et al. Deleterious mutations in the essential mRNA metabolism factor, hGle1, in amyotrophic lateral sclerosis. Hum Mol Genet. 2015;24:1363–73. https://doi.org/10.1093/hmg/ddu545.
282. Nousiainen HO, Kestilä M, Pakkasjärvi N, Honkala H, Kuure S, Tallila J, et al. Mutations in mRNA export mediator GLE1 result in a fetal motoneuron disease. Nat Genet. 2008;40:155–7. https://doi.org/10.1038/ng.2007.65.
283. Folkmann AW, Noble KN, Cole CN, Wente SR. Dbp5, Gle1-IP6 and Nup159: a working model for mRNP export. Nucleus. 2011;2:540–8. https://doi.org/10.4161/nucl.2.6.17881.
284. Murphy R, Watkins JL, Wente SR. GLE2, a Saccharomyces cerevisiae homologue of the Schizosaccharomyces pombe export factor RAE1, is required for nuclear pore complex structure and function. Mol Biol Cell. 1996;7:1921–37. Available: https://www.ncbi.nlm.nih.gov/pubmed/8970155
285. Bharathi A, Ghosh A, Whalen WA, Yoon JH, Pu R, Dasso M, et al. The human RAE1 gene is a functional homologue of Schizosaccharomyces pombe rae1 gene involved in nuclear export of Poly(A)+ RNA. Gene. 1997;198:251–8. Available: https://www.ncbi.nlm.nih.gov/pubmed/9370289
286. Freibaum BD, Lu Y, Lopez-Gonzalez R, Kim NC, Almeida S, Lee K-H, et al. GGGGCC repeat expansion in C9orf72 compromises nucleocytoplasmic transport. Nature. 2015;525:129–33. https://doi.org/10.1038/nature14974.
287. Jovičić A, Mertens J, Boeynaems S, Bogaert E, Chai N, Yamada SB, et al. Modifiers of C9orf72 dipeptide repeat toxicity connect nucleocytoplasmic transport defects to FTD/ALS. Nat Neurosci. 2015;18:1226–9. https://doi.org/10.1038/nn.4085.
288. Zhang K, Donnelly CJ, Haeusler AR, Grima JC, Machamer JB, Steinwald P, et al. The C9orf72 repeat expansion disrupts nucleocytoplasmic transport. Nature. 2015;525:56–61. https://doi.org/10.1038/nature14973.
289. Garcia-Lopez A, Monferrer L, Garcia-Alcover I, Vicente-Crespo M, Alvarez-Abril MC, Artero RD. Genetic and chemical modifiers of a CUG toxicity model in Drosophila. PLoS One. 2008;3:e1595. https://doi.org/10.1371/journal.pone.0001595.
290. Sun X, Li PP, Zhu S, Cohen R, Marque LO, Ross CA, et al. Nuclear retention of full-length HTT RNA is mediated by splicing factors MBNL1 and U2AF65. Sci Rep. 2015;5:12521. https://doi.org/10.1038/srep12521.
291. Grima JC, Daigle JG, Arbez N, Cunningham KC, Zhang K, Ochaba J, et al. Mutant Huntingtin disrupts the nuclear pore complex. Neuron. 2017;94:93–107.e6. https://doi.org/10.1016/j.neuron.2017.03.023.
292. D'Angelo MA, Raices M, Panowski SH, Hetzer MW. Age-dependent deterioration of nuclear pore complexes causes a loss of nuclear integrity in postmitotic cells. Cell. 2009;136:284–95. https://doi.org/10.1016/j.cell.2008.11.037.
293. Mertens J, Paquola ACM, Ku M, Hatch E, Böhnke L, Ladjevardi S, et al. Directly reprogrammed human neurons retain aging-associated transcriptomic signatures and reveal age-related nucleocytoplasmic defects. Cell Stem Cell. 2015;17:705–18. https://doi.org/10.1016/j.stem.2015.09.001.
294. Shi KY, Mori E, Nizami ZF, Lin Y, Kato M, Xiang S, et al. Toxic PRn poly-dipeptides encoded by the C9orf72 repeat expansion block nuclear import and export. Proc Natl Acad Sci U S A. 2017;114:E1111–7. https://doi.org/10.1073/pnas.1620293114.
295. Mizielinska S, Grönke S, Niccoli T, Ridler CE, Clayton EL, Devoy A, et al. C9orf72 repeat expansions cause neurodegeneration in Drosophila through arginine-rich proteins. Science. 2014;345:1192–4. https://doi.org/10.1126/science.1256800.

296. Eggens VR, Barth PG, Niermeijer J-MF, Berg JN, Darin N, Dixit A, et al. EXOSC3 mutations in pontocerebellar hypoplasia type 1: novel mutations and genotype-phenotype correlations. Orphanet J Rare Dis. 2014;9:23. https://doi.org/10.1186/1750-1172-9-23.

297. Rudnik-Schöneborn S, Senderek J, Jen JC, Houge G, Seeman P, Puchmajerová A, et al. Pontocerebellar hypoplasia type 1: clinical spectrum and relevance of EXOSC3 mutations. Neurology. 2013;80:438–46. https://doi.org/10.1212/WNL.0b013e31827f0f66.

298. Zanni G, Scotton C, Passarelli C, Fang M, Barresi S, Dallapiccola B, et al. Exome sequencing in a family with intellectual disability, early onset spasticity, and cerebellar atrophy detects a novel mutation in EXOSC3. Neurogenetics. 2013;14:247–50. https://doi.org/10.1007/s10048-013-0371-z.

299. Boczonadi V, Müller JS, Pyle A, Munkley J, Dor T, Quartararo J, et al. EXOSC8 mutations alter mRNA metabolism and cause hypomyelination with spinal muscular atrophy and cerebellar hypoplasia. Nat Commun. 2014;5:4287. https://doi.org/10.1038/ncomms5287.

300. Pan Q, Shai O, Lee LJ, Frey BJ, Blencowe BJ. Deep surveying of alternative splicing complexity in the human transcriptome by high-throughput sequencing. Nat Genet. 2008;40:1413–5. https://doi.org/10.1038/ng.259.

301. Johnson MB, Kawasawa YI, Mason CE, Krsnik Z, Coppola G, Bogdanović D, et al. Functional and evolutionary insights into human brain development through global transcriptome analysis. Neuron. 2009;62:494–509. https://doi.org/10.1016/j.neuron.2009.03.027.

302. Yeo G, Holste D, Kreiman G, Burge CB. Variation in alternative splicing across human tissues. Genome Biol. 2004;5:R74. https://doi.org/10.1186/gb-2004-5-10-r74.

303. Faustino NA, Cooper TA. Pre-mRNA splicing and human disease. Genes Dev. 2003;17:419–37. https://doi.org/10.1101/gad.1048803.

304. Sassi C, Capozzo R, Gibbs R, Crews C, Zecca C, Arcuti S, et al. A novel splice-acceptor site mutation in GRN (c. 709-2 A>T) causes frontotemporal dementia spectrum in a large family from Southern Italy. J Alzheimers Dis. 2016;53:475–85. Available: http://content.iospress.com/articles/journal-of-alzheimers-disease/jad151170

305. Luzzi S, Colleoni L, Corbetta P, Baldinelli S, Fiori C, Girelli F, et al. Missense mutation in GRN gene affecting RNA splicing and plasma progranulin level in a family affected by frontotemporal lobar degeneration. Neurobiol Aging. 2017;54:214.e1–6. https://doi.org/10.1016/j.neurobiolaging.2017.02.008.

306. Mukherjee O, Wang J, Gitcho M, Chakraverty S, Taylor-Reinwald L, Shears S, et al. Molecular characterization of novel progranulin (GRN) mutations in frontotemporal dementia. Hum Mutat. 2008;29:512–21. https://doi.org/10.1002/humu.20681.

307. Guven G, Lohmann E, Bras J, Gibbs JR, Gurvit H, Bilgic B, et al. Mutation frequency of the major frontotemporal dementia genes, MAPT, GRN and C9ORF72 in a Turkish cohort of dementia patients. PLoS One. 2016;11:e0162592. https://doi.org/10.1371/journal.pone.0162592.

308. Humphrey J, Emmett W, Fratta P, Isaacs AM, Plagnol V. Quantitative analysis of cryptic splicing associated with TDP-43 depletion. BMC Med Genet. 2017;10:38. https://doi.org/10.1186/s12920-017-0274-1.

309. Tan Q, Yalamanchili HK, Park J, De Maio A, Lu H-C, Wan Y-W, et al. Extensive cryptic splicing upon loss of RBM17 and TDP43 in neurodegeneration models. Hum Mol Genet. 2016;25:5083–93. https://doi.org/10.1093/hmg/ddw337.

310. Jeong YH, Ling JP, Lin SZ, Donde AN, Braunstein KE, Majounie E, et al. Tdp-43 cryptic exons are highly variable between cell types. Mol Neurodegener. 2017;12:13. https://doi.org/10.1186/s13024-016-0144-x.

311. Shum EY, Jones SH, Shao A, Dumdie J, Krause MD, Chan W-K, et al. The antagonistic gene paralogs Upf3a and Upf3b govern nonsense-mediated RNA decay. Cell. 2016;165:382–95. https://doi.org/10.1016/j.cell.2016.02.046.

312. Barmada SJ, Ju S, Arjun A, Batarse A, Archbold HC, Peisach D, et al. Amelioration of toxicity in neuronal models of amyotrophic lateral sclerosis by hUPF1. Proc Natl Acad Sci U S A. 2015;112:7821–6. https://doi.org/10.1073/pnas.1509744112.

313. Lander ES, Linton LM, Birren B, Nusbaum C, Zody MC, Baldwin J, et al. Initial sequencing and analysis of the human genome. Nature. 2001;409:860–921. https://doi.org/10.1038/35057062.
314. Doucet AJ, Hulme AE, Sahinovic E, Kulpa DA, Moldovan JB, Kopera HC, et al. Characterization of LINE-1 ribonucleoprotein particles. PLoS Genet. 2010;6. https://doi.org/10.1371/journal.pgen.1001150.
315. Cordaux R, Hedges DJ, Herke SW, Batzer MA. Estimating the retrotransposition rate of human Alu elements. Gene. 2006;373:134–7. https://doi.org/10.1016/j.gene.2006.01.019.
316. Maxwell PH, Burhans WC, Curcio MJ. Retrotransposition is associated with genome instability during chronological aging. Proc Natl Acad Sci U S A. 2011;108:20376–81. https://doi.org/10.1073/pnas.1100271108.
317. Hua-Van A, Le Rouzic A, Boutin TS, Filée J, Capy P. The struggle for life of the genome's selfish architects. Biol Direct. 2011;6:19. https://doi.org/10.1186/1745-6150-6-19.
318. Ramírez MA, Pericuesta E, Fernandez-Gonzalez R, Moreira P, Pintado B, Transcriptional G-AA. post-transcriptional regulation of retrotransposons IAP and MuERV-L affect pluripotency of mice ES cells. Reprod Biol Endocrinol. 2006;4:55. https://doi.org/10.1186/1477-7827-4-55.
319. Aguilera A. The connection between transcription and genomic instability. EMBO J. 2002;21:195–201. Available: http://emboj.embopress.org/content/21/3/195.abstract
320. Gregersen LH, Schueler M, Munschauer M, Mastrobuoni G, Chen W, Kempa S, et al. MOV10 Is a 5′ to 3′ RNA helicase contributing to UPF1 mRNA target degradation by translocation along 3′ UTRs. Mol Cell. 2014;54:573–85. https://doi.org/10.1016/j.molcel.2014.03.017.
321. Moldovan JB, Moran JV. The zinc-finger antiviral protein ZAP inhibits LINE and Alu retrotransposition. PLoS Genet. 2015;11:e1005121. https://doi.org/10.1371/journal.pgen.1005121.
322. Saito K, Siomi MC. Small RNA-mediated quiescence of transposable elements in animals. Dev Cell. 2010;19:687–97. https://doi.org/10.1016/j.devcel.2010.10.011.
323. De Cecco M, Criscione SW, Peterson AL, Neretti N, Sedivy JM, Kreiling JA. Transposable elements become active and mobile in the genomes of aging mammalian somatic tissues. Aging. 2013;5:867–83. https://doi.org/10.18632/aging.100621.
324. Li W, Prazak L, Chatterjee N, Grüninger S, Krug L, Theodorou D, et al. Activation of transposable elements during aging and neuronal decline in Drosophila. Nat Neurosci. 2013;16:529–31. https://doi.org/10.1038/nn.3368.
325. Douville R, Liu J, Rothstein J, Nath A. Identification of active loci of a human endogenous retrovirus in neurons of patients with amyotrophic lateral sclerosis. Ann Neurol. 2011;69:141–51. https://doi.org/10.1002/ana.22149.
326. Greenwood AD, Vincendeau M, Schmädicke A-C, Montag J, Seifarth W, Motzkus D. Bovine spongiform encephalopathy infection alters endogenous retrovirus expression in distinct brain regions of cynomolgus macaques (Macaca fascicularis). Mol Neurodegener. 2011;6:44. https://doi.org/10.1186/1750-1326-6-44.
327. Tan H, Qurashi A, Poidevin M, Nelson DL, Li H, Jin P. Retrotransposon activation contributes to fragile X premutation rCGG-mediated neurodegeneration. Hum Mol Genet. 2012;21:57–65. https://doi.org/10.1093/hmg/ddr437.
328. Li W, Jin Y, Prazak L, Hammell M, Dubnau J. Transposable elements in TDP-43-mediated neurodegenerative disorders. PLoS One. 2012;7:e44099. https://doi.org/10.1371/journal.pone.0044099.
329. Krug L, Chatterjee N, Borges-Monroy R, Hearn S, Liao W-W, Morrill K, et al. Retrotransposon activation contributes to neurodegeneration in a Drosophila TDP-43 model of ALS. PLoS Genet. 2017;13:e1006635. https://doi.org/10.1371/journal.pgen.1006635.
330. Subramanian RP, Wildschutte JH, Russo C, Coffin JM. Identification, characterization, and comparative genomic distribution of the HERV-K (HML-2) group of human endogenous retroviruses. Retrovirology. 2011;8:90. https://doi.org/10.1186/1742-4690-8-90.

331. Li W, Lee M-H, Henderson L, Tyagi R, Bachani M, Steiner J, et al. Human endogenous retrovirus-K contributes to motor neuron disease. Sci Transl Med. 2015;7:307ra153. https://doi.org/10.1126/scitranslmed.aac8201.
332. Carroll B, Hewitt G, Korolchuk VI. Autophagy and ageing: implications for age-related neurodegenerative diseases. Essays Biochem. 2013;55:119–31. https://doi.org/10.1042/bse0550119.
333. Romano AD, Serviddio G, de Matthaeis A, Bellanti F, Vendemiale G. Oxidative stress and aging. J Nephrol. 2010;23(Suppl 15):S29–36. Available: https://www.ncbi.nlm.nih.gov/pubmed/20872368
334. Gensler HL, Bernstein H. DNA damage as the primary cause of aging. Q Rev Biol. 1981;56:279–303. Available: https://www.ncbi.nlm.nih.gov/pubmed/7031747

Chapter 6
RNP Assembly Defects in Spinal Muscular Atrophy

Phillip L. Price, Dmytro Morderer, and Wilfried Rossoll

Abstract Spinal muscular atrophy (SMA) is a motor neuron disease caused by mutations/deletions within the survival of motor neuron 1 (*SMN1*) gene that lead to a pathological reduction of SMN protein levels. SMN is part of a multiprotein complex, functioning as a molecular chaperone that facilitates the assembly of spliceosomal small nuclear ribonucleoproteins (snRNP). In addition to its role in spliceosome formation, SMN has also been found to interact with mRNA-binding proteins (mRBPs), and facilitate their assembly into mRNP transport granules. The association of protein and RNA in RNP complexes plays an important role in an extensive and diverse set of cellular processes that regulate neuronal growth, differentiation, and the maturation and plasticity of synapses. This review discusses the role of SMN in RNP assembly and localization, focusing on molecular defects that affect mRNA processing and may contribute to SMA pathology.

Keywords Spinal muscular atrophy (SMA) · Survival of motor neuron (SMN) · RNA-binding protein (RBP) · Ribonucleoprotein (RNP) · Molecular chaperone · RNA processing · RNA localization

6.1 SMA Clinical Background

Spinal muscular atrophy (SMA) is an autosomal recessive disorder characterized by the early-onset of skeletal muscle atrophy and a progressive degeneration of motor neurons in the anterior horn of the spinal cord [1]. Impairments in synaptic

P. L. Price
Department of Neuroscience, Mayo Clinic, Jacksonville, FL, USA

Department of Cell Biology, Emory University, Atlanta, GA, USA
e-mail: Price.Phillip@mayo.edu

D. Morderer · W. Rossoll (✉)
Department of Neuroscience, Mayo Clinic, Jacksonville, FL, USA
e-mail: Morderer.Dmytro@mayo.edu; Rossoll.Wilfried@mayo.edu

© Springer International Publishing AG, part of Springer Nature 2018
R. Sattler, C. J. Donnelly (eds.), *RNA Metabolism in Neurodegenerative Diseases*, Advances in Neurobiology 20,
https://doi.org/10.1007/978-3-319-89689-2_6

maturation, sensory-motor circuitry, and synaptic transmission at the neuromuscular junction (NMJ), followed by a dying-back axonopathy, precede muscle denervation and loss of α-motor neurons in the spinal cord. In classical SMA, proximal muscles are more severely affected than distal muscles.

SMA is the leading genetic cause of death in infancy [2]. Across ethnicities, an incidence of 1 in 8,000–20,000 has been estimated [3]. Disease classification is based on the age of onset and clinical severity, the most common classification scheme distinguishes between Types I–IV [4]. The most common form of SMA (Type I; SMA1) typically leads to muscle weakness within the first 6 months and death due to respiratory failure by the age of 2. Type II SMA patients present with signs of muscle weakness during the first 7–18 months. Affected children may crawl and sit unassisted, but often require support for standing and mobility. These patients typically have a life expectancy into early adulthood. Type III and IV SMA are milder forms of the disease, with patients having a normal life expectancy and displaying muscle weakness presenting in adolescence and adulthood. Type III SMA patients are able to stand unsupported and walk with moderate difficulty. Type IV (adult-onset) SMA patients are usually not diagnosed until early adulthood. These patients experience slowly progressing muscle weakness, primarily affecting the legs, hips, shoulders, arms.

6.2   SMA Is Caused by Reduced SMN Protein Levels

In >95% of cases, SMA is caused by homozygous deletions or compound heterozygous mutations in the survival of motor neuron 1 (*SMN1*) gene encoding the SMN protein [5]. Humans possess a nearly identical copy of this gene (*SMN2*), which carries a splice site mutation in exon 7. This C>T transition in *SMN2* promotes the exclusion of exon 7 from the full-length protein, leading to the expression of only 10–20% of full-length SMN protein, and 80–90% of a truncated SMN protein isoform (SMNΔ7) that is rapidly degraded [6–8]. Since SMA patients lack functional *SMN1* genes, SMN protein is expressed only from the *SMN2* gene, leading to reduced levels of full-length SMN protein. Thus, SMA is directly caused by a pathological reduction of functional SMN protein levels below a critical threshold, and disease severity is correlated with *SMN2* copy number [6, 7]. Therapeutic approaches have mainly focused on raising SMN protein levels via gene therapy [8], or via increasing the splicing efficiency of *SMN2* exon 7. Small molecules [9] as well as antisense oligonucleotides (ASOs) have been developed as splicing modifiers of *SMN2* [10], leading to the introduction of the ASO Nusinersen as the first drug approved by the US Food and Drug Administration for the treatment of SMA [11–13]. The high cost for the treatment and the need for administering ASOs via lumbar puncture have created practical challenges that raise important ethical questions [14]. A continued effort will be required to provide more effective, affordable, and accessible treatment options.

6.3   SMN Protein Deficiency Primarily Affects Synapses in the Motor Circuitry

During early development, the assembly and stabilization of highly organized synaptic structures is essential for the maturation and function of the central nervous system. In neurodegenerative diseases, structural and functional abnormalities in synaptic connections often precede neuronal loss and cell death, and are thought to account for early clinical deficits [15–17]. The underlying cause for the vulnerability of motor neurons to reduced SMN protein levels remains unclear. Aside from *Drosophila*, *C. elegans*, and zebrafish SMA animal models, several mouse models with different severity have been engineered to closely recapitulate pathological hallmarks observed in human patients [18].

Unlike humans, mice only possess one gene encoding SMN, and a complete knockout results in early embryonic lethality [19]. To recapitulate the disease phenotype of human SMA, the introduction of the human *SMN2* transgene into the background of a homozygous deletion of murine *Smn1* has allowed for the creation of severe mouse models of SMA [20, 21]. Further mouse models have been generated to represent less severe forms of SMA as important preclinical models for therapy development [22–24].

Although born with a normal number of motor neurons, severe SMA mice experience a 35–40% loss of spinal cord and lower-brainstem motor neurons by day 5 [20]. As summarized in Fig. 6.1, pre-synaptic deficits include the aggregation of neurofilaments in the presynaptic terminal, poor terminal arborization, irregular distribution and positioning of synaptic vesicles, and reduced neurotransmission [25]. In addition, multiple studies in SMA mouse models have observed significant impairments in mitochondrial function and axonal transport, including increased oxidative stress levels and organelle fragmentation [26, 27].

Deleterious effects on the maturation and maintenance of NMJs support the characterization of SMA as an NMJ synaptopathy [28]. As the most commonly studied mouse model of SMA, SMNΔ7 mice carry a homozygous deletion of the murine *Smn1* gene and contain two transgenic constructs, one containing a single copy of the human *SMN2* gene locus, and a second encoding the human *SMN2* promoter driving expression of human *SMN2* cDNA lacking exon 7 (*SMNΔ7*) [22]. These mice have an average life span of 17.7 days, and display several similar phenotypes observed in SMA patients. Although most NMJs remain innervated until late in the disease time course of SMNΔ7 mice, thorough explorations of the synapse electrophysiology and ultra-structure revealed a significant decrease in synaptic vesicle density and release probability [29]. These deficits were found to be associated with a delayed maturation of NMJ terminals and myofibers, and together indicate that NMJ synaptic dysfunction precedes degeneration of the motor axon and finally the loss of motor neurons in severe SMA mouse models [29]. Reduced subsynaptic clefts and lack of synaptic vesicles at the NMJ and abnormal preterminal accumulation of vesicles have also been observed in SMA1 patients [30, 31].

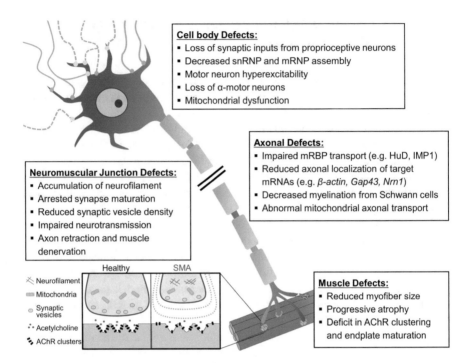

Cell body Defects:
- Loss of synaptic inputs from proprioceptive neurons
- Decreased snRNP and mRNP assembly
- Motor neuron hyperexcitability
- Loss of α-motor neurons
- Mitochondrial dysfunction

Axonal Defects:
- Impaired mRBP transport (e.g. HuD, IMP1)
- Reduced axonal localization of target mRNAs (e.g. *β-actin, Gap43, Nrn1*)
- Decreased myelination from Schwann cells
- Abnormal mitochondrial axonal transport

Neuromuscular Junction Defects:
- Accumulation of neurofilament
- Arrested synapse maturation
- Reduced synaptic vesicle density
- Impaired neurotransmission
- Axon retraction and muscle denervation

Healthy   SMA

Neurofilament
Mitochondria
Synaptic vesicles
Acetylcholine
AChR clusters

Muscle Defects:
- Reduced myofiber size
- Progressive atrophy
- Deficit in AChR clustering and endplate maturation

Fig. 6.1 Cellular Defects in SMA motor neurons. Schematic of a spinal motor neuron, highlighting morphological and molecular defects within different cellular compartments

Related to NMJ function, myotubes in SMA1 fetuses can display a significant retardation in growth and maturation [32]. One of the principal prenatal defects observed in mouse models and human SMA1 patients was an arrest in acetylcholine receptor clustering into 'pretzel'-shaped structures during postsynaptic endplate maturation, compromising the structural and functional integrity of the NMJ [30]. In SMA mice, abnormal molecular composition, disruptions in normal satellite cell differentiation, and reductions in myofiber size, have been described in skeletal muscle [22, 33, 34]. Agrin, a protein best known for its role in organizing acetylcholine receptors at the NMJ, is misspliced and reduced in motor neurons of SMA mice [35]. Rescue of the Z+ Agrin isoform prevented the development of several pathological phenotypes, and improved mean survival by 40% [35]. Taken together, this research points to abnormal function and maturation of the NMJ as key contributors to SMA pathogenesis, and as a potential target for therapy [36].

However, notable defects are also present within other cell-types that relay information, support motor neuron function and viability, and contribute to motor circuitry. In SMNΔ7 mice, loss of proprioceptive sensory synaptic input onto spinal motor neurons has been observed in embryonic mice, suggesting that the disruption of the spinal motor circuitry at multiple levels is an early phenotype contributing to motor dysfunction (Fig. 6.1) [37, 38]. Defects in astrocyte activity and myelination may also influence the severity of SMA [39–42]. SMA astrocytes display significant deficits in stimulating neurite outgrowth and differentiation of motor neurons, but may also display potential toxic gain-of-function properties [43, 44].

The generation of conditional SMA mouse models with promoter-driven depletion or rescue of SMN expression has made it possible to study the pathological effects of selectively reduced SMN levels in specific cell-types and tissues. Using this approach has demonstrated that depletion or restoration of SMN in motor neurons (*Hb9-Cre*; *ChAT-Cre*) significantly alters the functional synaptic output and excitability of the motor unit and retention of sensory-motor synapses [45, 46]. Nevertheless, restoration of SMN solely in motor neurons provides little to no improvement in life span, likely due to abnormal cardiac innervation by the autonomic nervous system in this severe SMA mouse model [45]. Multiple studies using a muscle-specific Cre driver (*Myf5-Cre*; *HSA-Cre*) to restore normal SMN in muscle of SMA mice showed that both replacement or depletion of SMN in muscle had little to no phenotypic effect on the mice [47–49]. The greatest improvement in survival and function is the result of SMN restoration throughout the entire nervous system (*Nestin-Cre* and *ChAT-Cre*; *PrP-Cre*), affecting neurons and glia alike and largely rescuing SMA phenotypes and life span [45, 47]. Although lower motor neurons and their circuitry are the primary targets of SMA pathology, mounting evidence suggests that SMN deficiency may contribute to defects in multiple tissues and across additional peripheral organs [49, 50].

In summary, these studies highlight the necessity of SMN protein during development, and demonstrate the physiological consequences of insufficient levels of SMN on function and survival of various cell-types. While motor neuron degeneration and loss of central synapses and NMJs in the motor circuitry are the primary targets of SMA pathology, restoration of SMN in multiple cell-types may be necessary for a complete rescue of the SMA phenotype. The advent of effective therapies targeting the CNS may lead to the development of multi-organ impairment in surviving SMA patients, requiring systemic delivery of therapies [51].

6.4   SMA Is Caused by Reduced RNP Assembly

SMN is an evolutionarily conserved and ubiquitously expressed protein with an essential role in RNA processing. Complete loss of SMN is lethal in all organisms and depends on maternal contribution across different species, highlighting its importance to cell development and survival [52]. The temporal expression of SMN protein levels is developmentally regulated, with the highest expression levels during the embryonic period and a gradual decrease into the early postnatal period in mice and humans [53]. SMN granules are also present in the axons and growth cones of developing and regenerating motor neurons, and at the postsynaptic endplate of the neuromuscular junctions [54–56]. Active bi-directional fast axonal transport of SMN has been demonstrated in primary forebrain and motor neurons [57, 58].

As its best characterized molecular function, SMN facilitates the assembly of small nuclear ribonuclear proteins (snRNPs), bringing together specific sets of protein and RNA molecules that form the building blocks for spliceosome formation and pre-messenger RNA splicing [59]. More recent studies from multiple laboratories have demonstrated that SMN plays a broader role in the assembly of various RNP complexes with divergent roles in RNA processing, including mRNA splicing, turnover, and trafficking [60]. In contrast to late onset neurodegenerative diseases that are often characterized by the accumulation of RNA-binding proteins into pathological aggregates [61], SMA is set apart by an SMN-dependent deficiency in the formation of RNPs, and is therefore best described as an RNP hypo-assembly disease [62, 63].

SMN associates with eight proteins (Gemins 2–8 and Unrip) to form a complex that is present in the cytoplasm and in discrete nuclear bodies called "gems," for Gemini of Cajal bodies (or coiled bodies) [64–66]. The self-oligomerization of SMN and subsequent formation of the macromolecular SMN complex requires the evolutionarily conserved YG-box. Located at the carboxy-terminus, the YG-box provides a structural basis for the SMN complex to form higher-order complexes ranging from 20S to 80S [69, 70]. A subunit of the SMN complex that includes SMN and Gemin2 recognizes Sm proteins, and assists in the ATP-dependent assembly of the heptameric Sm core complex [67–69]. Spliceosomal Sm proteins belong to a large family of Sm and Sm-like (LSm) proteins that share a conserved Sm motif necessary for protein-protein interaction, and are essential for snRNP biogenesis. Symmetrical dimethylation of a subset of Sm proteins by the protein arginine N-methyltransferase 5 (PRMT5) complex enhances their affinity for the conserved Tudor domain within the SMN protein [70, 71]. Gemin3 is a DEAD-box RNA-dependent RNA helicase and ATPase [72]. Gemin5 recognizes and interacts with large, ~50–60 nucleotide sequences or, snRNP codes, on specific spliceosomal U snRNAs [73, 74]. Gemin6 and Gemin7 are thought to possess an Sm protein-like structure, facilitating the recruitment of Sm proteins into snRNPs [75].

6.5   The SMN Complex Is an Assembly Machine for Spliceosomal snRNPs

Although SMN and the associated Gemin 2–8 proteins increase the efficiency and specificity of snRNP complex assembly, they do not become part of the final structure, thus acting as a molecular chaperone [60]. Sm proteins have an intrinsic ability to associate with snRNAs in vitro, forming snRNP complexes with little regard to RNA specificity. The presence of the SMN complex restricts illicit associations of Sm proteins with erroneous RNAs, and promotes the recognition of snRNAs. The assembly of Sm proteins and binding of specific RNA requires a coordinated interaction between the SMN complex and the PRMT5 complex [76]. The PRMT5 complex consists of PRMT5, pICln, and WD45 (Mep50), and pre-assembles specific sets of Sm proteins via the pICln subunit [68, 77]. pICln is displaced from these recruited Sm proteins by the SMN complex, which promotes the transfer of Sm proteins from an intermediate RNP complex onto snRNA to form U snRNPs [59, 78, 79]. Therefore, the SMN complex functions as an assemblysome that regulates snRNP biogenesis, structure, and function [80]. Recognition and binding of splice sites require the association of several small nuclear RNAs (snRNAs) and Sm proteins. Typically, uridine-rich snRNAs (U1, U2, U4, U5 U6, U11, U12, U4atac, and U6atac) are assembled with a set of seven Sm proteins (Sm B/B', D1, D2, D3, E, F and G) into different classes of heptameric snRNP core complexes that are essential to the catalytic activity of the spliceosome [69, 81]. The U2-dependent major spliceosome comprised of U1, U2, U4/6, and U5, is the predominant machinery responsible for the accurate removal of canonical "GT-AG" introns from most eukaryotic transcripts, whereas the U12-dependent minor spliceosomal complex comprised of U11, U12, U4atac/U6atac, and U5, removes rare "AT-AC" introns. Despite U12 introns representing only <1% of all human introns, the U12-dependent spliceosome is essential for the viability and development of many multicellular organisms, including humans [82]. U12-type introns have been identified mainly in genes with a role in DNA replication and repair, transcription, RNA processing, and translation, but can also be found in genes related to vesicular transport, cytoskeletal organization and assembly, and voltage-gated ion channel activity [83]. As discussed below, alterations to U12-dependent spliceosomal activity may have particularly deleterious effects on the morphology and physiology of neurons. Of note, a mutation in the gene encoding U12 snRNA has been identified as the potential cause of early onset cerebellar ataxia in one pedigree [84], whereas mutations of core spliceosomal factors are typically associated with severe developmental disorders [85, 86].

Aside from its role in spliceosomal snRNP assembly, SMN has also been shown to be involved in the assembly of related RNP complexes with diverse roles in RNA metabolism [60]. Unlike the U2 and U12 complexes, U7 snRNPs function not in splicing, but in the unique 3'-end processing of replication-dependent histone mRNAs that comprise the most abundant class of intronless and non-polyadenylated transcripts in metazoans [87]. Facilitated by the SMN complex, U7 snRNA associates with Sm-like (LSm) proteins LSm10 and LSm11 instead of SmD1 and SmD2, to form the heptameric Sm core characteristic of snRNPs complexes [88]. Interactions between U7 snRNA and the stem-loop-binding protein (SLBP) mediate the recruitment and positioning of the trans-acting factors that cleave histone pre-mRNA [87]. It remains to be seen whether SMN also plays a role in the assembly of the structurally related but functionally distinct LSm2–8 and LSm1–7 complexes, which play a role in pre-mRNA processing and mRNA decay [89].

While it is well established that the SMN complex promotes snRNP assembly, and ultimately spliceosome formation, several questions regarding the arrangement and association of SMN with Sm proteins remain. Further examination into the structural arrangement and functions of these complexes in vivo and their regulation by cellular signaling pathways are necessary to fully understand the physiological relevance of these complexes in development and disease.

6.6 SMA Deficiency Causes Widespread Splicing Defects

The extensively examined role of SMN in snRNP biogenesis and pre-mRNA splicing led researchers to hypothesize that SMA phenotypes are the result of SMN-
dependent alterations in snRNP biogenesis and splicing, and that SMA can be
described as a general splicing disease [90]. Evidence supporting this hypothesis is
substantial, yet incomplete. A potential direct link between defective snRNP assembly activity and SMA phenotypes was provided by a study showing that injection of
purified U snRNPs could rescue embryonic arrest and SMA-like axon degeneration
caused by a reduction of SMN or Gemin2 in zebrafish embryos [91], although later
studies in zebrafish have not found low Gemin2 levels to cause specific motor axon
defects, arguing for a separate role for SMN in the SMA disease process that is
snRNP independent [92]. Moreover, studies have shown a reduction in SMN-
dependent snRNP activity in SMA patient tissue and animal models, demonstrating
a correlation between snRNP activity and disease severity, but no selectivity for
vulnerable cell types or tissues was found [81, 90, 93].

As previously described, the SMN complex facilitates snRNP assembly of the
major (U2-dependent) and minor (U12-dependent) spliceosomes, as well as the U7
histone processing complex. Accordingly, several studies exploring how SMN deficiency influences the assembly and activity of each pathway have provided insight
into the relationship between SMN-dependent snRNP activity and SMA phenotypes. Interestingly, changes in snRNP assembly in mouse models of SMA primarily affect the (U12 dependent) minor spliceosome pathway [93]. Caused by a
deficiency in SMN, an inability of U11 snRNP to accumulate and form the U12
spliceosome machinery results in increased U12 intron retention, exon skipping,
and aberrant splicing events. In a *Drosophila* model of SMA, mis-splicing of the
U12 intron-containing gene *Stasimon* correlated with motor neuron pathology [94].
While overexpression of Stasimon in a *Drosophila* model of SMA rescued axonal
pathfinding and outgrowth defects in motor neurons, it failed to restore normal viability and locomotion [94]. Table 6.1 provides a list of selected mRNAs, which are
affected by SMN-dependent splicing alterations and have also been suggested to
contribute to SMA pathology. SMN depletion may also affect U7 histone mRNA
processing. Due to an accumulation of U7 pre-snRNA, U7 snRNP steady-state levels are significantly reduced in SMN-deficient cell lines and SMA mouse tissue,
decreasing the post-transcriptional regulation of histone mRNA and resulting in the
accumulation of uncleaved, 3′-end-extended histones [88].

Although these experiments emphasize the physiological relevance of snRNP
core assembly, it remains unclear whether SMN-dependent alterations to snRNP
biogenesis can account for the full spectrum of pathology observed in SMA patients
and disease models. Studies in *Drosophila Smn* null mutant larvae showed no appre-
ciable defects in the splicing of mRNAs containing minor-class introns, despite
significant reductions in minor-class spliceosomal snRNAs [95]. These findings
suggest that SMN's role in snRNP biogenesis can be uncoupled from its effect on
viability and locomotion. A comparison of snRNP-dependent and SMN-specific

Table 6.1 Mis-spliced transcripts in models of SMA

Gene	Description	Species	Experimental condition	Cell line/tissue	Reference	Relevance to SMA pathology
SMN2	Splicing-deficient gene copy encoding SMN	Human	SMA patient	iPS cells	[175]	SMN-deficiency affects SMN2 splicing in a feedback loop that further reduces SMN protein levels[175]
			SMN1 knockdown	HEK293		
TMEM41b/Stasimon	Transmembrane protein	Drosophila	smn loss-of-function	Larvae	[94]	Stasimon knockdown in cholinergic neurons increases EPSP amplitude in NMJ, restored by Stasimon expression. Stasimon expression rescues motor neuron defects in Zebrafish SMA model [94]
		Mouse	SMN1 knockdown	NIH3T3		
			Moderate SMA model $(SMN2^{+/+};SMN\Delta7^{+/+};Smn^{-/-})$ [22]	Spinal cord, L1 DRG		
Nrxn2a	Pre-synaptic membrane protein	Zebrafish	SMN knockdown	Embryo	[176]	Knock-down of Nrxn2a results in motor axon defects [176]
		Mouse	Severe SMA model $(Smn^{-/-}/SMN2)$ [20]	Spinal cord		
Rit1	G protein	Mouse	SMN1 knockdown	NSC-34	[177]	Alternatively spliced Rit1 isoform reduces neuritic length in NSC-34 cells [177]
			Severe SMA model $(Smn^{-/-}/SMN2)$ [21]	Lumbar spinal cord		
Agrn	Heparin sulfate proteoglycan; organizer of the NMJ	Mouse	Moderate SMA model $(SMN2^{+/+};SMN\Delta7^{+/+};Smn^{-/-})$ [22]	Motor neurons	[178]	SMN depletion leads to exclusion of Z exons from Agrn mRNA [178]. It was previously established that Agrn Z^+ isoforms are important for maturation of postsynaptic termini in NMJ [179]. Expression of Agrn Z^+ in motor neurons of SMA mice mitigates NMJ defects [35]
			Moderate SMA model $(SMN2^{+/+};SMN\Delta7^{+/+};Smn^{-/-})$ [22]	Motor neurons	[35]	

(continued)

Table 6.1 (continued)

Gene	Description	Species	Experimental condition	Cell line/tissue	Reference	Relevance to SMA pathology
Uba1	Ubiquitin-like modifier activating enzyme 1	Mouse	Severe SMA model (*Smn*[-/-]/*SMN2*) [20, 21]	Spinal cord	[180]	*Uba1* knockdown disrupts axonal growth and branching, and leads to increase in β-catenin level in *Zebrafish*. β-catenin levels are also increased in SMA mouse models, and inhibition of β-catenin signaling ameliorates neuromuscular pathology in SMA mice [180]
Anxa2	Ca[2+]-binding actin regulating protein	Mouse	Moderate SMA model (*SMN2*[+/+];*SMNΔ7*[+/+];*Smn*[-/-]) [22]	Spinal cord	[90]	In addition to reported alterations in splicing, *Anxa2* was also shown to be overexpressed in SMA in several proteomic studies [181–183], and *Anxa2* mRNA is mislocalized from neurites in SMN-deficient NSC-34 cells [128]
Cacna1a *Cacna1b* *Cacna1c* *Cacna1e* *Cacna1h*	Voltage-gated Ca[2+] channel subunits	Mouse	Severe SMA model (*Smn*[-/-]/*SMN2*) [21]	Spinal cord	[184]	In addition to reported alterations in splicing, reduction of Cacna1b in axonal growth cones of motor neurons from another severe SMA model mice [20], accompanied with alterations in their excitability, has been shown [138]

RNA changes in SMA models suggests that defects in snRNP supply are unlikely to be the primary drivers of SMA pathophysiology, at least in *Drosophila* [96]. Despite impairment of snRNP synthesis, endogenous snRNP and snRNA levels were found to be unaltered in SMA1 patient-derived fibroblasts, a chicken cell line, and a severe *Drosophila* mutant, all of which had severely reduced SMN levels [81, 93, 97]. Moreover, despite a significant difference in lifespan between the severe and SMNΔ7 models (~9 days), there was no difference in snRNP assembly activity, suggesting that the difference in disease severity is caused by differential effects on an additional function of the SMN protein [93]. It should be noted that snRNAs that are not associated with Sm cores are unstable, so snRNA levels are similar to snRNP levels [98].

Taken together, this research strongly suggests that while the direct effects of SMN deficiency on altered snRNP assembly and splicing are likely to contribute to SMA phenotypes, it fails to fully explain motor neuron susceptibility and the full spectrum of phenotypes observed in SMA pathology.

6.7 SMN Acts as a Molecular Chaperone for mRNP Assembly

Aside from its role in the assembly of noncoding RNAs into snRNP complexes, SMN has also been implicated in transport and local translation of mRNAs to the distal end of neurites [54, 55, 99–106]. It is based on observations from several labs that SMN associates with multiple mRNA-binding proteins (mRBPs) and regulates the localization of specific transcripts and RNA-binding proteins into axons [103, 104, 106–109]. Localized mRNA create micro-environments of newly synthesized proteins in specific subcellular compartments, promoting autonomous control of local proteomes and stimulus driven adaptive responses [110–112]. Both developing and adult axons contain complex transcriptomes that support the formation and maintenance of neural circuits in vivo [113, 114]. mRNA localization is most commonly achieved through the association of its 3′ untranslated region (3′UTR) with mRBPs, to form messenger ribonucleoprotein (mRNP) transport granules [115]. This assembly into mRNPs serves as a major regulator of multiple steps of mRNA processing, including nuclear export, intracellular trafficking, turnover, and translation. Regulatory sequences within the 3′UTR serve as platforms for the assembly of mRNPs [116] and act as cis-acting localization sequences or "zipcodes" that govern the precise spatiotemporal expression of the transcripts [117, 118]. Directed by these zipcode sequences, mRNPs associate with molecular adaptors and motor proteins to form transport granules, which translocate along microtubules and actin filaments to specific microdomains within the cell. In developing and regenerating axons, RNPs containing growth-promoting mRNA transcripts localize to the growth-cone and allow the cell to navigate and respond to environmental factors [110–112].

Indications of possible splicing-unrelated functions of SMN first came out from studies of its localization in neuronal cell cultures. Besides its principal localization in the cytoplasm and within nuclear gems [119], SMN was also found in axons and dendrites of motor neurons from rat spinal cord sections in association with cyto-skeletal components, suggesting potential motor-driven transport [120]. It was also shown that SMN is only partially colocalized with Gemin2 in cytoplasm of mouse cultured embryonic motor neurons, indicating that functions of some SMN sub-populations are not associated with snRNP biogenesis [121]. Cytoplasmic localiza-tion of SMN was shown to be exon 7-dependent, which points to possible role of SMN mislocalization in SMA pathology [57]. In addition, SMN was shown to be a part of granules that are actively transported along cytoskeletal structures in neurites of chicken forebrain neurons and in axons of motor neurons [57, 58]. Furthermore, it has been found that overexpression of wild-type SMN was able to promote neurite growth in differentiated PC12 cells [104], and downregulation of SMN leads to axon growth defects in mouse cultured motor neurons [104], zebrafish motor neu-rons [122], *Xenopus* motor neurons [123], as well as to reduced neurite outgrowth in PC12 cells [124]. Although SMN partially colocalizes with Gemins in cytoplas-mic granules in axons and dendrites, the significance of this association is unclear [103, 125, 126].

To date, the most studied non-splicing function of SMN is its role in mRNA localization and transport in neurons. The first evidence for a role in mRNA traffick-ing was the observed reduction of *β-actin* mRNA in axonal growth cones of cul-tured motor neurons isolated from a severe SMA mouse model [104]. Similar effects were also observed for the localization of the neurite-outgrowth promoting *neuritin 1* or *candidate plasticity-related gene 15* (*Nrn1/Cpg15*) transcript in corti-cal neuron neurites upon SMN knockdown [109] and *growth-associated protein 43* (*Gap43*) mRNA in SMN-deficient motor neurons [55]. In addition, a general reduc-tion of poly-A mRNA abundance in axons of anti-*Smn1* shRNA treated mouse motor neurons and a reduction in axonal protein synthesis in *Smn1* shRNA treated cortical neurons have been reported [55, 103]. A transcriptomic microarray analysis using microfluidic chambers to divide axonal and somatodendritic compartments of cultured mouse motor neurons revealed 1189 downregulated probe sets in axonal compartment upon *Smn1* knockdown [127]. Interestingly, while *Smn1* knockdown led primarily to the upregulation of transcripts in the somatodendritic compartment, the overwhelming majority of significantly altered axonal RNAs were downregu-lated, indicating that these changes resulted from a deficiency in axonal targeting rather than from changes in gene expression rate. The characterization of SMN-associated mRNAs in NSC-34 cells found that transcripts encoding annexin A2 (*Anxa2*) and Cytochrome c oxidase subunit 4 isoform (*Cox4i2*) colocalized with SMN in neurites of differentiated NSC-34 cells and were depleted in neurites upon *Smn1* knockdown [128]. A list of mRNAs that are known to be axonally transported or de-stabilized in an SMN-dependent manner is shown in Table 6.2.

Since SMN does not contain a canonical RNA-binding domain, and unlike Gemin5 and unrip has not been identified as an mRNA-binding protein in large-scale UV-crosslinking experiments [129], it is currently not clear whether its role in

Table 6.2 mRNA components of SMN-dependent mRNPs

Transcript	Species	Experimental condition	Cell line/ tissue	Reference	Type of evidence
β-actin	Mouse	Severe SMA mouse model [20]	Motor neurons	[104]	SMN deficiency reduces *β-actin* mRNA localization in distal axons
		Severe SMA mouse model [22]	DRG sensory neurons	[185]	*β-actin* mRNA is reduced in growth cones of cultured sensory neurons from Smn-deficient embryos
		Smn knockdown	MN-1	[132]	SMN is required for *β-actin* mRNA targeting to RNA granules
	Rat	Transfection with hnRNP R expression constructs	PC12	[104]	Interaction between SMN and hnRNP R modulate *β-actin* mRNA localization in neuritic growth cones
		–	Cortical neurons	[109]	Co-precipitates with SMN
Gap43	Mouse	Severe SMA mouse model [22]	Motor neurons	[55]	*Gap43* mRNA is reduced in axons and growth cones
		Smn knockdown	Motor neurons	[55]	*Gap43* mRNA is reduced in axons and growth cones
		Smn knockdown	MN-1	[132]	SMN is required for recruitment of *Gap43* mRNA to RNA granules
	Zebrafish	*SMN* and *HuD* mutants	Motor neurons	[136]	*Gap43* mRNA levels are decreased in motor neurons in HuD-dependent manner
	Rat	–	Cortical neurons	[109]	Co-precipitates with SMN
Nrn1/Cpg15	Rat	*Smn* knockdown	Cortical neurons	[109]	Co-precipitates with SMN. SMN knockdown affects *Nrn1* mRNA levels in both soma and neurites
Anxa2	Mouse	*Smn* knockdown	NSC-34	[128]	Associates with SMN complex and is reduced in axons upon SMN knockdown
Cox4i2	Mouse	*Smn* knockdown	NSC-34	[128]	Associates with SMN complex and is reduced in axons upon SMN knockdown
Tau	Mouse	*Smn* knockdown	MN-1	[132]	SMN is required for recruitment of *Tau* mRNA to RNA granules
p21	Mouse	Mild SMA mouse model [186]	Spinal cord	[108]	SMN depletion increases *p21* transcript stability

Table 6.3 mRBP components of SMN-dependent mRNPs

Name	Species	Tissue/cell line	Reference	Other supporting evidence
hnRNP R	Human	HEK293	[107]	Interaction with SMN is required for association between *β-actin* mRNA and hnRNP R [104]
	Mouse	Motor neurons, spinal cord extracts	[54]	
KSRP/ FBP2/ MARTA1	Mouse	N2a, spinal cord	[108]	p21 mRNA that is targeted for degradation by KSRP is upregulated in SMA tissues [108]
HuD/ ELAVL4	Rat	Cortical neurons	[109]	SMN is required for HuD targeting into RNA granules [132]. The HuD target mRNA *Gap43* is decreased in motor neurons from *Smn* mutant zebrafish [136]
	Mouse	Spinal cord		
		Motor neurons	[103]	
		MN-1	[132]	
	Zebrafish	Motor neurons	[136]	
IMP1/ZBP1	Rat	Brain	[106]	SMN facilitates association of IMP1 with *β-actin* mRNA [63]
	Mouse	Motor neurons		
SBP2	Human	HEK293	[187]	Levels of several SBP2-dependent selenoprotein mRNAs are reduced in spinal cords from SMA mice [135]

mRNA localization involves its direct interaction with mRNA or is mediated via associated mRBPs. SMN was shown to associate with a large number of mRBPs, including hnRNP R and Q [107, 130], FMRP [131], HuD/ELAVL4 [103, 109, 132], IMP1/ZBP1 [106], KSRP/FBP2/MARTA1 [108], TDP-43 [133], and FUS/TLS [134]. For several of these mRBPs, their association with mRNAs has been shown to be SMN-dependent (Table 6.3). In most cases these interactions are mediated by the Tudor domain of SMN [103, 106, 108, 132]. Some of these mRBPs ensure neuritic mRNA localization, acting in concert with SMN. hnRNP R was shown to modulate *β-actin* mRNA localization in differentiated PC12 neurites, and its SMN-interacting domain was required for this activity [104]. Moreover, hnRNP R directly interacts with *β-actin* mRNA, and SMN facilitates this interaction [104]. Along with their target mRNA, axonal localization of corresponding mRBPs, such as HuD and IMP1, is also affected by SMN deficiency [103, 106]. While the molecular function of SMN in mRNA localization is not as well defined as its role in the snRNP assembly, these results indicated a potentially related role of SMN as an organizer of protein-mRNA complexes (mRNPs) that are then transported within neurites. In support of this hypothesis, SMN was shown to mediate recruitment of HuD and its target mRNAs *Gap43* and *Tau* to RNA granules in differentiated MN-1 cells [132]. Finally, the binding of IMP1 to the *β-actin* 3′UTR was shown to be impaired in motor neurons from an SMA mouse model, and IMP1-containing mRNP granules were largely reduced in size in SMA1 patient fibroblasts [63]. These data indicate that the function of SMN as a molecular chaperone for RNP assembly [60] is not limited to snRNPs, but also includes other RNP types, including the assembly of mRNP transport granules [63].

More detailed mechanistic aspects of SMN activity in mRNP assembly have yet to be revealed. In particular, it would be interesting to determine if the same components of SMN complex that promote snRNP assembly also act in mRNP formation. The observation that selenocysteine insertion sequence (SECIS)-binding protein 2 (SBP2), which is an mRBP for selenoprotein mRNA, directly interacts with the SMN-complex proteins Gemins 3, 4, 7 and 8, suggests their involvement in the assembly of selenoprotein mRNPs [135]. However, their role in the formation of transport mRNP granules in motor neurons still has to be determined.

6.8   mRNP Assembly Defects Can Contribute to SMA Pathology

It is presently not known whether the defects of mRNP assembly contribute to neurodegeneration in SMA patients, but there are several lines of evidence supporting this hypothesis. As stated above, mRNP granules in fibroblasts from SMA patients are reduced in size, indicating that mRNP assembly is indeed impaired in SMA [63]. Furthermore, these granules show decreased association with the cytoskeleton, indicating that mRBP transport defects may occur in SMA [63]. In addition, downregulation of SMN-dependent mRBPs resemble the effects of SMN downregulation. HuD knockout in Zebrafish results in decreased branching of motor axons, and HuD expression in motor neurons from SMN-mutant Zebrafish rescues its defects [136]. Knockdown of hnRNP R also leads to defects in axonal growth in Zebrafish and mouse motor neurons, resulting in defective clustering of voltage-gated Ca^{2+} channels in axonal growth cones, similar to defects described in motor neurons from SMA mouse models [137, 138]. On the other hand, increased expression of both IMP1 and HuD, which are mislocalized in cellular SMA models, restore *Gap43* mRNA and protein levels in growth cones and rescues axon outgrowth defects in SMN-deficient motor neurons [55]. Taken together, these data indicate a potential role for impaired mRNP formation in SMA pathology.

mRNA localization in neurites has been identified as a key process that determines the enrichment of neuritic compartments for specific proteins [139]. Therefore, disruption of mRNA transport by defective mRNP formation has the potential to dramatically alter the neuritic proteome. It is well established that local translation of *β-actin* mRNA regulates directed growth of axonal growth cones in response to guidance cues [112, 140], and interaction between IMP1 and *β-actin* mRNA is required for these responses [140]. Since association between IMP1 and *β-actin* mRNA is decreased in SMA [63], deficiency in IMP1 mRNP formation may contribute to axonal defects observed in SMA models. Similarly, *Gap43* mRNA that is targeted to axons by association with mRBPs IMP1 and HuD [141] is reduced in axons and growth cones of SMN-deficient motor neurons [55]. Although local translation of *Gap43* is required for axon elongation, *β-actin* translation is more important for axonal branching [142]. Interestingly, defects in both elongation and branching of axons upon SMN depletion were reported in zebrafish in vivo [122]. Another example for an SMN-dependently localized mRNA is *Nrn1*, which can rescue axonal defects in a zebrafish SMA model [109]. Since neuritin is known to regulate synapse stability [143], lack of this protein due to insufficient mRNA transport could affect NMJs, where the earliest SMA-associated structural changes occur [28]. Of note, limited amount of IMP1 in DRG neurons restricts axonal localization of *Gap43* and *β-actin* mRNA, while limited availability of HuD induces competition between *Gap43* and *Nrn1* mRNA for HuD binding and axonal localization [144, 145]. It is possible that this phenomenon provides the basis for the regulation of specific mRNA in axons depending on neuronal developmental stage or specific conditions, such as response to neuronal injury. Therefore, it would be of interest to establish how SMN depletion affects relative abundances of specific mRNAs and mRBPs in axons and growth cones. Another question is how SMN regulates association of mRNA with distinct mRBPs. While it is well known that binding of HuD provides stabilization of its mRNA targets, such as *Gap43* [146], association with the SMN-interacting mRBP KSRP leads to increased *Gap43* mRNA decay [147].

mRNPs have the ability to form higher-order cytoplasmic mRNP granules in response to certain environmental conditions. Cellular stress can result in the assembly of mRNPs and stalled translation pre-initiation complexes into cytoplasmic stress granules [148]. The sequestration of translationally stalled housekeeping mRNAs and enhanced expression of stress-response factors, such as molecular chaperones, is believed to be protective for cell survival under stress conditions. SMN has been shown to facilitate stress granule formation [149], whereas its downregulation inhibits cellular stress response [150]. Thus, a reduced capacity for stress granule formation could make cells more vulnerable to environmental conditions. One of the described functions of stress granules is the prevention of apoptosis by sequestration of certain molecules, such as MTK1 and mTORC1 kinases, and inhibition of apoptotic signaling [151, 152]. If SMA is associated with a deficiency in stress granule formation, this could contribute to apoptotic mechanisms in neurodegeneration and cell death [153].

6.9   Other mRNA-Processing Functions of SMN

Aside from its role in snRNP and mRNP assembly, there is evidence for the involvement of SMN at other stages of the mRNA life cycle, including transcription and translation (Fig. 6.2). The first evidence for the involvement of SMN in transcription came from the finding that it interacts with bovine papillomavirus transcriptional activator E2 and stimulates E2-dependent transcription [154]. Subsequent studies identified additional SMN interactors involved in transcription, including the tumor

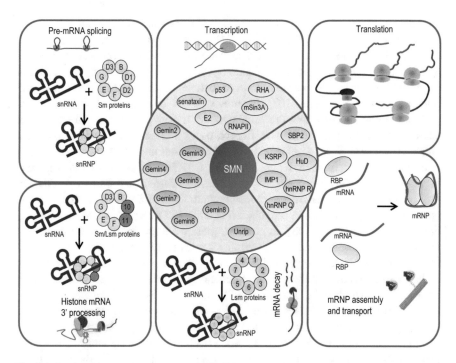

Fig. 6.2 Molecular Functions of SMN in mRNA processing. SMN can associate with a large selection of proteins to regulate snRNP assembly (splicing, histone mRNA processing, mRNA decay), transcription, translation, and mRNP assembly (mRNA transport and local translation)

suppressor and transcriptional activator p53 [155] and the transcription corepressor mSin3A [156]. Moreover, artificial recruitment of SMN to promotor regions resulted in repression of transcription [156]. SMN was found to associate with key components of transcription machinery, such as RNA helicase A and RNA polymerase II. Overexpression of truncated SMNΔN27 results in transcription inhibition and accumulation of these components in gems and coiled bodies [157]. SMN was also shown to interact with the DNA/RNA helicase senataxin [158] and facilitate the association of senataxin and the C-terminal domain (CTD) of RNA polymerase II in a manner that was dependent on CTD symmetric dimethylation [159]. Formation of this complex is required for resolving DNA-RNA loops (R-loops) and proper transcription termination [159]. It has been shown that SMN knockdown in SH-SY5Y cells leads to increased R-loop formation and DNA damage [160]. Of note, senataxin mutations are a rare cause of proximal spinal muscular atrophy [161], juvenile amyotrophic lateral sclerosis [162], and ataxia-ocular apraxia 2 [163], indicating a possible link between the regulation of transcription termination via R-loops and neurodegeneration across different neurodegenerative diseases (for reviews, see [164, 165]).

The role of SMN in mRNP assembly and transport implies that *SMN* mutations can cause defects in local translation due to inefficient localization of mRNA to their destination sites. Indeed, defects in axonal translation in motor neurons from a mouse model of severe SMA ($Smn^{-/-}$;$SMN2$) and in cortical neurons upon SMN knockdown were reported [55, 166]. In addition, there is accumulating evidence that SMN can directly regulate translation. SMN can associate with the translation machinery, and it has been found in polyribosome fractions purified by ultracentrifugation from MN-1 cells [167]. Moreover, SMN can anchor ribosomes to the plasma membrane, since the ribosomal content in plasma membrane fractions was depleted in SMA patient-derived fibroblasts or normal fibroblasts upon SMN knockdown [168]. SMN deficiency in MN-1 cells does not affect overall translation rates but leads to increased translation of CARM1 arginine methyltransferase mRNA, and possibly other specific mRNAs, via currently unknown mechanism [167]. Another study demonstrated that low amounts of SMN in cortical neurons reduce protein synthesis by upregulation of miR-183 microRNA and downregulation of mTOR pathway [105]. In contrast, there were no significant differences in mTOR activation status and protein synthesis rate upon SMN knockdown in human fibroblasts under steady-state conditions. However, when membrane protrusion formation was stimulated in these cells, a decrease in translation rate was observed in *SMN* knockdown fibroblasts, and this difference was mTOR-dependent [168].

An SMN-dependent defect in translation has also been suggested by polysome profiling experiments, which indicated a reduction in the polysome peak of profiles from late-symptomatic SMA mouse tissue [169]. RNA-seq data analysis identified genes associated with translation-related processes as significantly dysregulated in SMA motor neurons, providing evidence for a role of SMN in the regulation of ribosome biogenesis and translational activity [169].

SMN-mediated regulation of translation is an emerging field that needs further studies to elucidate its molecular mechanism and potential role in SMA pathology, but it may very well be related to a function for SMN in assembling mRNA and associated proteins, similar to its role in snRNP and mRNP assembly.

6.10   Open Questions and Future Perspectives

While diverse functions of SMN in regulating different aspects of mRNA processing are well established, the relative contributions of these SMN-dependent pathways and molecular mechanisms that result in disease pathology remain unclear. Studies examining the biological function of SMN are beginning to reveal the dynamic roles that SMN plays as a chaperone in snRNP assembly and splicing, mRNP assembly and mRNA trafficking, and as a regulator of post-transcriptional gene expression. Additional studies to discern which SMN function is critically affected in SMA will be crucial to our understanding of SMN biology and SMA pathology.

Although there is substantial evidence suggesting that low levels of SMN result in motor neuron dysfunction and loss, little is known about the pathological consequences of SMN depletion in other cell-types. It now appears likely that other neurons in the spinal cord motor circuitry, glia, myofibers, and other tissues outside the CNS may contribute to the pathophysiology of SMA [45–50]. Importantly, the recent introduction of Spinraza™ and the anticipated addition of gene therapy (AVXS-101) as a treatment for SMA is expected to attenuate aggressive aspects of disease pathology in the CNS of SMA I patients, potentially leading to a more chronic disease, and necessitating the characterization of otherwise masked disease phenotypes in peripheral organs and tissues.

As a regulator of snRNP complex assembly, SMN deficiency is known to cause widespread changes in splicing and gene expression in various cellular and animal models of SMA. However, the question if and how a defect in the canonical house-keeping function of SMN in snRNP assembly directly causes the neurodevelop-mental and neurodegenerative processes that lead to SMA pathogenesis remains. While the discovery of *stasimon* as a potential regulator of motor neuron circuitry provided the first link between SMN-dependent splicing variations and the develop-ment of SMA pathology [94], future studies will be needed to examine and charac-terize the role of stasimon in human patients. As listed in Table 6.1, research into identified SMN-dependent splicing isoforms will be required to substantiate a direct link between splicing defects and SMA phenotypes. A thorough characterization of aberrant pre-mRNA processing in motor neurons of SMA mouse models and patients will continue to expand our understanding of the down-stream conse-quences of SMN deficiency that may explain motor neuron susceptibility and dis-ease pathology.

Research from several groups has begun to elucidate the noncanonical functions of SMN in regulating mRNP assembly and trafficking, as well as local and general translation [101, 170]. Understanding SMN-dependent interactions, assembly, and localization of mRNP complexes could elucidate how extensive cell polarity and trafficking demands characteristic of motor neurons contribute to disease vulnera-bility. Axonal localization defects are prevalent in multiple neurological disorders including amyotrophic lateral sclerosis (ALS), Alzheimer disease (AD), Huntington disease (HD), and Fragile X Syndrome (FXS), with recent studies emphasizing the role of local protein synthesis in regulating synaptic transmission and axon mainte-nance, and its relevance for human disease [113, 171–173].

While the precise molecular mechanism of SMN-mediated snRNP assembly is well characterized (reviewed in [60]), much less is known about the exact molecular processes that govern mRNP assembly. SMN was shown to interact with several mRBPs using its Tudor domain, and similar to Sm and Lsm proteins, these interac-tions are triggered by arginine methylation of mRBPs [108, 132]. Indeed, work from several labs has shown that SMN associates with mRBPs known to regulate the axonal localization and synthesis of growth-promoting mRNAs. With a growing body of research linking SMN deficiency to reduced RNP assembly and transport, several questions are beginning to arise regarding the molecular mechanisms and the nature of the SMN interactome. Questions such as: How extensive is SMN's involvement in mRNP assembly and trafficking? By what molecular mechanisms does SMN mediate the association of mRBPs with target mRNAs? While current models for the formation of mRNPs suggest that the interaction of mRNAs bearing distinct localization elements with mRBPs is sufficient to initiate mRNP assembly [174], a recent study has identified SMN as a chaperone for the assembly of mRNP granules, at least for those containing the IMP1 protein [63]. Do other core components of the SMN complex (Gemins and unrip) contribute to noncanonical functions of SMN? These questions are currently unanswered, and should be addressed in future studies. Another important task is to identify a potential link between hypo-assembly of mRNPs and the neurodegeneration of motor neurons in SMA. Comparing the molecular composition of mRNPs in normal cells and in SMA models may offer clues into the components that are necessary to achieve normal biological function. A recent transcriptomic study identified a large number of mRNAs that are mislocalized in axons upon SMN knockdown in cultured mouse motor neurons in vitro [127], and it remains to be seen if similar changes can be identified in vivo.

Selective disruption and rescue of diverse SMN-dependent RNA processing functions, as summarized in Fig. 6.2, should allow us to assess their contribution to SMA pathology. The identification of pivotal pathways and molecules in RNP assembly and transport will expand our understanding of the underlying biology that contributes to organismal development and regeneration, and potentially offer novel strategies to treat and rescue degenerative phenotypes in a wide variety of neurological disorders and disease. While current efforts for the treatment of SMA are mainly focused on raising SMN protein levels, a thorough understanding of SMN's role in the SMA disease process may lead to the identification of additional targets for therapy.

References

1. Kolb SJ, Kissel JT. Spinal muscular atrophy: a timely review. Arch Neurol. 2011;68(8):979–84.
2. Prior TW, Snyder PJ, Rink BD, Pearl DK, Pyatt RE, Mihal DC, et al. Newborn and carrier screening for spinal muscular atrophy. Am J Med Genet A. 2010;152A(7):1608–16.
3. Verhaart IEC, Robertson A, Wilson IJ, Aartsma-Rus A, Cameron S, Jones CC, et al. Prevalence, incidence and carrier frequency of 5q-linked spinal muscular atrophy - a literature review. Orphanet J Rare Dis. 2017;12(1):124.
4. Zerres K, Wirth B, Rudnik-Schöneborn S. Spinal muscular atrophy—clinical and genetic correlations. Neuromuscul Disord. 1997;7(3):202–7.
5. Lefebvre S, Burglen L, Reboullet S, Clermont O, Burlet P, Viollet L, et al. Identification and characterization of a spinal muscular atrophy-determining gene. Cell. 1995;80(1):155–65.
6. McAndrew PE, Parsons DW, Simard LR, Rochette C, Ray PN, Mendell JR, et al. Identification of proximal spinal muscular atrophy carriers and patients by analysis of SMNT and SMNC gene copy number. Am J Hum Genet. 1997;60(6):1411–22.
7. Campbell L, Potter A, Ignatius J, Dubowitz V, Davies K. Genomic variation and gene conversion in spinal muscular atrophy: implications for disease process and clinical phenotype. Am J Hum Genet. 1997;61(1):40–50.
8. Foust KD, Nurre E, Montgomery CL, Hernandez A, Chan CM, Kaspar BK. Intravascular AAV9 preferentially targets neonatal-neurons and adult-astrocytes in CNS. Nat Biotechnol. 2009;27(1):59–65.

9. Naryshkin NA, Weetall M, Dakka A, Narasimhan J, Zhao X, Feng Z, et al. Motor neuron disease. SMN2 splicing modifiers improve motor function and longevity in mice with spinal muscular atrophy. Science (New York, NY). 2014;345(6197):688–93.
10. Hua Y, Sahashi K, Hung G, Rigo F, Passini MA, Bennett CF, et al. Antisense correction of SMN2 splicing in the CNS rescues necrosis in a type III SMA mouse model. Genes Dev. 2010;24(15):1634–44.
11. Rigo F, Chun SJ, Norris DA, Hung G, Lee S, Matson J, et al. Pharmacology of a central nervous system delivered 2′-O-methoxyethyl-modified survival of motor neuron splicing oligonucleotide in mice and nonhuman primates. J Pharmacol Exp Ther. 2014;350(1):46–55.
12. Singh NN, Howell MD, Androphy EJ, Singh RN. How the discovery of ISS-N1 led to the first medical therapy for spinal muscular atrophy. Gene Ther. 2017;24(9):520–6.
13. Hoy SM. Nusinersen: first global approval. Drugs. 2017;77(4):473–9.
14. Burgart AM, Magnus D, Tabor HK, Paquette ED, Frader J, Glover JJ, et al. Ethical challenges confronted when providing nusinersen treatment for spinal muscular atrophy. JAMA Pediatr. 2018;172(2):188–92.
15. Day M, Wang Z, Ding J, An X, Ingham CA, Shering AF, et al. Selective elimination of glutamatergic synapses on striatopallidal neurons in Parkinson disease models. Nat Neurosci. 2006;9(2):251–9.
16. Shankar GM, Li S, Mehta TH, Garcia-Munoz A, Shepardson NE, Smith I, et al. Amyloidbeta protein dimers isolated directly from Alzheimer's brains impair synaptic plasticity and memory. Nat Med. 2008;14(8):837–42.
17. Fischer LR, Culver DG, Tennant P, Davis AA, Wang M, Castellano-Sanchez A, et al. Amyotrophic lateral sclerosis is a distal axonopathy: evidence in mice and man. Exp Neurol. 2004;185(2):232–40.
18. Edens BM, Ajroud-Driss S, Ma L, Ma Y-C. Molecular mechanisms and animal models of spinal muscular atrophy. Biochim Biophys Acta Mol Basis Dis. 2015;1852(4):685–92.
19. Schrank B, Gotz R, Gunnersen JM, Ure JM, Toyka KV, Smith AG, et al. Inactivation of the survival motor neuron gene, a candidate gene for human spinal muscular atrophy, leads to massive cell death in early mouse embryos. Proc Natl Acad Sci U S A. 1997;94(18):9920–5.
20. Monani UR, Sendtner M, Coovert DD, Parsons DW, Andreassi C, Le TT, et al. The human centromeric survival motor neuron gene (SMN2) rescues embryonic lethality in Smn(−/−) mice and results in a mouse with spinal muscular atrophy. Hum Mol Genet. 2000;9(3):333–9.
21. Hsieh-Li HM, Chang JG, Jong YJ, Wu MH, Wang NM, Tsai CH, et al. A mouse model for spinal muscular atrophy. Nat Genet. 2000;24(1):66–70.
22. Le TT, Pham LT, Butchbach ME, Zhang HL, Monani UR, Coovert DD, et al. SMNDelta7, the major product of the centromeric survival motor neuron (SMN2) gene, extends survival in mice with spinal muscular atrophy and associates with full-length SMN. Hum Mol Genet. 2005;14(6):845–57.
23. Bowerman M, Murray LM, Beauvais A, Pinheiro B, Kothary R. A critical smn threshold in mice dictates onset of an intermediate spinal muscular atrophy phenotype associated with a distinct neuromuscular junction pathology. Neuromuscul Disord. 2012;22(3):263–76.
24. Hammond SM, Gogliotti RG, Rao V, Beauvais A, Kothary R, DiDonato CJ. Mouse survival motor neuron alleles that mimic SMN2 splicing and are inducible rescue embryonic lethality early in development but not late. PLoS One. 2010;5(12):e15887.
25. Cifuentes-Diaz C, Nicole S, Velasco ME, Borra-Cebrian C, Panozzo C, Frugier T, et al. Neurofilament accumulation at the motor endplate and lack of axonal sprouting in a spinal muscular atrophy mouse model. Hum Mol Genet. 2002;11(12):1439–47.
26. Xu CC, Denton KR, Wang ZB, Zhang X, Li XJ. Abnormal mitochondrial transport and morphology as early pathological changes in human models of spinal muscular atrophy. Dis Model Mech. 2016;9(1):39–49.
27. Miller N, Shi H, Zelikovich AS, Ma Y-C. Motor neuron mitochondrial dysfunction in spinal muscular atrophy. Hum Mol Genet. 2016;25(16):3395–406.

28. Kariya S, Park GH, Maeno-Hikichi Y, Leykekhman O, Lutz C, Arkovitz MS, et al. Reduced SMN protein impairs maturation of the neuromuscular junctions in mouse models of spinal muscular atrophy. Hum Mol Genet. 2008;17(16):2552–69.
29. Kong L, Wang X, Choe DW, Polley M, Burnett BG, Bosch-Marce M, et al. Impaired synaptic vesicle release and immaturity of neuromuscular junctions in spinal muscular atrophy mice. J Neurosci. 2009;29(3):842–51.
30. Martinez-Hernandez R, Bernal S, Also-Rallo E, Alias L, Barcelo MJ, Hereu M, et al. Synaptic defects in type I spinal muscular atrophy in human development. J Pathol. 2013;229(1):49–61.
31. Diers A, Kaczinski M, Grohmann K, Hubner C, Stoltenburg-Didinger G. The ultrastructure of peripheral nerve, motor end-plate and skeletal muscle in patients suffering from spinal muscular atrophy with respiratory distress type 1 (SMARD1). Acta Neuropathol. 2005;110(3):289–97.
32. Martinez-Hernandez R, Soler-Botija C, Also E, Alias L, Caselles L, Gich I, et al. The developmental pattern of myotubes in spinal muscular atrophy indicates prenatal delay of muscle maturation. J Neuropathol Exp Neurol. 2009;68(5):474–81.
33. Lee YI, Mikesh M, Smith I, Rimer M, Thompson W. Muscles in a mouse model of spinal muscular atrophy show profound defects in neuromuscular development even in the absence of failure in neuromuscular transmission or loss of motor neurons. Dev Biol. 2011;356(2):432–44.
34. Hayhurst M, Wagner AK, Cerletti M, Wagers AJ, Rubin LL. A cell-autonomous defect in skeletal muscle satellite cells expressing low levels of survival of motor neuron protein. Dev Biol. 2012;368(2):323–34.
35. Kim JK, Caine C, Awano T, Herbst R, Monani UR. Motor neuronal repletion of the NMJ organizer, Agrin, modulates the severity of the spinal muscular atrophy disease phenotype in model mice. Hum Mol Genet. 2017;26(13):2377–85.
36. Boido M, Vercelli A. Neuromuscular junctions as key contributors and therapeutic targets in spinal muscular atrophy. Front Neuroanat. 2016;10:6.
37. Ling KKY, Lin M-Y, Zingg B, Feng Z, Ko C-P. Synaptic defects in the spinal and neuromuscular circuitry in a mouse model of spinal muscular atrophy. PLoS One. 2010;5(11):e15457.
38. Mentis GZ, Blivis D, Liu W, Drobac E, Crowder ME, Kong L, et al. Early functional impairment of sensory-motor connectivity in a mouse model of spinal muscular atrophy. Neuron. 2011;69(3):453–67.
39. Rindt H, Feng Z, Mazzasette C, Glascock JJ, Valdivia D, Pyles N, et al. Astrocytes influence the severity of spinal muscular atrophy. Hum Mol Genet. 2015;24(14):4094–102.
40. Zhou C, Feng Z, Ko CP. Defects in motoneuron-astrocyte interactions in spinal muscular atrophy. J Neurosci. 2016;36(8):2543–53.
41. McGivern JV, Patitucci TN, Nord JA, Barabas M-EA, Stucky CL, Ebert AD. Spinal muscular atrophy astrocytes exhibit abnormal calcium regulation and reduced growth factor production. Glia. 2013;61(9):1418–28.
42. Hunter G, Aghamaleky Sarvestany A, Roche SL, Symes RC, Gillingwater TH. SMN-dependent intrinsic defects in Schwann cells in mouse models of spinal muscular atrophy. Hum Mol Genet. 2014;23(9):2235–50.
43. Martin JE, Nguyen TT, Grunseich C, Nofziger JH, Lee PR, Fields D, et al. Decreased motor neuron support by SMA astrocytes due to diminished MCP1 secretion. J Neurosci. 2017;37(21):5309–18.
44. Sison SL, Patitucci TN, Seminary ER, Villalon E, Lorson CL, Ebert AD. Astrocyte-produced miR-146a as a mediator of motor neuron loss in spinal muscular atrophy. Hum Mol Genet. 2017;26(17):3409–20.
45. Gogliotti RG, Quinlan KA, Barlow CB, Heier CR, Heckman CJ, Didonato CJ. Motor neuron rescue in spinal muscular atrophy mice demonstrates that sensory-motor defects are a consequence, not a cause, of motor neuron dysfunction. J Neurosci. 2012;32(11):3818–29.
46. McGovern VL, Iyer CC, Arnold WD, Gombash SE, Zaworski PG, Blatnik AJ, et al. SMN expression is required in motor neurons to rescue electrophysiological deficits in the SMNΔ7 mouse model of SMA. Hum Mol Genet. 2015;24(19):5524–41.

47. Gavrilina TO, McGovern VL, Workman E, Crawford TO, Gogliotti RG, DiDonato CJ, et al. Neuronal SMN expression corrects spinal muscular atrophy in severe SMA mice while muscle-specific SMN expression has no phenotypic effect. Hum Mol Genet. 2008;17(8):1063–75.
48. Iyer CC, McGovern VL, Murray JD, Gombash SE, Zaworski PG, Foust KD, et al. Low levels of Survival Motor Neuron protein are sufficient for normal muscle function in the SMNDelta7 mouse model of SMA. Hum Mol Genet. 2015;24(21):6160–73.
49. Shababi M, Lorson CL, Rudnik-Schoneborn SS. Spinal muscular atrophy: a motor neuron disorder or a multi-organ disease? J Anat. 2014;224(1):15–28.
50. Nash LA, Burns JK, Chardon JW, Kothary R, Parks RJ. Spinal muscular atrophy: more than a disease of motor neurons? Curr Mol Med. 2016;16(9):779–92.
51. Wirth B, Barkats M, Martinat C, Sendtner M, Gillingwater TH. Moving towards treatments for spinal muscular atrophy: hopes and limits. Expert Opin Emerg Drugs. 2015;20(3):353–6.
52. Burghes AHM, Beattie CE. Spinal muscular atrophy: why do low levels of SMN make motor neurons sick? Nat Rev Neurosci. 2009;10(8):597–609.
53. Jablonka S, Sendtner M. Developmental regulation of SMN expression: pathophysiological implications and perspectives for therapy development in spinal muscular atrophy. Gene Ther. 2017;24(9):506–13.
54. Dombert B, Sivadasan R, Simon CM, Jablonka S, Sendtner M. Presynaptic localization of Smn and hnRNP R in axon terminals of embryonic and postnatal mouse motoneurons. PLoS One. 2014;9(10):e110846.
55. Fallini C, Donlin-Asp PG, Rouanet JP, Bassell GJ, Rossoll W. Deficiency of the survival of motor neuron protein impairs mRNA localization and local translation in the growth cone of motor neurons. J Neurosci. 2016;36(13):3811–20.
56. Hao le T, Duy PQ, Jontes JD, Beattie CE. Motoneuron development influences dorsal root ganglia survival and Schwann cell development in a vertebrate model of spinal muscular atrophy. Hum Mol Genet. 2015;24(2):346–60.
57. Zhang HL, Pan F, Hong D, Shenoy SM, Singer RH, Bassell GJ. Active transport of the survival motor neuron protein and the role of exon-7 in cytoplasmic localization. J Neurosci. 2003;23(16):6627–37.
58. Fallini C, Bassell GJ, Rossoll W. High-efficiency transfection of cultured primary motor neurons to study protein localization, trafficking, and function. Mol Neurodegener. 2010;5:17.
59. Gruss OJ, Meduri R, Schilling M, Fischer U. UsnRNP biogenesis: mechanisms and regulation. Chromosoma. 2017;126(5):577–93.
60. Li DK, Tisdale S, Lotti F, Pellizzoni L. SMN control of RNP assembly: from post-transcriptional gene regulation to motor neuron disease. Semin Cell Dev Biol. 2014;32:22–9.
61. Ramaswami M, Taylor JP, Parker R. Altered ribostasis: RNA-protein granules in degenerative disorders. Cell. 2013;154(4):727–36.
62. Shukla S, Parker R. Hypo- and hyper-assembly diseases of RNA–protein complexes. Trends Mol Med. 2016;22(7):615–28.
63. Donlin-Asp PG, Fallini C, Campos J, Chou CC, Merritt ME, Phan HC, et al. The survival of motor neuron protein acts as a molecular chaperone for mRNP assembly. Cell Rep. 2017;18(7):1660–73.
64. Pellizzoni L. Chaperoning ribonucleoprotein biogenesis in health and disease. EMBO Rep. 2007;8(4):340–5.
65. Cauchi RJ. SMN and Gemins: 'we are family' ... or are we?: insights into the partnership between Gemins and the spinal muscular atrophy disease protein SMN. Bioessays. 2010;32(12):1077–89.
66. Otter S, Grimmler M, Neuenkirchen N, Chari A, Sickmann A, Fischer U. A comprehensive interaction map of the human survival of motor neuron (SMN) complex. J Biol Chem. 2007;282(8):5825–33.
67. Fischer U, Liu Q, Dreyfuss G. The SMN-SIP1 complex has an essential role in spliceosomal snRNP biogenesis. Cell. 1997;90(6):1023–9.
68. Meister G, Buhler D, Pillai R, Lottspeich F, Fischer U. A multiprotein complex mediates the ATP-dependent assembly of spliceosomal U snRNPs. Nat Cell Biol. 2001;3(11):945–9.

69. Pellizzoni L, Yong J, Dreyfuss G. Essential role for the SMN complex in the specificity of snRNP assembly. Science. 2002;298(5599):1775–9.
70. Brahms H, Meheus L, de Brabandere V, Fischer U, Luhrmann R. Symmetrical dimethylation of arginine residues in spliceosomal Sm protein B/B′ and the Sm-like protein LSm4, and their interaction with the SMN protein. RNA. 2001;7(11):1531–42.
71. Friesen WJ, Massenet S, Paushkin S, Wyce A, Dreyfuss G. SMN, the product of the spinal muscular atrophy gene, binds preferentially to dimethylarginine-containing protein targets. Mol Cell. 2001;7(5):1111–7.
72. Charroux B, Pellizzoni L, Perkinson RA, Shevchenko A, Mann M, Dreyfuss G. Gemin3: a novel DEAD box protein that interacts with SMN, the spinal muscular atrophy gene product, and is a component of gems. J Cell Biol. 1999;147(6):1181–94.
73. Battle DJ, Lau CK, Wan L, Deng H, Lotti F, Dreyfuss G. The Gemin5 protein of the SMN complex identifies snRNAs. Mol Cell. 2006;23(2):273–9.
74. Lau CK, Bachorik JL, Dreyfuss G. Gemin5-snRNA interaction reveals an RNA binding function for WD repeat domains. Nat Struct Mol Biol. 2009;16(5):486–91.
75. Ma Y, Dostie J, Dreyfuss G, Van Duyne GD. The Gemin6-Gemin7 heterodimer from the survival of motor neurons complex has an Sm protein-like structure. Structure. 2005;13(6):883–92.
76. Meister G, Eggert C, Fischer U. SMN-mediated assembly of RNPs: a complex story. Trends Cell Biol. 2002;12(10):472–8.
77. Friesen WJ, Paushkin S, Wyce A, Massenet S, Pesiridis GS, Van Duyne G, et al. The methylosome, a 20S complex containing JBP1 and pICln, produces dimethylarginine-modified Sm proteins. Mol Cell Biol. 2001;21(24):8289–300.
78. Meister G, Eggert C, Bühler D, Brahms H, Kambach C, Fischer U. Methylation of Sm proteins by a complex containing PRMT5 and the putative U snRNP assembly factor pICln. Curr Biol. 2001;11(24):1990–4.
79. Chari A, Golas MM, Klingenhäger M, Neuenkirchen N, Sander B, Englbrecht C, et al. An assembly chaperone collaborates with the SMN complex to generate spliceosomal SnRNPs. Cell. 2008;135(3):497–509.
80. Paushkin S, Gubitz AK, Massenet S, Dreyfuss G. The SMN complex, an assemblyosome of ribonucleoproteins. Curr Opin Cell Biol. 2002;14(3):305–12.
81. Wan L, Battle DJ, Yong J, Gubitz AK, Kolb SJ, Wang J, et al. The survival of motor neurons protein determines the capacity for snRNP assembly: biochemical deficiency in spinal muscular atrophy. Mol Cell Biol. 2005;25(13):5543–51.
82. Patel AA, Steitz JA. Splicing double: insights from the second spliceosome. Nat Rev Mol Cell Biol. 2003;4(12):960–70.
83. Turunen JJ, Niemelä EH, Verma B, Frilander MJ. The significant other: splicing by the minor spliceosome. Wiley Interdiscip Rev RNA. 2013;4(1):61–76.
84. Elsaid MF, Chalhoub N, Ben-Omran T, Kumar P, Kamel H, Ibrahim K, et al. Mutation in noncoding RNA RNU12 causes early onset cerebellar ataxia. Ann Neurol. 2017;81(1):68–78.
85. Bacrot S, Doyard M, Huber C, Alibeu O, Feldhahn N, Lehalle D, et al. Mutations in SNRPB, encoding components of the core splicing machinery, cause cerebro-costo-mandibular syndrome. Hum Mutat. 2015;36(2):187–90.
86. Singh RK, Cooper TA. Pre-mRNA splicing in disease and therapeutics. Trends Mol Med. 2012;18(8):472–82.
87. Marzluff WF, Wagner EJ, Duronio RJ. Metabolism and regulation of canonical histone mRNAs: life without a poly(A) tail. Nat Rev Genet. 2008;9(11):843–54.
88. Tisdale S, Lotti F, Saieva L, Van Meerbeke JP, Crawford TO, Sumner CJ, et al. SMN is essential for the biogenesis of U7 snRNP and 3′-end formation of histone mRNAs. Cell Rep. 2013;5(5). https://doi.org/10.1016/jcelrep2013.11.012.
89. Vindry C, Marnef A, Broomhead H, Twyffels L, Ozgur S, Stoecklin G, et al. Dual RNA processing roles of Pat1b via cytoplasmic Lsm1-7 and nuclear Lsm2-8 complexes. Cell Rep. 2017;20(5):1187–200.

90. Zhang Z, Lotti F, Dittmar K, Younis I, Wan L, Kasim M, et al. SMN deficiency causes tissue-specific perturbations in the repertoire of snRNAs and widespread defects in splicing. Cell. 2008;133(4):585–600.

91. Winkler C, Eggert C, Gradl D, Meister G, Giegerich M, Wedlich D, et al. Reduced U snRNP assembly causes motor axon degeneration in an animal model for spinal muscular atrophy. Genes Dev. 2005;19(19):2320–30.

92. McWhorter ML, Boon KL, Horan ES, Burghes AH, Beattie CE. The SMN binding protein Gemin2 is not involved in motor axon outgrowth. Dev Neurobiol. 2008;68(2):182–94.

93. Gabanella F, Butchbach ME, Saieva L, Carissimi C, Burghes AH, Pellizzoni L. Ribonucleoprotein assembly defects correlate with spinal muscular atrophy severity and preferentially affect a subset of spliceosomal snRNPs. PLoS One. 2007;2(9):e921.

94. Lotti F, Imlach WL, Saieva L, Beck ES, Hao le T, Li DK, et al. An SMN-dependent U12 splicing event essential for motor circuit function. Cell. 2012;151(2):440–54.

95. Praveen K, Wen Y, Matera AG. A Drosophila model of spinal muscular atrophy uncouples snRNP biogenesis functions of survival motor neuron from locomotion and viability defects. Cell Rep. 2012;1(6):624–31.

96. Garcia EL, Wen Y, Praveen K, Matera AG. Transcriptomic comparison of Drosophila snRNP biogenesis mutants reveals mutant-specific changes in pre-mRNA processing: implications for spinal muscular atrophy. RNA (New York, NY). 2016;22(8):1215–27.

97. Rajendra TK, Gonsalvez GB, Walker MP, Shpargel KB, Salz HK, Matera AG. A Drosophila melanogaster model of spinal muscular atrophy reveals a function for SMN in striated muscle. J Cell Biol. 2007;176(6):831–41.

98. Sauterer RA, Feeney RJ, Zieve GW. Cytoplasmic assembly of snRNP particles from stored proteins and newly transcribed snRNA's in L929 mouse fibroblasts. Exp Cell Res. 1988;176(2):344–59.

99. Fallini C, Bassell GJ, Rossoll W. Spinal muscular atrophy: the role of SMN in axonal mRNA regulation. Brain Res. 2012;1462:81–92.

100. Briese M, Esmaeili B, Sattelle DB. Is spinal muscular atrophy the result of defects in motor neuron processes? Bioessays. 2005;27(9):946–57.

101. Donlin-Asp PG, Bassell GJ, Rossoll W. A role for the survival of motor neuron protein in mRNP assembly and transport. Curr Opin Neurobiol. 2016;39:53–61.

102. Rossoll W, Bassell GJ. Spinal muscular atrophy and a model for survival of motor neuron protein function in axonal ribonucleoprotein complexes. Results Probl Cell Differ. 2009;48:289–326.

103. Fallini C, Zhang H, Su Y, Silani V, Singer RH, Rossoll W, et al. The survival of motor neuron (SMN) protein interacts with the mRNA-binding protein HuD and regulates localization of poly(A) mRNA in primary motor neuron axons. J Neurosci. 2011;31(10):3914–25.

104. Rossoll W, Jablonka S, Andreassi C, Kroning AK, Karle K, Monani UR, et al. Smn, the spinal muscular atrophy-determining gene product, modulates axon growth and localization of beta-actin mRNA in growth cones of motoneurons. J Cell Biol. 2003;163(4):801–12.

105. Kye MJ, Niederst ED, Wertz MH, Goncalves Ido C, Akten B, Dover KZ, et al. SMN regulates axonal local translation via miR-183/mTOR pathway. Hum Mol Genet. 2014;23(23):6318–31.

106. Fallini C, Rouanet JP, Donlin-Asp PG, Guo P, Zhang H, Singer RH, et al. Dynamics of survival of motor neuron (SMN) protein interaction with the mRNA-binding protein IMP1 facilitates its trafficking into motor neuron axons. Dev Neurobiol. 2014;74(3):319–32.

107. Rossoll W, Kroning AK, Ohndorf UM, Steegborn C, Jablonka S, Sendtner M. Specific interaction of Smn, the spinal muscular atrophy determining gene product, with hnRNP-R and gry-rbp/hnRNP-Q: a role for Smn in RNA processing in motor axons? Hum Mol Genet. 2002;11(1):93–105.

108. Tadesse H, Deschenes-Furry J, Boisvenue S, Cote J. KH-type splicing regulatory protein interacts with survival motor neuron protein and is misregulated in spinal muscular atrophy. Hum Mol Genet. 2008;17(4):506–24.

109. Akten B, Kye MJ, Hao le T, Wertz MH, Singh S, Nie D, et al. Interaction of survival of motor neuron (SMN) and HuD proteins with mRNA cpg15 rescues motor neuron axonal deficits. Proc Natl Acad Sci U S A. 2011;108(25):10337–42.

110. Wu KY, Hengst U, Cox LJ, Macosko EZ, Jeromin A, Urquhart ER, et al. Local translation of RhoA regulates growth cone collapse. Nature. 2005;436(7053):1020–4.

111. Campbell DS, Holt CE. Chemotropic responses of retinal growth cones mediated by rapid local protein synthesis and degradation. Neuron. 2001;32(6):1013–26.

112. Leung KM, van Horck FP, Lin AC, Allison R, Standart N, Holt CE. Asymmetrical beta-actin mRNA translation in growth cones mediates attractive turning to netrin-1. Nat Neurosci. 2006;9(10):1247–56.

113. Shigeoka T, Jung H, Jung J, Turner-Bridger B, Ohk J, Lin Julie Q, et al. Dynamic axonal translation in developing and mature visual circuits. Cell. 2016;166(1):181–92.

114. Wong HH, Lin JQ, Strohl F, Roque CG, Cioni JM, Cagnetta R, et al. RNA docking and local translation regulate site-specific axon remodeling in vivo. Neuron. 2017;95(4):852–68 e8.

115. Andreassi C, Riccio A. To localize or not to localize: mRNA fate is in 3′UTR ends. Trends Cell Biol. 2009;19(9):465–74.

116. Berkovits BD, Mayr C. Alternative 3′ UTRs act as scaffolds to regulate membrane protein localization. Nature. 2015;522(7556):363–7.

117. Kiebler MA, Bassell GJ. Neuronal RNA granules: movers and makers. Neuron. 2006;51(6):685–90.

118. Vuppalanchi D, Coleman J, Yoo S, Merianda TT, Yadhati AG, Hossain J, et al. Conserved 3′-untranslated region sequences direct subcellular localization of chaperone protein mRNAs in neurons. J Biol Chem. 2010;285(23):18025–38.

119. Liu Q, Dreyfuss G. A novel nuclear structure containing the survival of motor neurons protein. EMBO J. 1996;15(14):3555–65.

120. Pagliardini S, Giavazzi A, Setola V, Lizier C, Di Luca M, DeBiasi S, et al. Subcellular localization and axonal transport of the survival motor neuron (SMN) protein in the developing rat spinal cord. Hum Mol Genet. 2000;9(1):47–56.

121. Jablonka S, Bandilla M, Wiese S, Buhler D, Wirth B, Sendtner M, et al. Co-regulation of survival of motor neuron (SMN) protein and its interactor SIP1 during development and in spinal muscular atrophy. Hum Mol Genet. 2001;10(5):497–505.

122. McWhorter ML, Monani UR, Burghes AH, Beattie CE. Knockdown of the survival motor neuron (Smn) protein in zebrafish causes defects in motor axon outgrowth and pathfinding. J Cell Biol. 2003;162(5):919–31.

123. Ymlahi-Ouazzani Q, O JB, Paillard E, Ballagny C, Chesneau A, Jadaud A, et al. Reduced levels of survival motor neuron protein leads to aberrant motoneuron growth in a Xenopus model of muscular atrophy. Neurogenetics. 2010;11(1):27–40.

124. van Bergeijk J, Rydel-Konecke K, Grothe C, Claus P. The spinal muscular atrophy gene product regulates neurite outgrowth: importance of the C terminus. FASEB J. 2007;21(7):1492–502.

125. Zhang H, Xing L, Rossoll W, Wichterle H, Singer RH, Bassell GJ. Multiprotein complexes of the survival of motor neuron protein SMN with Gemins traffic to neuronal processes and growth cones of motor neurons. J Neurosci. 2006;26(33):8622–32.

126. Todd AG, Morse R, Shaw DJ, Stebbings H, Young PJ. Analysis of SMN-neurite granules: core Cajal body components are absent from SMN-cytoplasmic complexes. Biochem Biophys Res Commun. 2010;397(3):479–85.

127. Saal L, Briese M, Kneitz S, Glinka M, Sendtner M. Subcellular transcriptome alterations in a cell culture model of spinal muscular atrophy point to widespread defects in axonal growth and presynaptic differentiation. RNA. 2014;20(11):1789–802.

128. Rage F, Boulisfane N, Rihan K, Neel H, Gostan T, Bertrand E, et al. Genome-wide identification of mRNAs associated with the protein SMN whose depletion decreases their axonal localization. RNA. 2013;19(12):1755–66.

129. Castello A, Fischer B, Frese Christian K, Horos R, Alleaume A-M, Foehr S, et al. Comprehensive identification of RNA-binding domains in human cells. Mol Cell. 2016;63(4):696–710.

130. Mourelatos Z, Abel L, Yong J, Kataoka N, Dreyfuss G. SMN interacts with a novel family of hnRNP and spliceosomal proteins. EMBO J. 2001;20(19):5443–52.
131. Piazzon N, Rage F, Schlotter F, Moine H, Branlant C, Massenet S. In vitro and in cellulo evidences for association of the survival of motor neuron complex with the fragile X mental retardation protein. J Biol Chem. 2008;283(9):5598–610.
132. Hubers L, Valderrama-Carvajal H, Laframboise J, Timbers J, Sanchez G, Cote J. HuD interacts with survival motor neuron protein and can rescue spinal muscular atrophy-like neuronal defects. Hum Mol Genet. 2011;20(3):553–79.
133. Wang IF, Reddy NM, Shen CK. Higher order arrangement of the eukaryotic nuclear bodies. Proc Natl Acad Sci U S A. 2002;99(21):13583–8.
134. Yamazaki T, Chen S, Yu Y, Yan B, Haertlein TC, Carrasco MA, et al. FUS-SMN protein interactions link the motor neuron diseases ALS and SMA. Cell Rep. 2012;2(4):799–806.
135. Gribling-Burrer AS, Leichter M, Wurth L, Huttin A, Schlotter F, Troffer-Charlier N, et al. SECIS-binding protein 2 interacts with the SMN complex and the methylosome for selenoprotein mRNP assembly and translation. Nucleic Acids Res. 2017;45(9):5399–413.
136. Hao LT, Duy PQ, An M, Talbot J, Iyer CC, Wolman M, et al. HuD and the Survival Motor Neuron protein interact in motoneurons and are essential for motoneuron development, function and mRNA regulation. J Neurosci. 2017;37(48):11559–71.
137. Glinka M, Herrmann T, Funk N, Havlicek S, Rossoll W, Winkler C, et al. The heterogeneous nuclear ribonucleoprotein-R is necessary for axonal beta-actin mRNA translocation in spinal motor neurons. Hum Mol Genet. 2010;19(10):1951–66.
138. Jablonka S, Beck M, Lechner BD, Mayer C, Sendtner M. Defective Ca2+ channel clustering in axon terminals disturbs excitability in motoneurons in spinal muscular atrophy. J Cell Biol. 2007;179(1):139–49.
139. Zappulo A, van den Bruck D, Ciolli Mattioli C, Franke V, Imami K, McShane E, et al. RNA localization is a key determinant of neurite-enriched proteome. Nat Commun. 2017;8(1):583.
140. Yao J, Sasaki Y, Wen Z, Bassell GJ, Zheng JQ. An essential role for beta-actin mRNA localization and translation in Ca2+-dependent growth cone guidance. Nat Neurosci. 2006;9(10):1265–73.
141. Yoo S, Kim HH, Kim P, Donnelly CJ, Kalinski AL, Vuppalanchi D, et al. A HuD-ZBP1 ribonucleoprotein complex localizes GAP-43 mRNA into axons through its 3′ untranslated region AU-rich regulatory element. J Neurochem. 2013;126(6):792–804.
142. Donnelly CJ, Park M, Spillane M, Yoo S, Pacheco A, Gomes C, et al. Axonally synthesized β-actin and GAP-43 proteins support distinct modes of axonal growth. J Neurosci. 2013;33(8):3311–22.
143. Fujino T, Leslie JH, Eavri R, Chen JL, Lin WC, Flanders GH, et al. CPG15 regulates synapse stability in the developing and adult brain. Genes Dev. 2011;25(24):2674–85.
144. Donnelly CJ, Willis DE, Xu M, Tep C, Jiang C, Yoo S, et al. Limited availability of ZBP1 restricts axonal mRNA localization and nerve regeneration capacity. EMBO J. 2011;30(22):4665–77.
145. Gomes C, Lee SJ, Gardiner AS, Smith T, Sahoo PK, Patel P, et al. Axonal localization of neuritin/CPG15 mRNA is limited by competition for HuD binding. J Cell Sci. 2017;130(21):3650–62.
146. Beckel-Mitchener AC, Miera A, Keller R, Perrone-Bizzozero NI. Poly(A) tail lengthdependent stabilization of GAP-43 mRNA by the RNA-binding protein HuD. J Biol Chem. 2002;277(31):27996–8002.
147. Bird CW, Gardiner AS, Bolognani F, Tanner DC, Chen C-Y, Lin W-J, et al. KSRP modulation of GAP-43 mRNA stability restricts axonal outgrowth in embryonic hippocampal neurons. PLoS One. 2013;8(11):e79255.
148. Sheinberger J, Shav-Tal Y. mRNPs meet stress granules. FEBS Lett. 2017;591(17):2534–42.
149. Hua Y, Zhou J. Survival motor neuron protein facilitates assembly of stress granules. FEBS Lett. 2004;572(1–3):69–74.

150. Zou T, Yang X, Pan D, Huang J, Sahin M, Zhou J. SMN deficiency reduces cellular ability to form stress granules, sensitizing cells to stress. Cell Mol Neurobiol. 2011;31(4):541–50.

151. Arimoto K, Fukuda H, Imajoh-Ohmi S, Saito H, Takekawa M. Formation of stress granules inhibits apoptosis by suppressing stress-responsive MAPK pathways. Nat Cell Biol. 2008;10(11):1324–32.

152. Thedieck K, Holzwarth B, Prentzell MT, Boehlke C, Klasener K, Ruf S, et al. Inhibition of mTORC1 by astrin and stress granules prevents apoptosis in cancer cells. Cell. 2013;154(4):859–74.

153. Gallotta I, Mazzarella N, Donato A, Esposito A, Chaplin JC, Castro S, et al. Neuron-specific knock-down of SMN1 causes neuron degeneration and death through an apoptotic mechanism. Hum Mol Genet. 2016;25(12):2564–77.

154. Strasswimmer J, Lorson CL, Breiding DE, Chen JJ, Le T, Burghes AH, et al. Identification of survival motor neuron as a transcriptional activator-binding protein. Hum Mol Genet. 1999;8(7):1219–26.

155. Young PJ, Day PM, Zhou J, Androphy EJ, Morris GE, Lorson CL. A direct interaction between the survival motor neuron protein and p53 and its relationship to spinal muscular atrophy. J Biol Chem. 2002;277(4):2852–9.

156. Zou J, Barahmand-Pour F, Blackburn ML, Matsui Y, Chansky HA, Yang L. Survival motor neuron (SMN) protein interacts with transcription corepressor mSin3A. J Biol Chem. 2004;279(15):14922–8.

157. Pellizzoni L, Charroux B, Rappsilber J, Mann M, Dreyfuss G. A functional interaction between the survival motor neuron complex and RNA polymerase II. J Cell Biol. 2001;152(1):75–85.

158. Suraweera A, Lim Y, Woods R, Birrell GW, Nasim T, Becherel OJ, et al. Functional role for senataxin, defective in ataxia oculomotor apraxia type 2, in transcriptional regulation. Hum Mol Genet. 2009;18(18):3384–96.

159. Zhao DY, Gish G, Braunschweig U, Li Y, Ni Z, Schmitges FW, et al. SMN and symmetric arginine dimethylation of RNA polymerase II C-terminal domain control termination. Nature. 2016;529(7584):48–53.

160. Jangi M, Fleet C, Cullen P, Gupta SV, Mekhoubad S, Chiao E, et al. SMN deficiency in severe models of spinal muscular atrophy causes widespread intron retention and DNA damage. Proc Natl Acad Sci U S A. 2017;114(12):E2347–E56.

161. Rudnik-Schoneborn S, Arning L, Epplen JT, Zerres K. SETX gene mutation in a family diagnosed autosomal dominant proximal spinal muscular atrophy. Neuromuscul Disord. 2012;22(3):258–62.

162. Chen YZ, Bennett CL, Huynh HM, Blair IP, Puls I, Irobi J, et al. DNA/RNA helicase gene mutations in a form of juvenile amyotrophic lateral sclerosis (ALS4). Am J Hum Genet. 2004;74(6):1128–35.

163. Moreira MC, Klur S, Watanabe M, Nemeth AH, Le Ber I, Moniz JC, et al. Senataxin, the ortholog of a yeast RNA helicase, is mutant in ataxia-ocular apraxia 2. Nat Genet. 2004;36(3):225–7.

164. Salvi JS, Mekhail K. R-loops highlight the nucleus in ALS. Nucleus. 2015;6(1):23–9.

165. Gama-Carvalho M, L Garcia-Vaquero M, R Pinto F, Besse F, Weis J, Voigt A, et al. Linking amyotrophic lateral sclerosis and spinal muscular atrophy through RNA-transcriptome homeostasis: a genomics perspective. J Neurochem. 2017;141(1):12–30.

166. Rathod R, Havlicek S, Frank N, Blum R, Sendtner M. Laminin induced local axonal translation of beta-actin mRNA is impaired in SMN-deficient motoneurons. Histochem Cell Biol. 2012;138(5):737–48.

167. Sanchez G, Dury AY, Murray LM, Biondi O, Tadesse H, El Fatimy R, et al. A novel function for the survival motoneuron protein as a translational regulator. Hum Mol Genet. 2013;22(4):668–84.

168. Gabanella F, Pisani C, Borreca A, Farioli-Vecchioli S, Ciotti MT, Ingegnere T, et al. SMN affects membrane remodelling and anchoring of the protein synthesis machinery. J Cell Sci. 2016;129(4):804–16.

169. Bernabò P, Tebaldi T, Groen EJN, Lane FM, Perenthaler E, Mattedi F, et al. In vivo transla-tome profiling in spinal muscular atrophy reveals a role for SMN protein in ribosome biology. Cell Rep. 2017;21(4):953–65.
170. Donlin-Asp PG, Rossoll W, Bassell GJ. Spatially and temporally regulating translation via mRNA-binding proteins in cellular and neuronal function. FEBS Lett. 2017;591(11):1508–25.
171. Costa CJ, Willis DE. To the end of the line: axonal mRNA transport and local translation in health and neurodegenerative disease. Dev Neurobiol. 2018;78(3):209–20.
172. Batista AF, Hengst U. Intra-axonal protein synthesis in development and beyond. Int J Dev Neurosci. 2016;55:140–9.
173. Coyne AN, Zaepfel BL, Zarnescu DC. Failure to deliver and translate-new insights into RNA dysregulation in ALS. Front Cell Neurosci. 2017;11:243.
174. Hutten S, Sharangdhar T, Kiebler M. Unmasking the messenger. RNA Biol. 2014;11(8):992–7.
175. Jodelka FM, Ebert AD, Duelli DM, Hastings ML. A feedback loop regulates splicing of the spinal muscular atrophy-modifying gene, SMN2. Hum Mol Genet. 2010;19(24):4906–17.
176. See K, Yadav P, Giegerich M, Cheong PS, Graf M, Vyas H, et al. SMN deficiency alters Nrxn2 expression and splicing in zebrafish and mouse models of spinal muscular atrophy. Hum Mol Genet. 2014;23(7):1754–70.
177. Custer SK, Gilson TD, Li H, Todd AG, Astroski JW, Lin H, et al. Altered mRNA splicing in SMN-depleted motor neuron-like cells. PLoS One. 2016;11(10):e0163954.
178. Zhang Z, Pinto AM, Wan L, Wang W, Berg MG, Oliva I, et al. Dysregulation of synaptogen-esis genes antecedes motor neuron pathology in spinal muscular atrophy. Proc Natl Acad Sci U S A. 2013;110(48):19348–53.
179. Burgess RW, Nguyen QT, Son YJ, Lichtman JW, Sanes JR. Alternatively spliced isoforms of nerve- and muscle-derived agrin: their roles at the neuromuscular junction. Neuron. 1999;23(1):33–44.
180. Wishart TM, Mutsaers CA, Riessland M, Reimer MM, Hunter G, Hannam ML, et al. Dysregulation of ubiquitin homeostasis and beta-catenin signaling promote spinal muscular atrophy. J Clin Invest. 2014;124(4):1821–34.
181. Mutsaers CA, Lamont DJ, Hunter G, Wishart TM, Gillingwater TH. Label-free proteomics identifies Calreticulin and GRP75/Mortalin as peripherally accessible protein biomarkers for spinal muscular atrophy. Genome Med. 2013;5(10):95.
182. Sarvestany AA, Hunter G, Tavendale A, Lamont DJ, Hurtado ML, Graham LC, et al. Label-free quantitative proteomic profiling identifies disruption of ubiquitin homeosta-sis as a key driver of Schwann cell defects in spinal muscular atrophy. J Proteome Res. 2014;13(11):4546–57.
183. Fuller HR, Mandefro B, Shirran SL, Gross AR, Kaus AS, Botting CH, et al. Spinal muscular atrophy patient iPSC-derived motor neurons have reduced expression of proteins important in neuronal development. Front Cell Neurosci. 2016;9:506.
184. Doktor TK, Hua Y, Andersen HS, Broner S, Liu YH, Wieckowska A, et al. RNA-sequencing of a mouse-model of spinal muscular atrophy reveals tissue-wide changes in splicing of U12-dependent introns. Nucleic Acids Res. 2017;45(1):395–416.
185. Jablonka S, Karle K, Sandner B, Andreassi C, von Au K, Sendtner M. Distinct and overlap-ping alterations in motor and sensory neurons in a mouse model of spinal muscular atrophy. Hum Mol Genet. 2006;15(3):511–8.
186. Jablonka S, Schrank B, Kralewski M, Rossoll W, Sendtner M. Reduced survival motor neu-ron (Smn) gene dose in mice leads to motor neuron degeneration: an animal model for spinal muscular atrophy type III. Hum Mol Genet. 2000;9(3):341–6.
187. Wurth L, Gribling-Burrer AS, Verheggen C, Leichter M, Takeuchi A, Baudrey S, et al. Hypermethylated-capped selenoprotein mRNAs in mammals. Nucleic Acids Res. 2014;42(13):8663–77.

Chapter 7
Stress Granules and ALS: A Case of Causation or Correlation?

Nikita Fernandes, Nichole Eshleman, and J. Ross Buchan

Abstract Amyotrophic Lateral Sclerosis (ALS) is a fatal neurodegenerative disease characterized by cytoplasmic protein aggregates within motor neurons. These aggregates are linked to ALS pathogenesis. Recent evidence has suggested that stress granules may aid the formation of ALS protein aggregates. Here, we summarize current understanding of stress granules, focusing on assembly and clearance. We also assess the evidence linking alterations in stress granule formation and dynamics to ALS protein aggregates and disease pathology.

Keywords Stress granules · ALS · TDP-43 · FUS · SOD1 · mRNA · Autophagy · C9ORF72 · Chaperones · Cytoskeleton

7.1 Introduction to ALS

Amyotrophic lateral sclerosis (ALS) is a fatal neurodegenerative disease characterized by premature degeneration of upper and lower motor neurons, typically in mid-adult life. Death usually results within 2–5 years due to paralysis and respiratory failure. Approximately 90% of ALS cases are sporadic, with 10% familial [1].

Mutations in >30 genes are linked to ALS onset (Table 7.1), several of which are linked to another neurodegenerative disease, Frontotemporal dementia (FTD). FTD results from neuronal atrophy within frontal and temporal cortices that causes cognitive, behavioral, and language defects [66]. Many patients diagnosed with ALS or FTD exhibit symptoms of the other disease. Given this, and commonalities at the genetic and cellular levels, ALS and FTD are often considered different facets of a neurodegenerative disease continuum [1]. Here, we focus on ALS, though much discussed is also relevant to FTD.

ALS is characterized by cytoplasmic aggregates within affected neurons, often termed "inclusion bodies." In 97% of ALS cases (including all sporadic cases),

N. Fernandes · N. Eshleman · J. R. Buchan (✉)
Department of Molecular and Cellular Biology, University of Arizona, Tucson, AZ, USA
e-mail: rbuchan@email.arizona.edu

© Springer International Publishing AG, part of Springer Nature 2018
R. Sattler, C. J. Donnelly (eds.), *RNA Metabolism in Neurodegenerative Diseases*, Advances in Neurobiology 20,
https://doi.org/10.1007/978-3-319-89689-2_7

Table 7.1 Evidence for ALS-linked mutations affecting SG dynamics

Gene	Location of mutation(s)	Function	In SGs?	Evidence ALS mutations affect SGs		References
				Mutations that increase SG persistence or formation	Mutations that decrease SG persistence or formation	
RNA metabolism						
TDP-43	C-term IDR [2]	Transcription/splicing/ translation regulation/ transport	Y	Mutant ↑ number [3] and size of SG+ inclusions [4]	Mutant (endogenous) in IDR ↓ SG assembly; mutant outside IDR no effect [5]	Mackenzie [2] Yesucevitz [3] Dewey [4] McDonald [5]
FUS	Dispersed [2]	Transcription/splicing/ translation regulation/ transport/DNA damage Repair	Y	Mutant FUS delays SG assembly, once formed ↑ number and size of SG+ inclusions [6, 7]; another mutant ↓ release of FUS from SGs [6]	Mutant FUS ↑ SG disassembly [7]	Mackenzie [2] Baron [7] Ryu [6]
TIA1	IDR [8]	Major SG component/ splicing/translation regulation	Y	Mutant delays SG disassembly [8]	Unknown	Mackenzie [8]
hnRNP A2B1/ hnRNPA1	IDR [9]	Splicing/translation regulation/stability/ transport	Y	Mutant ↑ recruitment of mutant protein into SGs [10]	Unknown	Kapeli [9] Kim [10]
ATXN2	PolyQ expansion [11]	Translation/endocytosis	Y	Involved in SG assembly and recruitment of TDP-43 to SGs [12]	Unknown	Elden [11] Becker [12]
ANG	Dispersed [13]	Angiogenesis	Y	Angiogenin-generated tiRNAs ↑ SG assembly [14]	Mutant ↓ SG assembly [15]	Padhi [13] Lyons [14] Thiyagarajan [15]
ELP3	KAT domain [16]	Translation elongation, histone acetylation	N	Unknown	Unknown	Glatt [16]

Gene	Location of mutation(s)	Function	Evidence ALS mutations affect SGs			References
			In SGs?	Mutations that increase SG persistence or formation	Mutations that decrease SG persistence or formation	
SETX	N-term protein interaction domain, Helicase domain [17]	Predicted helicase	N	Unknown	Unknown	Bennett [17]
MATR3	Dispersed [18]	Nuclear matrix protein	N	Mutant nuclear. Few cells MATR3+ SGs [19]	Unknown	Boehringer [18] Gallego-Iradi [19]
SMN1	Abnormal copy numbers [20, 21]	Biogenesis of snRNPs	Y	WT involved in SG assembly [22]	Unknown	Blauw [20] Corcia [21] Hua [22]
TAF15	C-term Zn finger and RGG domains [9]	Transcription regulation	Y	Unknown	Unknown	Kapeli [9]
EWSR1	Around Zn finger motif, N-term QGSY domain [9]	RNA-binding protein	Y	Unknown	Unknown	Kapeli [9]
SS18L1	Dispersed, 1 CBP binding motif deletion [23]	Transcription regulation, subunit of chromatin remodeling complex	Y	Unknown	Unknown	Kukharsky [23]
GLE1	Dispersed, 1 mutant has C-term extension [24]	RNA export	Y	Regulates SG assembly [25]. Mutants localize to SGs and no effect on SG dynamics [26]	Unknown	Kaneb [24] Aditi [25] Aditi [26]

(continued)

Table 7.1 (continued)

Gene	Location of mutation(s)	Function	In SGs?	Evidence ALS mutations affect SGs		References
				Mutations that increase SG persistence or formation	Mutations that decrease SG persistence or formation	
Protein turnover/vesicular trafficking						
C9ORF72	Intronic hexanucleotide repeat expansion [27]	DENN protein with Rab GEF activity, endocytosis, autophagy	Y	DPRs ↑ stress, ↓ G3BP1 dynamics in SGs, change composition of SGs [28–30]	Unknown	DeJesus-Hernandez [27] Maharjan [28] Boeynaems [30] Lee [29]
VCP	N-term [31]	Proteasomal degradation, autophagy	Y	Disease mutants induce constitutive SGs [32]	Unknown	Ayaki [31] Buchan [32]
UBQLN2	Proline repeat motif (Pxx) in central domain [33]	Proteasomal degradation, autophagy, endosomal trafficking	N	Unknown	Unknown	Deng [33]
p62	Dispersed [34]	Ubiquitination, selective autophagy adaptor	Y	p62 localizes to SGs, KD ↓ SG disassembly [35],OE of p62 reduces TDP-43 [36]	Unknown	Fecto [34] Guo [35] Brady [36]
OPTN	Coiled coil and ubiquitin binding domains, deletion of some exons [37]	Selective autophagy adaptor	N	Unknown	Unknown	Swarup [37]
ALS2	Truncations [11, 38]	Vesicle trafficking (Rab GEF)	N	Unknown	Unknown	Yang [38] Hadano [11]
CHMP2B	I29V, T104N or Q206H [39]	Vesicle trafficking (endocytosis and autophagy)	N		Unknown	Cox [39]
FIG4	Dispersed [40]	Vesicle trafficking	N	Unknown	Unknown	Chow [40]

| | | | | Evidence ALS mutations affect SGs | | |
Gene	Location of mutation(s)	Function	In SGs?	Mutations that increase SG persistence or formation	Mutations that decrease SG persistence or formation	References
VAPB	Major sperm protein domain [41]	Vesicle trafficking, facilitates UPR	N	Unknown	Unknown	Nishimura [41]
TBK1	Dispersed [42]	Regulator of selective autophagy	N		Unknown	Freischmidt [42]
CCNF	Dispersed [43]	Ubiquitination	N		Unknown	Williams [43]
Chaperones						
SIGR1	Ligand binding motif [44]	ER chaperone	N	Mutant induce constitutive SGs [45]	Unknown	Watanabe [44] Dreser [45]
Cytoskeleton						
PFN1	N-term [46]	Actin cytoskeletal dynamics	Y	Mutants induce constitutive SGs, ↓ SG disassembly and ↓ recruitment of Pfn1 to SGs [47]	Unknown	Wu [46] Figley [47]
DCTN1	Dispersed [48, 49]	Interacts with dynein facilitating transport of membrane organelles/ vesicles	Y	Unknown	Unknown	Münch [48] Vilariño-Güell [49]
NEFH	C-term [50, 51]	Structural neurofilament, axonal transport	N	Unknown	Unknown	Al-Chalabi [50] Skvortsova [51]
SPG11	N-term, C-term [52]	DNA damage repair, cytoskeleton stability, synaptic vesicle transport	N	Unknown	Unknown	Couthouis [52]

(continued)

Table 7.1 (continued)

Gene	Location of mutation(s)	Function	In SGs?	Evidence ALS mutations affect SGs			References
				Mutations that increase SG persistence or formation	Mutations that decrease SG persistence or formation	References
MAPT	Dinucleotide polymorphism [53]	Microtubule assembly	Y	Unknown	Unknown	Münch [53]
TUBA4A	Dispersed, many C-term [54, 55]	Alpha tubulin	Y	Unknown	Unknown	Smith [54] Perrone [55]
PRPH	Rod domain [56, 57]	Intermediate filament	N	Unknown	Unknown	Corrado [56] Gros-Louis [57]
TRPM7	C-term [58]	Ion channel, Ser/Thr protein kinase	N	Unknown	Unknown	Hermosura [58]
DNA damage repair						
NEK1	Kinase domain, basic domain and coiled coil domain [59]	DNA damage repair	N	Unknown	Unknown	Brenner [59]
C21ORF2	Unclear	DNA damage repair, celia formation, mitochondrial function	N		Unknown	
Other						
SOD1	Dispersed [60]	Superoxide metabolism	N	Mutants ↓ dynamics of SOD1 + SGs [61]	Mutants delay SG assembly [62]	Cleveland [60] Mateju [61] Gal [62]
ERBB4	C-term [63]	Receptor Tyr Protein Kinase	N	Unknown	Unknown	Takahashi [63]
CHCHD10	Dispersed [64]	Associated with MICOS	N	Mutants induce constitutive SGs, colocalization of TDP-43 with SGs [65]	Unknown	Cozzolino [64] Woo [65]

Acronyms: C-term = C-terminal, N-term = N-terminal, IDR = Intrinsically disordered region, SGs = stress granules, DENN = differentially expressed in neoplastic vs. normal cells, MICOS = mitochondrial contact site and cristae organizing system

these aggregates are enriched for 43 kDa TAR DNA-binding protein (TDP-43) [1, 67, 68]. TDP-43 is typically hyper-phosphorylated, ubiquitinated, and N-terminally truncated within these aggregates; phosphorylation may be mediated by Casein Kinase δ1 [69], whereas cleavage is due to Caspase-3 activity [1, 67, 68, 70]. Under normal conditions, TDP-43 is a nuclear RNA-binding protein that regulates many steps of mRNA metabolism including transcription, splicing, export, translation, and mRNA stability [71]. In familial ALS cases lacking TDP-43 pathology, aggregates of either Fused in Sarcoma (*FUS*) [71, 72], another normally nuclear RNA-binding protein whose functions overlap with TDP-43, or Superoxide dismutase (*SOD1*), an antioxidant enzyme, are observed. Mutations in *FUS* and *SOD1* typically drive formation of these aggregates [73–75].

TDP-43, FUS, and SOD1 aggregates may lead to both a toxic gain of function and a loss of function. Many excellent reviews have addressed this [1, 76–79]. However, perturbations in mRNA metabolism caused by these aggregates are often suggested as an underlying cause of ALS pathology [1, 78, 80–84]. This reflects the localization of other RNA-binding proteins (see below) and mRNA [85, 86] within these aggregates, and the fact that TDP-43 and FUS toxicity require RNA-binding activity in numerous model systems [87–90]. Preventing formation or facilitating removal of these aggregates is considered a promising therapeutic approach, hence considerable research is now focused in these areas.

Much interest has focused on whether ALS cytoplasmic aggregates are linked to perturbations of endogenous mRNA-protein (mRNP) granules, particularly stress granules (SGs). It has been hypothesized that SGs may facilitate TDP-43/FUS aggregation, and/or that perturbations of SG dynamics may contribute to the ALS disease mechanism [78, 81–83]. Here, we introduce the reader to SGs, before assessing evidence that SGs may be a component of ALS pathogenesis.

7.2   Introduction to Stress Granules

Various mRNP granules exist in eukaryotic biology, including SGs, P-bodies, and neuronal transport granules (NTGs) [91]. While distinguished by composition, morphology, and cellular context, they are all dynamic, self-assembling structures that lack a limiting membrane. All mRNP granules harbor non-translating mRNPs, including proteins that regulate mRNA translation, localization, and stability.

SGs are cytoplasmic mRNP granules that are conserved throughout eukaryotes. They usually form only during cellular stress, when translation of most mRNAs is repressed. SGs contain polyadenylated mRNA, translation initiation factors, small ribosomal subunits, various RNA-binding proteins (RBPs), and several cell signaling proteins [82, 92–95]. SGs are typically dynamic, with most components examined via kinetic microscopy methods exhibiting residency times of <30s. However, dynamics of SG components can change over time or under different growth conditions, and immobile populations of mRNAs and proteins are also commonly observed, suggestive of a storage role for SGs [94].

SGs may function as "triage" sites for mRNAs, in which some mRNAs may be stored, while others are returned to translation or targeted for decay [93]. This may occur in P-bodies, with which SGs can physically dock and may exchange mRNP components [96, 97]. mRNP localization within SGs is selective, as stress-responsive mRNAs are often excluded from SGs during stress responses [98]. Finally, SGs can regulate signaling pathways by virtue of sequestration of kinases from specific substrate proteins [94], and also have roles in viral defense [99].

7.3   SG Assembly and Disassembly

SG assembly can occur rapidly following cellular stress (<10 min in yeast; <15 min in human cells—[97, 100, 101], and often involves proteins whose importance in assembly varies in a stress-specific manner [95, 102]. However, key principles have emerged.

7.3.1   Non-translating mRNA Is Required for SG Assembly

SG assembly requires non-translating mRNAs. SGs normally exhibit an inverse relationship with translation, requiring translation repression in order to induce SG assembly [92, 103, 104]. Drugs that arrest ribosomes on mRNAs (e.g., cycloheximide) prevent SG assembly if administered before or coincident with a SG-inducing stress, and cause disassembly of already-formed SGs [103, 105]. This suggests that mRNAs undergo regular exchange between SGs and polysomes, and once mRNAs are trapped in polysomes, they cannot nucleate SG assembly. Conversely, SG assembly is enhanced by drugs that dissociate ribosomes from mRNAs (e.g., puromycin), thus increasing the non-translating mRNA pool [103, 106]. Finally, the absence of 60S ribosomal subunits from SGs, and assays to identify sites of active protein synthesis, indicate that SGs are translationally silent [104, 107, 108].

In principle, noncoding RNA molecules (ncRNAs) could help nucleate SG assembly. Many ncRNAs localize in SGs including small ribosomal subunit RNA [108], tRNA fragments ("tiRNAs") [109], miRNAs [110], and long ncRNAs [111]. Interestingly, ncRNAs also co-purify with P-bodies, albeit mRNAs are preferentially enriched [112]. However, no current evidence suggests a direct scaffolding role for ncRNAs in SG assembly, through a precedent exists for paraspeckles [113].

7.3.2   Protein-Protein and Protein-RNA Interactions Drive SG Assembly

Proteins that aid SG assembly often self-interact and form oligomeric complexes. Such interactions may occur via defined structural domains, as with G3BP1 [114], or via intrinsically disordered regions (IDRs; also termed "low-complexity" or "prion-like" depending on composition), as with TIA1 [115]. Both proteins are considered key to SG assembly. IDRs often aid protein localization in SGs [116–118] and/or facilitate SG assembly through formation of amyloid-like structures [95, 115, 118, 119].

Protein-RNA interactions also drive SG assembly. Approximately half of SG-localizing proteins bind RNA [120], some via IDRs [118, 119], and others via classically recognized RNA-binding domains. RNA binding is key for the ability of several proteins to facilitate SG assembly and/or localize in SGs [114–116, 121].

7.3.3   Liquid-Liquid Phase Separation May Facilitate SG Assembly

SG assembly likely occurs in part via a liquid-liquid phase separation (LLPS) process. The phase separation model posits that SGs, and other RNP granules, are driven by both homo and heterotypic protein-protein and protein-RNA interactions once critical local concentrations of interactors are attained [96, 118, 122–126]. Molecules with high valency and binding affinity likely play the most prominent roles in such phase separation. In principle, RNA-RNA interactions may also facilitate SG assembly via an LLPS process. Indeed, RNAs containing G_4C_2 repeat expansions observed in ALS-mutant alleles of *C9ORF72* (Table 7.1) can undergo LLPS [127]. The role of phase separation in SG assembly has been expertly discussed elsewhere [95, 122], thus only key findings are now summarized.

Several full-length SG proteins can phase separate into homotypic liquid-like droplets in vitro, including FUS [124] and hnRNPA1 [118, 119]. Furthermore, IDR domains from various SG proteins, including TDP-43 and TIA1, can also phase separate in vitro into liquid-like droplets [119, 128]. Such phase separation is promoted by low salt, low temperature, crowding agents, and particularly the addition of RNA [118, 119, 129]. Interestingly, these liquid-like droplets can mature over time into less dynamic bodies, sometimes termed hydrogels, or form fibrillar aggregates [117–119, 124]. Rates of phase separation, hydrogel formation, and aggregation can be accelerated if the proteins in question harbor ALS-associated mutations (e.g., FUS, hnRNPA1 [118, 124, 130]).

7.3.4   SG Assembly: More Than Just a Passing Phase?

Although many SG proteins can phase separate, several observations suggest that other processes affect in vivo SG assembly. First, super-resolution microscopy indicates that human cell SGs possess substructure, consisting of SG "cores" that are protein dense, and a liquid-like protein shell that surrounds the cores [120]. SG cores can be detected and purified as soon as SGs become microscopically visible, both in yeast and human cells [101, 120]. Second, SG cores are relatively stable, and do not dissolve in dilute lysate as would be predicted for LLPS droplets. Third, SG assembly in human cells is inhibited at low temperature, contrary to enhanced LLPS and/or hydrogel formation in vitro [117, 118, 126]. Finally, ATP depletion strongly inhibits SG assembly, and reduces the mobility and internal dynamics of already-formed SGs [120]. This suggests the involvement of energy-driven processes in regulating SG formation.

7.3.5 ATP-Driven Machines Affect SG Assembly and Disassembly

Sensitivity of SGs to ATP depletion is likely due to inhibition of numerous ATP-driven chaperones and helicases, which localize within SGs [32, 115, 131, 132], and are purified in SG cores [120]. Genetic analysis of these ATPases indicates contrasting effects on SG assembly and disassembly. For example, inhibition of Hsp70 chaperones and associated cofactors causes SG persistence following alleviation of various stresses [131, 133–136], indicating that Hsp70 aids SG disassembly. In contrast, inhibition of the yeast mini chromosome maintenance (MCM) and RuvB-like helicase complexes causes faster SG disassembly, indicating a role in SG persistence [120]. Inhibition of the Chaperonin-containing T-complex (CCT complex) accelerates SG assembly and increases SG numbers following stress [120], indicating that CCT inhibits SG assembly. An ATPase-deficient allele of yeast Ded1, an RNA helicase, causes strong accumulation of constitutive SGs, within which mRNPs are stalled in translation re-entry [132]. Finally, under certain stresses, the AAA-ATPase Vasolin-containing protein (VCP; Cdc48 in yeast), a "ubiquitin segregase" implicated in proteasomal and autophagic turnover [137], also facilitates SG clearance at least partly via an autophagic mechanism (discussed later [32]). Thus, SGs dynamics likely depend on constant activity from ATP-dependent machines, which may facilitate assembly, disassembly and prevent conversion of SGs proteins into non-dynamic aggregates.

7.3.6 Posttranslational Modifications Affect SG Dynamics

Modification of SG proteins often affects their ability to help assemble or disassemble SGs, as well as their localization within SGs [138]. For example, G3BP1 oligomerization, which facilitates SG assembly, is inhibited by a Ras-dependent phosphorylation event [114]. This specifically favors binding to G3BP1 by USP10 at the expense of Caprin1, which limits G3BP1-driven SG assembly [139]. Additionally, SG disassembly is aided by phosphorylation of Grb7 by focal adhesion kinase. Grb7 phosphorylation weakens interactions between SG proteins, such as HuR and TIA1, and their binding to specific mRNAs [140]. Finally, localization of cold-inducible RNA-binding protein and calreticulin to SGs requires methylation and arginylation, respectively [141, 142]. Other modifications implicated in SG dynamics include O-glcnacylation [143], acetylation, ubiquitination [144], neddylation [145], and poly-(ADP) ribosylation [146].

Modifications also affect LLPS processes. In vitro, phosphorylation of the IDR of FUS limits its retention in pre-formed hydrogels of non-phosphorylated FUS [147]. In addition, in vitro droplets of the RNA helicase DDX4 are destabilized by methylation [126].

7.3.7 Role of the Cytoskeleton in SG Dynamics

SG formation is strongly linked to microtubules. Microtubule depolymerizing drugs or knockdown of Dynein microtubule motor proteins limit SG assembly, often resulting in the formation of miniature SGs [144, 148–151]. Localization of specific mRNAs to SGs also requires microtubules [152]. Microtubule disruption or knockdown of Kinesin microtubule motor proteins also impairs SG disassembly [149, 153]. SGs move along microtubules and undergo fusion and fission events, and SG mobility is strongly decreased when microtubules are depolymerized. However, once formed, SGs do not require microtubules to persist [150, 153].

In contrast, a role for actin in SG dynamics is unclear. While one study of actin disruption using cytochalasin B resulted in smaller SGs [149], another using latrunculin B leads to a slight increase in SG size [148], while other latrunculin B studies have reported no effect [144, 153]. Disruption of actin also has no effect on SG disassembly, nor do SGs show localization with actin filaments [153].

7.3.8 Additional Mechanisms of SG Disassembly

SG disassembly can be enacted either by disrupting interactions between proteins and mRNAs that sustain SG formation, or by degrading SG components. The simplest means of SG disassembly is the return of repressed mRNAs within SGs to translation. Indeed, bulk translation levels recover as SGs disassemble following stress alleviation [103, 131, 133], a process often aided by chaperones [133, 134]. However, translational recovery without complete SG disassembly is observed following microtubule disruption [149].

SG components can also be cleared by an autophagic mechanism termed "Granulophagy" [32]. This occurs in yeast and human cell lines in response to certain stresses, or following perturbation (in yeast) of cytoplasmic mRNA decay [32]. Granulophagy involves a diverse set of effectors. In yeast, the Hsp40 Sis1, together with Hsp70, facilitates autophagic targeting of SG components. In contrast, another Hsp40, Ydj1, facilitates SG disassembly with Hsp70 by promoting return of mRNAs to translation [134]. In human cells, the selective autophagy protein p62/SQSTM1 localizes within SGs, and p62 knockdown slows SG clearance [35]. Additionally, Syk kinase facilitates autophagic clearance of SGs in a manner dependent on its catalytic activity and its ability to localize within SGs [154]. Inhibiting autophagy genetically in both yeast, human cells [32], and neurons [6] leads to constitutive SG accumulation in the absence of stress, suggesting a basal level of autophagy helps prevent aberrant SG persistence.

In principle, SGs could also be cleared by the action of the proteasome and mRNA decay enzymes. Although SGs are induced by proteasomal inhibition [133], and G3BP1 itself harbors an endonuclease domain [114], no evidence for a role of either process in SG disassembly is known.

7.4 Hypothesized Role for SGs in Driving TDP-43/FUS/ SOD1 Aggregation

A popular model is that SGs may facilitate formation of cytoplasmic aggregates in ALS. Specifically, mutations or cellular conditions that increase SG persistence, either due to excessive SG assembly or impaired SG clearance, increase the chance of SG-localized TDP-43, FUS, or SOD1 stochastically undergoing conversion to a toxic aggregate [78, 81–84, 155]. Following this, other SG components could remain associated with these aggregates, or SGs may dissolve and thus serve as transient nucleators of TDP-43/FUS/SOD1 aggregation. A third possibility is that TDP-43/FUS/SOD1 aggregation occurs independently of SGs entirely (Fig. 7.1). We examine these possibilities below.

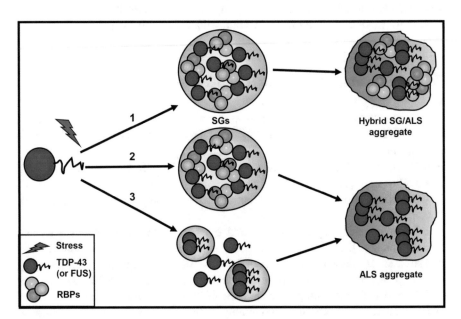

Fig. 7.1 Models for aggregate formation in ALS. (1) After stress induction, SGs form, and TDP 43 (or FUS/SOD1) is recruited. Over time, protein aggregation reduces SG dynamics, leading to formation of persistent SG/ALS aggregate hybrid state, which may sequester numerous mRNP components. (2) As in model 1, except that with time, SG components disassemble, leaving ALS aggregates behind. Thus, SGs nucleate aggregation, but are dispensable for aggregate persistence. (3) After stress induction, ALS protein aggregates begin to form, completely independent of SGs. Over time, ALS aggregates increase and persist

7.5 Localization of TDP-43, FUS, and SOD1 in SGs

If SGs facilitate TDP-43/FUS/SOD1 aggregation, these proteins should localize to SGs, at least transiently, if not throughout the disease. Indeed TDP-43, FUS, and SOD1 can all colocalize with SGs to varying degrees.

7.5.1 TDP-43 Localization to SGs

TDP-43 localizes to SGs in many cellular contexts, and under numerous stress conditions. These include oxidative (arsenite), osmotic (sorbitol), ER (thapsigargin), heat, serum deprivation, proteasome inhibition (MG132), and mitochondrial stress (paraquat) [3–5, 156–158]. Some caveats of these studies, common to many works in the field, include the unclear physiological relevance of these stresses in ALS-afflicted neurons, and that some (not all—[4, 156, 158]) of these studies utilize over-expression of TDP-43 and/or SG proteins, which may drive artifactual SG assembly or TDP-43 aggregation.

However, TDP-43 colocalization in SGs is not always observed. In other studies, arsenite-stressed HEK293 cells [4] and neuroblastoma cells subject to ER (Thapsigargin) or oxidative (SIN-1, Arginine) stress show no TDP-43 localization despite SG induction [159, 160]. SGs lacking TDP-43 are also observed following oxidative (Hydrogen peroxide) and proteasome inhibition (Epoxomicin) stress in multiple cell lines [159].

In several studies, TDP-43 aggregates in ALS and FTD patient spinal cord and brain tissue colocalize with SG markers including TIA1, eIF3, and PABP1 [3, 82, 116, 161]. In other similar studies, such colocalization is not seen [3]. The reason for this discrepancy is unclear, but could reflect differences in the cause or progression of ALS between different patient samples.

7.5.2 FUS and SOD1 Localization to SGs

Endogenous FUS localization to SGs has been reported in HeLa and HEK293 cells following arsenite or sorbitol stress [162, 163]. In contrast, most studies using various cell lines, including rat primary neurons and spinal cord neural cells, show no re-localization of FUS from the nucleus under numerous stresses [116, 163–165]. However, ALS-mutant alleles of FUS often localize to SGs under many stress conditions [116, 162]. Furthermore, spinal cord and hippocampal tissue from familial FUS-mutant ALS and FTD patients with WT FUS exhibits colocalization with the SG marker PABP-1 [166].

SOD1 localization in SGs is poorly studied. However, ALS-mutant SOD1 colocalizes with G3BP1-positive SGs following heat shock in HeLa cells while WT SOD1 does not [61]. Additionally, ALS-mutant SOD1 aggregates in patient-derived fibroblasts and in motor neurons of *SOD1* transgenic mice colocalized with G3BP1-positive SGs, whereas WT SOD1 did not [62].

In conclusion, there are many instances of colocalization of TDP-43 in cell line and ALS patient tissue, though exceptions are seen. Localization of FUS, and particularly SOD1 to SGs, is less studied, but can occur, particularly with ALS-mutant alleles.

7.6 Is SG Assembly Required for TDP-43, FUS, or SOD1 Aggregation?

The simplest test of whether SG assembly is required for TDP-43, FUS, or SOD1 aggregation would be to impair SG assembly, and then examine if aggregation still occurs under a condition of interest. Surprisingly, this question remains poorly addressed, perhaps because completely blocking SG assembly can be challenging due to redundant assembly mechanisms. However, in arsenite-stressed U2OS cells, Ataxin-2 knockdown, which slows SG assembly rate, is accompanied by fewer cells exhibiting SG colocalizing TDP-43 aggregates over the same time period, compared to WT cells [12]. This suggests that SG assembly may affect TDP-43 aggregation rates.

TDP-43 aggregates can sometimes persist following dissolution of SGs. Specifically, treatment of HeLa cells with Paraquat for 24 h leads to formation of SGs. However, 6 h after stress removal, SGs are mostly gone, whereas endogenous TDP-43 aggregates remain relatively unchanged [158]. Similarly, disassembly of paraquat-induced SGs by a 6 h cycloheximide treatment leaves a significant fraction of endogenous TDP-43 aggregates unaffected [158]. This partially contradicts another study in which TDP-43-GFP aggregates that form due to over-expression in neuroblastoma cells are fully cleared by 1 h of cycloheximide treatment. However expression of a TDP-43-GFP 25 kDa C-terminal fragment, which mimics TDP-43 fragments in ALS aggregates, also generates cycloheximide resistant TDP-43 aggregates [3]. Similarly, we have observed that TDP-43-GFP aggregates in yeast, while initially mostly SG localized, are cycloheximide resistant, unlike yeast SGs [167].

In summary, due to a lack of data concerning the effects of blocking SG assembly on TDP-43, FUS, and SOD1 aggregation, we cannot decisively say whether SG assembly is required for TDP-43/FUS/SOD1 aggregation. The failure to detect colocalization of SG markers with TDP-43 inclusions in some patient tissues could indicate SG-independent assembly, but does not rule out SGs as transient nucleation sites for TDP-43 aggregation.

7.7 Do ALS Mutations Always Perturb SG Formation and Dynamics?

Many ALS-linked mutations occur in SG localizing proteins and/or proteins that affect SG assembly, disassembly, or internal dynamics (i.e., rate at which SG components enter and exit SGs). For example, a commonly held view is that ALS-linked

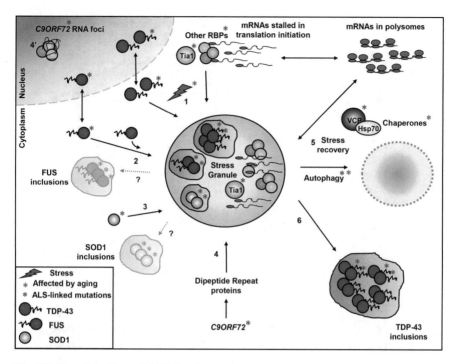

Fig. 7.2 Potential effects of ALS-linked mutations on SG-mediated aggregate formation. (1) Stress induces formation of SGs composed of mRNAs stalled in translation initiation and RBPs, many of which are mutated in ALS. Stress also leads to cytoplasmic localization of WT and/or aggregate-prone TDP-43 mutants and their recruitment to SGs. (2) FUS mutations cause cytoplasmic mis-localization of FUS which, upon stress, is also recruited to SGs. The fate of FUS aggregates, whether they remain within or become independent of SGs, is unclear. (3) SOD1 mutants are recruited to SGs, the fate of which is yet unknown. (4) *C9ORF72* repeat expansions generate dipeptide repeat proteins that induce SG formation and favor their transition to a less dynamic state. (4′) *C9ORF72* repeat expansions also lead to nuclear RNA foci capable of sequestering SG RBPs. (5) Stress recovery leads to SG disassembly via return of mRNAs into translation, or SG clearance via autophagy. (6) TDP-43 aggregates may persist following SG clearance. Alternatively, failure to clear SGs might lead to persistent SG positive TDP-43 inclusions. Aging leads to an increase in oxidative stress and decrease in protein quality control. Additional genetic or stress insults may ultimately drive pathological SG persistence and onset of ALS

mutant RBPs are more aggregation prone than their WT counterparts, and that this facilitates either faster SG assembly, slower SG disassembly, and/or reduced SG internal dynamics [78, 84]. For other ALS-associated genes, connections to SGs are less apparent. The effect of multiple ALS-associated mutations upon SGs is discussed below (also see Fig. 7.2), with select mutant-specific summaries presented in Table 7.2.

Table 7.2 Details of ALS mutants for which effects on SGs have been analyzed

Gene	Type of expression	Allele	Effect on SG localization/dynamics	References
TDP-43			WT and mutants localize to SGs	
	WT		Facilitates SG assembly	MacDonald [5] Aulas [168]
	OE	G294A (IDR)	Mutation does not increase aggregation potential [169], ↑ percent cells with SG+ TDP-43 inclusions [3]	Johnson [169] Yesucevitz [3]
	OE	A315T (IDR)	↑ percent cells with SG+ TDP-43 inclusions	Yesucevitz [3]
	OE	Q331K (IDR)	Mutation increases aggregation potential [169], ↑ percent cells with SG+ TDP-43 inclusions [3]	Johnson [169] Yesucevitz [3]
	OE	Q343R (IDR)	↑ percent cells with SG+ TDP-43 inclusions	Yesucevitz [3]
	OE	G294A (IDR)	↑ TDP-43 (SG?) granule size	Dewey [4]
	OE	A315T (IDR)	↑ TDP-43 (SG?) granule size	Dewey [4]
	OE	G348C (IDR)	↑ TDP-43 (SG?) granule size and faster assembly	Dewey [4]
	OE	N390S (IDR)	↑ TDP-43 (SG?) granule size	Dewey [4]
	Endo	R361S (IDR)	↓ percent cells with SGs	McDonald [5]
	Endo	D169G (outside IDR)	No effect	McDonald [5]
FUS			Mutants localize to SGs better	
	WT		No effect on SG assembly	Aulas [168]
	OE	R495X (truncation of NLS)	Delays SG assembly, once formed ↑ number and size, ↑ SG disassembly	Baron [7]
	OE	H517Q (mild mis-localization)	No effect	Baron [7]
	OE	R521C (in NLS)	↑ percent cells with FUS+ SGs, ↓ release of FUS from SGs on recovery	Ryu [6]
TIA1			WT and mutants localize to SGs	
	WT		Key SG assembly protein	Gilks [115]
	OE	P362L (IDR)	↓ SG disassembly, TDP-43 recruited to mutant TIA1 SGs	Mackenzie [8]

(continued)

Table 7.2 (continued)

Gene	Type of expression	Allele	Effect on SG localization/dynamics	References
	OE	A381T (IDR)	↓ SG disassembly, TDP-43 recruited to mutant TIA1 SGs	Mackenzie [8]
	OE	E384K (IDR)	↓ SG disassembly, TDP-43 recruited to mutant TIA1 SGs	Mackenzie [8]
hnRNPA2			WT and mutant localize to SGs	
	WT		Unknown effect on SG formation	
	OE	D290V (IDR)	↑ recruitment of mutant protein to SGs	Kim [10]
ANG			WT or mutant not known to localize to SGs	
	WT		Facilitates SG assembly	Emara [109]
	OE	K40I	Slight ↓ in SG formation	Thiyagarajan [15]
	OE	C39W	No effect	Thiyagarajan [15]
MATR3			WT does not localize to SGs, mutants weakly localize to SGs	
	WT		Unknown effect on SG assembly	
	OE	F115C	Few cells had MATR3 localizing to SGs	Gallego-Iradi [19]
SS18L1			WT and mutants localize to SGs	
	WT		Unknown effect on SG assembly	
	OE	Q388X	No effect on SG localization	Kukharsky [23]
	OE	I123M	No effect on SG localization	Kukharsky [23]
	OE	△222–224	No effect on SG localization	Kukharsky [23]
	OE	A264T	No effect on SG localization	Kukharsky [23]
GLE1			WT and mutants localize to SGs	
	WT		Facilitates SG assembly	Aditi [25]

(continued)

Table 7.2 (continued)

Gene	Type of expression	Allele	Effect on SG localization/ dynamics	References
	Rescue	Mutation causes novel C-term protein—hGle1-IVS14-2A > C	No effect on SG assembly	Aditi [26]
	OE	Mutation causes novel C-term protein—hGle1-IVS14-2A > C	↑ SG size like OE of WT, also formed some independent Gle1 aggregates	Aditi [26]
C9ORF72			WT and mutants localize to SGs	
	WT		Facilitates SG assembly	Maharjan [28]
	OE	GGGGCCx30 added at 5′ end of C9ORF72	Induces constitutive SGs	Maharjan [28]
	OE	GGGGCCx60 added 5′ end of C9ORF72	Induces constitutive SGs	Maharjan [28]
	OE	PR100	DPR localized to SGs, induces constitutive SGs, ↓ SG internal dynamics	Boeynaems [30]
	OE	PA100	No effect	Boeynaems [30]
	OE	GR50	DPR localized to SGs, induces constitutive SGs, ↓ SG internal dynamics	Lee [29]
	OE	PR50	DPR did not localize to SGs, induces constitutive SGs, ↓ SG internal dynamics	Lee [29]
VCP			WT and mutants localize to SGs	
	WT		Facilitates SG clearance	Buchan [32]
	OE	A232E	Induces constitutive TDP-43+ SGs	Buchan [32]
	OE	R155H	Induces constitutive TDP-43+ SGs	Buchan [32]
PFN1			WT and mutants localize to SGs to varying degrees	
	WT		Not essential to SG assembly/disassembly but OE induces constitutive SGs and ↓ SG disassembly	Figley [47]
	OE	C71G	Induces constitutive SGs, also forms separate Pfn1 aggregates; On stress, ↓ Pfn1 recruitment to SGs	Figley [47]

(continued)

Table 7.2 (continued)

Gene	Type of expression	Allele	Effect on SG localization/dynamics	References
	OE	M114T	Induces constitutive SGs, also forms separate Pfn1 aggregates; On stress, ↓ Pfn1 recruitment to SGs	Figley [47]
	OE	T109M	Induces constitutive SGs	Figley [47]
	OE	G118V	Various stresses, ↓ in Pfn1 recruitment to SGs	Figley [47]
	OE	E117G	↓ SG disassembly	Figley [47]
SOD1			Mutants localize to SGs, WT does not	
	WT		Unknown effect on SG assembly	
	OE	A4V	Delay in SG assembly	Gal [62]
	OE	A4V	Recruitment of mutant SOD1 ↓ SG internal dynamics (FRAP FUS and G3BP1)	Mateju [61]
CHCHD10			WT and mutants not localize to SGs	
	WT		Unknown effect on SG assembly	
	OE	R15L	Induces constitutive SGs, colocalization of TDP-43 with SGs	Woo [65]
	OE	S59L	Induces constitutive SGs, colocalization of TDP-43 with SGs	Woo [65]

Acronyms: SGs = stress granules, OE = overexpression, Endo = endogenous expression, IDR = Intrinsically disordered domain

7.7.1 ALS Mutations in RNA-Binding Proteins

TDP-43: Several ALS-associated TDP-43 mutations, most of which map to the C-terminal IDR, affect SG formation. For example, over-expression of mutant TDP-43 in neuroblastoma cells drives more numerous SG-localizing TDP-43 aggregates than in cells expressing WT TDP-43 [3]. In HEK293 cells, over-expression of mutant TDP-43 induces significantly larger SGs following sorbitol stress than cells expressing WT TDP-43; mutant TDP-43 also enters SGs faster than WT TDP-43 [4]. A common finding in these studies therefore is that TDP-43 ALS mutant alleles facilitate SG assembly better than WT TDP-43. This correlates with an increased aggregation propensity for many TDP-43 mutants in vitro [3, 116, 169, 170].

TDP-43 mutant alleles do not always facilitate SG assembly. In a study of endogenous ALS-linked TDP-43 mutations in ALS patient-derived lymphoblasts [5], arsenite-induced SG assembly was disrupted by a TDP-43 IDR mutant, but not by a

TDP-43 RNA-recognition motif mutant. SG assembly defects with the IDR mutant, or following TDP-43 knockdown, were attributed to a role in maintaining high G3BP1 mRNA levels [168]; similar results have been seen with another TDP-43 mutant [171].

FUS: ALS-associated mutations are dispersed throughout the FUS protein, though most studies focus on mutants that inactivate/truncate a C-terminal nuclear localization sequence (NLS). This causes strong cytoplasmic re-localization of FUS. In one study, a FUS NLS mutant delayed SG assembly in HEK293 cells, but once formed, SGs were larger and more numerous than in WT FUS-expressing cells [7]. Another study of mutant FUS in rodent cortical neurons did not observe this, but did observe stronger localization of mutant FUS in SGs versus WT FUS [6]. Mutant FUS localization in SGs correlates with cytoplasmic mis-localization [164, 165]. FUS mutants that exhibit only mild cytoplasmic mis-localization do not localize to or alter SGs [7]. Additionally, knockdown of endogenous WT FUS does not impair SG assembly [168].

Opposing findings regarding the effect of mutant FUS on SG disassembly and clearance have been observed. Expression of a truncated NLS FUS mutant in HEK293 cells caused increased SG dynamics (TIA1 and G3BP1 mobility increased), and more rapid SGs disassembly after arsenite stress compared to WT FUS-expressing cells [7]. In another study, SGs harboring NLS mutant FUS were impaired in clearance following arsenite stress, and were preferentially targeted by autophagy [6].

TIA1: An ALS-associated mutation was recently identified in the IDR domain of TIA1. This mutation drives TIA1 phase separation in vitro, slows SG disassembly following heat shock in HeLa cells, and decreases TDP-43 mobility in SGs [8]. TDP-43 insoluble aggregates also accumulated in the TIA1 mutant context, consistent with the idea that SG disassembly defects facilitate accumulation of TDP-43 aggregates [8].

hnRNPA2B1 and hnRNPA1: hnRNPA2B1 (A2B1) and hnRNPA1 (A1) localize in SGs [120, 172], and are mutated in a subset of ALS cases. Like TDP-43, these mutations occur in their IDR domains, which drives greater and faster assembly of A2B1 and A1 into fibrils in vitro compared to WT alleles [10]. Notably, mutant alleles can also "seed" fibrilization of their WT counterparts [10]. IDR mutations also induce A2-positive SGs in unstressed conditions and accelerate A2 incorporation into SGs. TDP-43 inclusions in Multi-System Proteinopathy, another degenerative disease, also colocalize with A2B1 and A1 [10]. The effect of WT or mutant A2B1 and A1 on SG disassembly remains unclear.

Ataxin-2: Polyglutamine expansions in Ataxin-2 increase ALS risk [87]. Consequences of polyglutamine-expanded Ataxin-2 include increased TDP-43 binding in an RNA-dependent manner, increased Ataxin-2 stability, and greater cytoplasmic mislocalization of TDP-43 in ALS patient-derived lymphoblasts following heat stress [87]. Additionally, knockdown experiments indicated that WT Ataxin-2 stimulates normal rates of arsenite-induced SG assembly and promotes recruitment of TDP-43 into SGs [12, 173]. Ataxin-2 knockdown also extends lifespan and reduces pathology in a TDP-43 transgenic mouse model [12]. Collectively, these data hint that mutant Ataxin-2 may promote TDP-43 aggregation, possibly within an SG context via altered SG dynamics.

In summary, analysis of the above-mentioned proteins generally supports the hypothesis that ALS-associated mutations (with some exceptions) increase SG persistence, either due to accelerated SG assembly or impaired disassembly. However, the effect on SGs of many RBPs mutated in ALS, and of most specific RBP ALS mutants, remain unknown (Tables 7.1 and 7.2). Analyses of such mutants at endogenous expression levels would help address if existing models are accurate.

7.7.2 ALS Mutations in Protein Quality Control Factors

Several genes implicated in protein quality control, particularly autophagy, are mutated in ALS (Table 7.1). SG clearance can occur via autophagy [6, 32, 35, 154], thus SG accumulation due to autophagic defects may facilitate aggregation of SG-localizing proteins such as TDP-43 [81]. However, only two ALS-associated genes (VCP, p62) implicated in autophagy have been studied for effects on SGs. It remains possible that protein quality control defects could affect TDP-43 turnover independently of SGs. Specific protein quality control genes and their known/possible effects on SGs are now discussed.

VCP: VCP localizes in SGs under various stresses, facilitates efficient SG clearance following heat-shock, participates in Granulophagy, and over-expression of ALS-associated VCP mutants induces constitutive SGs that harbor TDP-43 [32]. Interestingly, VCP is also implicated in SG assembly, with depletion leading to formation of SGs harboring misfolded proteins and 60S ribosomal subunits [174]. ALS-associated VCP mutants may be defective in SG clearance due to a failure to undergo efficient N-terminal SUMOylation [175]. This facilitates SG localization, as assessed by biochemical fractionation, and is required for formation of functional VCP hexamers [175].

Autophagy Factors: Several autophagy "receptors" that can selectively bind substrates and target them for autophagic turnover have been identified as ALS-linked genes [176]. These include p62, Optineurin and Ubiquilin-2, which also facilitates proteasomal turnover. To our knowledge, p62 is the only autophagy receptor that localizes in SGs and facilitates SG clearance [35]. Several other genes that promote autophagic functions, including *TBK1*, *CHMP2B*, *VAPB*, and *FIG4*, are mutated in ALS (Table 7.1), but effects on SG clearance are unknown.

Chaperones and Proteasomal turnover: No ALS-associated mutations in chaperones that localize within or directly act upon SGs have been identified. However, an E102Q mutation in the ER chaperone SigR1, which occurs in juvenile ALS [177], leads to impairment of autophagy, endocytosis and formation of SGs, within which SigR1 localizes [45]. Mechanistic understanding of these effects remains unclear. In addition, SOD1 aggregates can sequester Hsc70 chaperones (constitutive Hsp70s) [178], which impairs endocytosis, but could in principle affect SG disassembly.

Proteasomal inhibition induces SG assembly [133], and has been implicated in TDP-43 turnover [179, 180]. However, besides Ubiquilin-2, we are unaware of any ALS-associated mutants that perturb proteasomal activity and that affect SGs.

In summary, the evidence that defects in protein quality control lead to ALS protein aggregates due to accumulation of SGs remains relatively scant, and warrants further investigation.

7.7.3 ALS Mutations in Cytoskeletal Associated Proteins

Several cytoskeletal-related proteins are mutated in ALS. However, only Profilin, an actin-binding protein that regulates actin filament dynamics, has known effects on SGs. Specifically, Profilin ALS mutants [46] show poor recruitment to SGs, and lead to slower SG disassembly relative to WT Profilin-expressing cells [47]. As SG formation is generally independent of the actin cytoskeleton, mutant Profilin may affect other biological processes. Indeed Profilin genetically interacts with Dynein, and physically interacts with tubulin [47, 181], thus Profilin's mutant effects could be microtubule mediated. Alternatively, mutant Profilin protein also forms SG-distinct aggregates that persist through stress treatment and recovery [47] which may seed formation of TDP-43 aggregates [182].

Mutations in *TUBA4A* (alpha-tubulin subunit) are found in a rare fraction of ALS patients, which destabilizes microtubules [54]. In principle, this could affect SG assembly or disassembly. Other cytoskeletal ALS-linked genes include *DCTN1* (Dynactin subunit 1) [48], *NEFH* [50], and *SPG11* [183] (Table 7.1), though only dynactin subunit 1 has been detected in SGs [120].

In summary, despite a clear role for microtubules in SG assembly and disassembly, the effects of most cytoskeletal-associated ALS-mutant genes on SGs remain poorly characterized.

7.7.4 ALS-Linked C9ORF72 Repeat Expansions

C9ORF72 is the most commonly mutated gene in ALS (Table 7.1). An intronic hexanucleotide repeat expansion of G_4C_2 repeats occurs between exons 1a and 1b (WT range 2–23 repeats), which reduces transcription, protein levels and may cause loss of function [184–186]. Several WT functions and mechanisms of ALS-mutant associated toxicity have been proposed [1, 187–191], some of which affect SGs.

Two *C9ORF72* isoforms are expressed in human cells. The longer possesses a differentially expressed in normal and neoplastic cells (DENN) domain, a hallmark of guanine nucleotide exchange factors for Rab GTPases [192, 193]. Indeed C9ORF72 physically and functionally interacts with several Rab proteins, leading to proposed roles in endocytosis and autophagy [189, 191, 194–196]. Impairment of autophagy, or endocytosis [167] could contribute to impaired TDP-43 turnover and/

or delayed SG clearance. C9ORF72 also localizes in SGs and P-bodies, and stimulates G3BP1 and TIA1 protein levels based on knockdown data [28]. Over-expression of repeat-expanded *C9ORF72* alleles, unlike WT, also induces SG assembly in neuroblastoma and cortical neurons [28]. This may reflect gain-of-function mechanisms discussed below.

G_4C_2 repeats in *C9ORF72* are subject to noncanonical translation, termed repeat-associated non-ATG (RAN) translation. This generates dipeptide repeats (DPRs) from all reading frames of both sense and antisense transcripts [197, 198]. Arginine-containing DPR species are particularly toxic [199, 200], and interact with several SG-localizing proteins containing IDRs [29]. Over-expression of GFP-tagged DPRs, particularly Glycine-Arginine (GR) and Proline-Arginine (PR) dipeptides, causes translation repression and induces spontaneous SGs dependent on eIF2a phosphorylation and G3BP1 [29, 30]. Expression of GR (detectable in SGs) and PR dipeptides also reduces SG internal dynamics (G3BP1 mobility). GR and PR DPRs also enhance hnRNPA1 and TIA1 phase separation in vitro, and reduce the internal dynamics of such phase-separated bodies [29].

Finally, nuclear *C9ORF72* RNA foci accumulate in the brain and spinal cord of *C9ORF72* ALS patients [27, 198, 201]. Paralleling other "toxic-RNA" diseases such as Myotonic Dystrophy, repeat-expanded *C9ORF72* RNA interacts with and may sequester several RNA-binding proteins that could affect SGs. Partially over-lapping C9ORF72 interactomes have been described, with other proposed consequences of protein sequestration including defects in nuclear RNA processing, splicing, and nucleocytoplasmic transport (summarized in [202]). Transfection of G_4C_2 RNA into HeLa cells also induces SG assembly and translational repression [203]; whether such effects are direct or stem from RAN translation is unclear.

In summary, *C9ORF72* G_4C_2-derived DPRs may directly modulate SG assembly and dynamics. Roles for *C9ORF72* in autophagy, sequestration, or translation regulation of SG assembly proteins may also be important.

7.7.5   ALS-Linked SOD1 Mutants

SOD1 ALS mutations likely cause a toxic gain of function, as many SOD1 mutations have little or no effect on SOD1 antioxidant activity [204, 205]. In *SOD1* mutant transgenic mice, SOD1 aggregates in spinal cord motor neurons colocalize with G3BP1, as does mutant SOD1 in ALS patient-derived fibroblasts [62]. Additionally, mutant SOD1, but not WT, interacts with G3BP1 in an RNA-independent manner and delays SG assembly following hyperosomotic and arsenite stress when over-expressed in neuroblastoma cells [62]. Additionally, ALS-mutant SOD1 protein is preferentially recruited to SGs over WT SOD1 following heat-stress. This reduces SG internal dynamics (G3BP1 mobility) [61]. Thus, mutant forms of SOD1 can localize in SGs, and modulate SG dynamics, particularly via interactions with G3BP1.

The evidence for ALS mutations in RBPs and C9ORF72 in driving excessive SG assembly, and facilitating SG conversion to a less dynamic state, is reasonably extensive and compelling, although studies are not in uniform agreement (Tables 7.1 and 7.2). Evidence for how ALS mutations in cytoskeletal and protein quality control proteins affect SGs is at a more nascent stage, and the issue of how/if ALS-associated mutations affect SG disassembly is currently understudied.

7.8 Effect of Aging on ALS and SGs

About 90% of sporadic ALS cases have no clear genetic etiology [206]. Additionally, ALS onset typically occurs later in life, between 40 and 70 years of age (average age of onset 55). Interestingly, TDP-43 aggregation within certain brain regions increases during normal aging, albeit the severity and tissue distribution of such aggregates is less widespread than in ALS patients [207–209]. Thus, outstanding questions in the field include whether unidentified mutations, or combinations of mutations are driving disease, and to what extent environmental or age-associated factors affect disease onset [79]. Regarding the later, age-associated defects in protein clearance and accumulation of cellular stress are processes which could increase SG persistence.

7.8.1 Proteostasis in Neurons

Autophagy plays a key role in neuronal cell homeostasis. Supporting this, CNS-specific knockout of ATG5 and ATG7 in mice causes early-onset neurodegeneration, characterized by accumulation of ubiquitinated protein aggregates throughout the CNS, behavioral and motor defects, and premature death [210, 211]. Autophagy activity decreases with age in numerous tissues [212], including the brain as suggested by reduced mRNA and protein expression of core autophagy genes such as BECLIN-1, ATG5, and ATG7 [213–215]. Additionally, inducing autophagy facilitates aggregate clearance and improves cell/organism survival in many models of various neurodegenerative diseases, including ALS [216–218] (and see the therapeutics section 7.10.2).

Autophagic clearance of SGs can occur in neurons, with SGs harboring ALS-mutant FUS being preferentially associated with autophagosomes, versus SGs containing WT FUS [6]. Furthermore, SGs harboring ALS-mutant FUS particularly accumulated in ATG5 knockout mouse embryonic fibroblasts, and ATG7-knockdown neurons [6]. This suggests that SGs harboring ALS-mutant proteins may be particularly dependent on autophagic clearance.

Chaperone function is also thought to decline with age. Supporting this, the ability to increase chaperone levels in response to stress, particularly Hsp70 chaperones, decreases with age in multiple tissues and model systems [219]. Additionally, multiple chaperone proteins, particularly Hsp40s and ATP-dependent chaperones, decrease in aging human brains, and/or in brains from various neurodegenerative diseases [220]. Interestingly, motor neurons show a paucity of chaperone upregulation in response to stress, or following accumulation of ALS disease aggregates. Specifically, cultured rodent motor neurons show no induction of Hsp70 following heatshock stress, unlike glial cells, and also no upregulation of Hsp70 following expression of ALS-mutant SOD1 protein [221, 222]. Hsp70 and Hsp27 levels are also not elevated in the spinal cord tissue of familial or sporadic ALS patients versus healthy controls [222].

Taken together, a lack of a robust proteostatic mechanism in aged motor-neurons may increase the likelihood of SG persistence and TDP-43/FUS/SOD1 protein aggregation.

7.8.2   Oxidative Stress

The brain exhibits extremely high energy demands compared to other tissues, and consumes 20% of total oxygen within the human body despite accounting for only 2% of body weight [223]. This, coupled with a near complete reliance on mitochondrial-driven oxidative metabolism for energy, means that neurons, relative to other cells, are prone to accumulate reactive oxygen species (ROS) and thus oxidative stress. ROS accumulation generally increases in aging neural tissue, exacerbated by age-associated declines in mitochondrial function [224]. Oxidative stress in sporadic and familial ALS spinal cord and motor cortex tissue is typically even further elevated relative to aged healthy tissue controls [225–228]. Though currently unaddressed, this could induce SG assembly. Numerous other stresses that induce SGs are likely encountered by motor-neurons [229], though whether the levels of such stress would abnormally induce SGs is unclear.

Surprisingly, the effects of aging on SG formation are poorly studied. However, in *C. elegans*, accumulation of insoluble aggregates of SG proteins TIA1 and PABP in non-dynamic SG-like foci occurs in aged worms [230]. This did not occur in identically aged Insulin-like growth factor-1 mutant worms (*daf-2*), which exhibit a two- to threefold increase in lifespan, in part due to reduced protein metabolism and upregulated stress responsive genes [231, 232]. Additional study of SG formation in other aging models therefore seems highly warranted.

7.9   Why Does ALS Affect Motor-Neurons?

Why motor neurons are selectively subject to degeneration in ALS remains contentious, though several ideas based on motor neuron biology have been suggested [81]. First, as long lived post-mitotic cells, motor-neurons may acquire sub-lethal damage (e.g., misfolded protein aggregates, oxidized biomolecules) to a critical level over time. Cell types in other tissues may avoid this by undergoing cell division and/or being turned over and more efficiently replaced by stem cell populations. Second, motor-neurons are highly enriched in mRNP granules. Besides possessing SGs and P-bodies, the polarized morphology of motor neurons makes them extremely dependent on localized transport and storage of mRNAs in NTGs, which can harbor TDP-43 and FUS including under non-stress conditions [233, 234]. Thus, like SGs, NTGs could theoretically concentrate and increase the likelihood of TDP-43 and FUS forming aggregates. This may be particularly true of NTGs in proximal axonal regions near the cell soma, as super-resolution microscopy techniques indicate a higher concentration of TDP-43 in a more static state compared to TDP-43 found in distal axon NTGs [235]. Notably, ALS-mutant FUS and TDP-43 proteins also inhibit axonal translation and transport of their bound mRNAs in various model systems, which may underlie defects in axon outgrowth [130, 234–236] and neuron degeneration. Finally, the circuit-like nature of nerve tissue, and the ability of ALS-associated protein aggregates to spread between cells either via secretion or following cell lysis [237–239], may lead to more rapid dysfunction than in other tissues [81].

7.10   Are SGs a Promising Therapeutic Target for ALS?

Existing FDA-approved treatments for ALS currently have a limited or an unknown benefit in mitigating ALS symptoms and extending lifespan, and do not offer a cure [240]. While strategies involving antisense oligonucleotides (ASOs) that reduce SOD1 and C9ORF72 expression [241–243], and stem-cell based therapies show promise, we refer readers to other reviews on those topics [243, 244]. Below, we examine whether preventing SG assembly, or facilitating SG clearance, offers a viable therapeutic strategy in ALS.

7.10.1   Limiting SG Assembly

Under physiological conditions, SG assembly typically relies on phosphorylation of the eukaryotic initiation factor 2 (eIF2) α subunit. This results in general translation repression and increased availability of non-translating mRNA to nucleate SG assembly [104]. eIF2α phosphorylation inhibits the ability of eIF2B to promote exchange of GDP for GTP on eIF2, which is necessary for eIF2 to deliver initiator tRNA to the small ribosomal subunit during translation initiation [245, 246]. Four kinases phosphorylate eIF2α in human cells, responding to various stresses including ER protein folding stress (PERK), nutrient stress (GCN2), viral infection (PKR), and heme deprivation (HRI).

In a TDP-43 fly model, TDP-43 expression correlates with increased eIF2α phosphorylation, suggesting that global translation repression is accompanied by

SG formation [247]. Given this, genetic and pharmacological inhibition (GSK2606414; [248]) of PERK was assessed in the TDP-43 fly model. This resulted in reduced eIF2α phosphorylation, presumably limiting translation repression and SG assembly, and led to significant improvements in motor-neuron function in flies. Pharmacological inhibition of PERK also reduced TDP-43 toxicity in primary rat cortical neurons [248]. Unfortunately, GSK2606414 is toxic to pancreatic tissue in mice models. However, ISRIB, which also inhibits eIF2α phosphorylation-dependent SG assembly by rendering eIF2B largely insensitive to eIF2α phosphorylation [249, 250], lacks this problem, and is neuroprotective in mouse prion models [251]. ISRIB may therefore hold promise as an anti-SG assembly, ALS therapeutic agent.

A caveat of the above is that perturbing a global translation repression mechanism may have unexpected off-target effects, and/or limit the ability of cells to effectively adjust their transcriptomes during stress. Targeting of proteins or mechanisms that physically drive SG assembly could be an alternative approach. As previously described, slowing SG assembly by Ataxin-2 knockdown limits TDP-43 aggregation, extended lifespan and improved motor performance in TDP-43 transgenic mice [12]. Targeting other genes that affect SG assembly may be of future interest.

7.10.2   Enhancing SG Clearance

Determining mechanisms by which autophagy and chaperones regulate SG clearance and disassembly could identify new ways to selectively clear SGs that would be of therapeutic benefit. Selectivity may be important, as inducing autophagy non-selectively has produced mixed results in ALS models, sometimes suppressing ALS phenotypes [252, 253], and at other times exacerbating them [254, 255]. The effects on SGs in these studies were not examined. However, one study in rodent cortical neurons has examined how non-selective autophagy induction affects SG clearance rates. Specifically, clearance of SGs harboring ALS-mutant FUS was enhanced by autophagy induction with rapamycin. Clearance of these SGs also coincided with reduced neurite fragmentation and neuronal cell death attributed to ALS-mutant FUS expression [6]. Though correlative, this suggests that accelerating clearance of SGs may offer a viable therapeutic strategy, and is consistent with the notion of SGs as contributors to ALS pathology.

7.11   Summary and Future Directions

In our view, the data summarized above suggests that SGs are probably involved in formation of ALS protein aggregates (especially TDP-43). However, certain views in the field have little supporting data, and key experiments remain to be addressed.

Evidence suggesting that SGs facilitate formation of ALS protein aggregates include: (1) To varying degrees, WT and ALS-mutant TDP-43, FUS and SOD1 localize in SGs under stress, and sometimes colocalize with SG components in patient tissue. (2) Many proteins mutated in ALS also localize in SGs (Table 7.1). (3) By various mechanisms, some ALS-mutants alter SG assembly, disassembly or internal dynamics such that SGs/SG proteins become more persistent and static in nature (Table 7.2). (4) Manipulations that prevent SG assembly (impairing eIF2α-based translation repression; Ataxin-2 knockdown) or facilitate SG clearance (Autophagy upregulation) correlate with improvements in ALS models. However, these manipulations likely affect other cellular processes besides SG formation.

Useful future directions include determining if ALS protein aggregate formation is affected when SG assembly is inhibited or enhanced using robust, targeted means. Also, numerous ALS-associated mutants remain poorly studied regarding their effects on SGs (Tables 7.1 and 7.2). The extent to which SG disassembly and clearance are perturbed in ALS, particularly in cases involving protein quality control mutants, requires further study. Such work would better inform the potential of targeting SG formation for therapeutic purposes.

How ALS aggregates affect disease progression remains unclear. Efforts to purify ALS aggregates to unbiasedly identify what proteins and mRNAs are present within would shed light on the validity of the "perturbed mRNA metabolism" model, and perhaps suggest novel disease mechanisms. In addition, whether SGs affect truncation and modification of TDP-43, which may affect TDP-43 aggregation and toxicity, remains unclear. Finally, greater study of how aging impacts SG formation, particularly in a motor neuron context, may lead to a better general understanding of SG biology, and provide clues as to the age and tissue-specific patterns of pathology seen in ALS.

References

1. Ling S-C, Polymenidou M, Cleveland DW. Converging mechanisms in ALS and FTD: disrupted RNA and protein homeostasis. Neuron. 2013;79:416–38.
2. Mackenzie IRA, Rademakers R, Neumann M. TDP-43 and FUS in amyotrophic lateral sclerosis and frontotemporal dementia. Lancet Neurol. 2010;9:995–1007.
3. Liu-Yesucevitz L, Bilgutay A, Zhang Y-J, Vanderweyde T, Vanderwyde T, Citro A, et al. Tar DNA binding protein-43 (TDP-43) associates with stress granules: analysis of cultured cells and pathological brain tissue. PLoS One. 2010;e13250:5.
4. Dewey CM, Cenik B, Sephton CF, Dries DR, Mayer P, Good SK, et al. TDP-43 is directed to stress granules by sorbitol, a novel physiological osmotic and oxidative stressor. Mol Cell Biol. 2011;31:1098–108.
5. McDonald KK, Aulas A, Destroismaisons L, Pickles S, Beleac E, Camu W, et al. TAR DNA-binding protein 43 (TDP-43) regulates stress granule dynamics via differential regulation of G3BP and TIA-1. Hum Mol Genet. 2011;20:1400–10.
6. Ryu H-H, Jun M-H, Min K-J, Jang D-J, Lee Y-S, Kim HK, et al. Autophagy regulates amyotrophic lateral sclerosis-linked fused in sarcoma-positive stress granules in neurons. Neurobiol Aging. 2014;35:2822–31.

7. Baron DM, Kaushansky LJ, Ward CL, Sama RRK, Chian R-J, Boggio KJ, et al. Amyotrophic lateral sclerosis-linked FUS/TLS alters stress granule assembly and dynamics. Mol Neurodegener. 2013;8:30.
8. Mackenzie IR, Nicholson AM, Sarkar M, Messing J, Purice MD, Pottier C, et al. TIA1 mutations in amyotrophic lateral sclerosis and frontotemporal dementia promote phase separation and alter stress granule dynamics. Neuron. 2017;95:808–816.e9.
9. Kapeli K, Martinez FJ, Yeo GW. Genetic mutations in RNA-binding proteins and their roles in ALS. Hum Genet. 2017;136:1193–214.
10. Kim HJ, Kim NC, Wang Y-D, Scarborough EA, Moore J, Diaz Z, et al. Mutations in prion-like domains in hnRNPA2B1 and hnRNPA1 cause multisystem proteinopathy and ALS. Nature. 2013;495:467–73.
11. Hadano S, Hand CK, Osuga H, Yanagisawa Y, Otomo A, Devon RS, et al. A gene encoding a putative GTPase regulator is mutated in familial amyotrophic lateral sclerosis 2. Nat Genet. 2001;29:166–73.
12. Becker LA, Huang B, Bieri G, Ma R, Knowles DA, Jafar-Nejad P, et al. Therapeutic reduction of ataxin-2 extends lifespan and reduces pathology in TDP-43 mice. Nature. 2017;544:367–71.
13. Padhi AK, Jayaram B, Gomes J. Prediction of functional loss of human angiogenin mutants associated with ALS by molecular dynamics simulations. Sci Rep. 2013;3:1225.
14. Lyons SM, Achorn C, Kedersha NL, Anderson PJ, Ivanov P. YB-1 regulates tiRNA-induced Stress Granule formation but not translational repression. Nucleic Acids Res. 2016;44:6949–60.
15. Thiyagarajan N, Ferguson R, Subramanian V, Acharya KR. Structural and molecular insights into the mechanism of action of human angiogenin-ALS variants in neurons. Nat Commun. 2012;3:1114–21.
16. Glatt S, Zabel R, Kolaj-Robin O, Onuma OF, Baudin F, Graziadei A, et al. Structural basis for tRNA modification by Elp3 from Dehalococcoides mccartyi. Nat Struct Mol Biol. 2016;23:794–802.
17. Bennett C, La Spada A. Unwinding the role of senataxin in neurodegeneration. Discov Med. 2015;19(103):127–36.
18. Boehringer A, Garcia-Mansfield K, Singh G, Bakkar N, Pirrotte P, Bowser R. ALS associated mutations in matrin 3 alter protein-protein interactions and impede mRNA nuclear export. Sci Rep. 2017;7:1–14.
19. Gallego-Iradi MC, Clare AM, Brown HH, Janus C, Lewis J, Borchelt DR. Subcellular localization of Matrin 3 containing mutations associated with ALS and distal myopathy. PLoS One. 2015;10:1–15.
20. Blauw HM, Barnes CP, Van Vught PWJ, Van Rheenen W, Verheul M, Cuppen E, et al. SMN1 gene duplications are associated with sporadic ALS. Neurology. 2012;78:776–80.
21. Corcia P, Camu W, Halimi J-M, Vourc'h P, Antar C, Vedrine S, et al. SMN1 gene, but not SMN2, is a risk factor for sporadic ALS. Neurology. 2006;67:1147 LP–1150.
22. Hua Y, Zhou J. Survival motor neuron protein facilitates assembly of stress granules. FEBS Lett. 2004;572:69–74.
23. Kukharsky MS, Quintiero A, Matsumoto T, Matsukawa K, An H, Hashimoto T, et al. Calcium-responsive transactivator (CREST) protein shares a set of structural and functional traits with other proteins associated with amyotrophic lateral sclerosis. Mol Neurodegener. 2015;10:1–18.
24. Kaneb HM, Folkmann AW, Belzil VV, Jao LE, Leblond CS, Girard SL, et al. Deleterious mutations in the essential mRNA metabolism factor, hGle1, in amyotrophic lateral sclerosis. Hum Mol Genet. 2015;24:1363–73.
25. Aditi, Folkmann AW, Wente SR. Cytoplasmic hGle1A regulates stress granules by modulation of translation. Mol Biol Cell. 2015;26:1476–90.
26. Aditi, Glass L, Dawson TR, Wente SR. An amyotrophic lateral sclerosis-linked mutation in GLE1 alters the cellular pool of human Gle1 functional isoforms. Adv Biol Regul. 2016;62:25–36.

27. DeJesus-Hernandez M, Mackenzie IR, Boeve BF, Boxer AL, Baker M, Rutherford NJ, et al. Expanded GGGGCC hexanucleotide repeat in noncoding region of C9ORF72 causes chromosome 9p-linked FTD and ALS. Neuron. 2011;72:245–56.
28. Maharjan N, Künzli C, Buthey K, Saxena S. C9ORF72 regulates stress granule formation and its deficiency impairs stress granule assembly, hypersensitizing cells to stress. Mol Neurobiol. 2017;54:3062–77.
29. Lee K-H, Zhang P, Kim HJ, Mitrea DM, Sarkar M, Freibaum BD, et al. C9orf72 dipeptide repeats impair the assembly, dynamics, and function of membrane-less organelles. Cell. 2016;167:774–788.e17.
30. Boeynaems S, Bogaert E, Kovacs D, Konijnenberg A, Timmerman E, Volkov A, et al. Phase separation of C9orf72 dipeptide repeats perturbs stress granule dynamics. Mol Cell. 2017;65:1044–1055.e5.
31. Ayaki T, Ito H, Fukushima H, Inoue T, Kondo T, Ikemoto A, et al. Immunoreactivity of valosin-containing protein in sporadic amyotrophic lateral sclerosis and in a case of its novel mutant. Acta Neuropathol Commun. 2014;2:1–14.
32. Buchan JR, Kolaitis R-M, Taylor JP, Parker R. Eukaryotic stress granules are cleared by autophagy and Cdc48/VCP function. Cell. 2013;153:1461–74.
33. Deng H-X, Chen W, Hong S-T, Boycott KM, Gorrie GH, Siddique N, et al. Mutations in UBQLN2 cause dominant X-linked juvenile and adult-onset ALS and ALS/dementia. Nature. 2011;477:211–5.
34. Fecto F, Yan J, Vemula SP, Liu E, Yang Y, Chen W, et al. SQSTM1 mutations in familial and sporadic amyotrophic lateral sclerosis. Arch Neurol. 2011;68:1440–6.
35. Guo H, Chitiprolu M, Gagnon D, Meng L, Perez-Iratxeta C, Lagace D, et al. Autophagy supports genomic stability by degrading retrotransposon RNA. Nat Commun. 2014;5:5276.
36. Brady OA, Meng P, Zheng Y, Mao Y, Hu F. Regulation of TDP-43 aggregation by phosphorylation andp62/SQSTM1. J Neurochem. 2011;116:248–59.
37. Swarup G, Vaibhava V, Nagabhushana A. Functional defects caused by glaucoma–associated mutations in optineurin. Glaucoma Basic Clin Asp. 2017. https://doi.org/10.5772/52692.
38. Yang Y, Hentati A, Deng HX, Dabbagh O, Sasaki T, Hirano M, et al. The gene encoding alsin, a protein with three guanine-nucleotide exchange factor domains, is mutated in a form of recessive amyotrophic lateral sclerosis. Nat Genet. 2001;29:160–5.
39. Cox LE, Ferraiuolo L, Goodall EF, Heath PR, Higginbottom A, Mortiboys H, et al. Mutations in CHMP2B in lower motor neuron predominant amyotrophic lateral sclerosis (ALS). PLoS One. 2010;5:e9872.
40. Chow CY, Landers JE, Bergren SK, Sapp PC, Grant AE, Jones JM, et al. Deleterious variants of FIG4, a phosphoinositide phosphatase, in patients with ALS. Am J Hum Genet. 2009;84:85–8.
41. Nishimura AL, Mitne-Neto M, Silva HCA, Richieri-Costa A, Middleton S, Cascio D, et al. A mutation in the vesicle-trafficking protein VAPB causes late-onset spinal muscular atrophy and amyotrophic lateral sclerosis. Am J Hum Genet. 2004;75:822–31.
42. Freischmidt A, Wieland T, Richter B, Ruf W, Schaeffer V, Müller K, et al. Haploinsufficiency of TBK1 causes familial ALS and fronto-temporal dementia. Nat Neurosci. 2015;18:631–6.
43. Williams KL, Topp S, Yang S, Smith B, Fifita JA, Warraich ST, et al. CCNF mutations in amyotrophic lateral sclerosis and frontotemporal dementia. Nat Commun. 2016;7:11253.
44. Watanabe S, Ilieva H, Tamada H, Nomura H, Komine O, Endo F, et al. Mitochondria-associated membrane collapse is a common pathomechanism in SIGMAR1–and SOD1 -linked ALS. EMBO Mol Med. 2016;8:1421–37.
45. Dreser A, Vollrath JT, Sechi A, Johann S, Roos A, Yamoah A, et al. The ALS-linked E102Q mutation in Sigma receptor-1 leads to ER stress-mediated defects in protein homeostasis and dysregulation of RNA-binding proteins. Cell Death Differ. 2017;24:1655–71.
46. Wu C-H, Fallini C, Ticozzi N, Keagle PJ, Sapp PC, Piotrowska K, et al. Mutations in the profilin 1 gene cause familial amyotrophic lateral sclerosis. Nature. 2012;488:499–503.
47. Figley MD, Bieri G, Kolaitis R-M, Taylor JP, Gitler AD. Profilin 1 associates with stress granules and ALS-linked mutations alter stress granule dynamics. J Neurosci. 2014;34:8083–97.

48. Münch C, Sedlmeier R, Meyer T, Homberg V, Sperfeld AD, Kurt A, et al. Point mutations of the p150 subunit of dynactin (DCTN1) gene in ALS. Neurology. 2004;63:724–6.
49. Vilariño-Güell C, Wider C, Soto-Ortolaza AI, Cobb SA, Kachergus JM, Keeling BH, et al. Characterization of DCTN1 genetic variability in neurodegeneration. Neurology. 2009;72:2024–8.
50. Al-Chalabi A, Andersen PM, Nilsson P, Chioza B, Andersson JL, Russ C, et al. Deletions of the heavy neurofilament subunit tail in amyotrophic lateral sclerosis. Hum Mol Genet. 1999;8:157–64.
51. Skvortsova V, Shadrina M, Slominsky P, Levitsky G, Kondratieva E, Zherebtsova A, et al. Analysis of heavy neurofilament subunit gene polymorphism in Russian patients with sporadic motor neuron disease (MND). Eur J Hum Genet. 2004;12:241–4.
52. Couthouis J, Raphael AR, Daneshjou R, Gitler AD. Targeted exon capture and sequencing in sporadic amyotrophic lateral sclerosis. PLoS Genet. 2014;10(10):e1004704.
53. Münch C, Rosenbohm A, Sperfeld AD, Uttner I, Reske S, Krause BJ, et al. Heterozygous R1101K mutation of the DCTN1 gene in a family with ALS and FTD. Ann Neurol. 2005;58:777–80.
54. Smith BN, Ticozzi N, Fallini C, Gkazi AS, Topp S, Kenna KP, et al. Exome-wide rare variant analysis identifies TUBA4A mutations associated with familial ALS. Neuron. 2014;84:324–31.
55. Perrone F, Nguyen HP, Van Mossevelde S, Moisse M, Sieben A, Santens P, et al. Investigating the role of ALS genes CHCHD10 and TUBA4A in Belgian FTD-ALS spectrum patients. Neurobiol Aging. 2017;51:177.e9–177.e16.
56. Corrado L, Mazzini L, Oggioni GD, Luciano B, Godi M, Brusco A, et al. ATXN-2 CAG repeat expansions are interrupted in ALS patients. Hum Genet. 2011;130:575–80.
57. Gros-Louis F, Larivière R, Gowing G, Laurent S, Camu W, Bouchard JP, et al. A frame-shift deletion in peripherin gene associated with amyotrophic lateral sclerosis. J Biol Chem. 2004;279:45951–6.
58. Hermosura MC, Nayakanti H, Dorovkov MV, Calderon FR, Ryazanov AG, Haymer DS, et al. A TRPM7 variant shows altered sensitivity to magnesium that may contribute to the pathogenesis of two Guamanian neurodegenerative disorders. Proc Natl Acad Sci USA. 2005;102:11510–5.
59. Brenner D, Müller K, Wieland T, Weydt P, Böhm S, Lule D, et al. NEK1 mutations in familial amyotrophic lateral sclerosis. Brain. 2016;139:e28.
60. Cleveland DW, Rothstein JD. From charcot to lou gehrig. Nat Rev Neurosci. 2001;2:806–19.
61. Mateju D, Franzmann TM, Patel A, Kopach A, Boczek EE, Maharana S, et al. An aberrant phase transition of stress granules triggered by misfolded protein and prevented by chaperone function. EMBO J. 2017;36:1669–87.
62. Gal J, Kuang L, Barnett KR, Zhu BZ, Shissler SC, Korotkov KV, et al. ALS mutant SOD1 inter-acts with G3BP1 and affects stress granule dynamics. Acta Neuropathol. 2016;132:563–76.
63. Takahashi Y, Fukuda Y, Yoshimura J, Toyoda A, Kurppa K, Moritoyo H, et al. Erbb4 muta-tions that disrupt the neuregulin-erbb4 pathway cause amyotrophic lateral sclerosis type 19. Am J Hum Genet. 2013;93:900–5.
64. Cozzolino M, Rossi S, Mirra A, Carrì MT. Mitochondrial dynamism and the pathogenesis of Amyotrophic Lateral Sclerosis. Front Cell Neurosci. 2015;9:1–5.
65. Woo JAA, Liu T, Trotter C, Fang CC, De Narvaez E, Lepochat P, et al. Loss of function CHCHD10 mutations in cytoplasmic TDP-43 accumulation and synaptic integrity. Nat Commun. 2017;8:1–15.
66. Mackenzie IRA, Neumann M. Molecular neuropathology of frontotemporal dementia: insights into disease mechanisms from postmortem studies. J Neurochem. 2016;138:54–70.
67. Neumann M, Sampathu DM, Kwong LK, Truax AC, Micsenyi MC, Chou TT, et al. Ubiquitinated TDP-43 in frontotemporal lobar degeneration and amyotrophic lateral sclero-sis. Science. 2006;314(5796):130–3.

68. Arai T, Hasegawa M, Akiyama H, Ikeda K, Nonaka T, Mori H, et al. TDP-43 is a component of ubiquitin-positive tau-negative inclusions in frontotemporal lobar degeneration and amyotrophic lateral sclerosis. Biochem Biophys Res Commun. 2006;351:602–11.
69. Nonaka T, Suzuki G, Tanaka Y, Kametani F, Hirai S, Okado H, et al. Phosphorylation of TAR DNA-binding protein of 43 kDa (TDP-43) by truncated casein kinase 1δ triggers mislocalization and accumulation of TDP-43. J Biol Chem. 2016;291:5473–83.
70. Zhang Y-J, Xu Y-F, Cook C, Gendron TF, Roettges P, Link CD, et al. Aberrant cleavage of TDP-43 enhances aggregation and cellular toxicity. Proc Natl Acad Sci U S A. 2009;106:7607–12.
71. Lagier-Tourenne C, Polymenidou M, Cleveland DW. TDP-43 and FUS/TLS: emerging roles in RNA processing and neurodegeneration. Hum Mol Genet. 2010;19:R46–64.
72. Vance C, Rogelj B, Hortobagyi T, De Vos KJ, Nishimura AL, Sreedharan J, et al. Mutations in FUS, an RNA processing protein, cause familial amyotrophic lateral sclerosis type 6. Science. 2009;323:1208–11.
73. Mackenzie IRA, Bigio EH, Ince PG, Geser F, Neumann M, Cairns NJ, et al. Pathological TDP-43 distinguishes sporadic amyotrophic lateral sclerosis from amyotrophic lateral sclerosis withSOD1 mutations. Ann Neurol. 2007;61:427–34.
74. Tan C-F, Eguchi H, Tagawa A, Onodera O, Iwasaki T, Tsujino A, et al. TDP-43 immunoreactivity in neuronal inclusions in familial amyotrophic lateral sclerosis with or without SOD1 gene mutation. Acta Neuropathol. 2007;113:535–42.
75. Turner BJ, Bäumer D, Parkinson NJ, Scaber J, Ansorge O, Talbot K. TDP-43 expression in mouse models of amyotrophic lateral sclerosis and spinal muscular atrophy. BMC Neurosci. 2008;9:104.
76. Lee EB, Lee VM-Y, Trojanowski JQ. Gains or losses: molecular mechanisms of TDP43-mediated neurodegeneration. Nat Rev Neurosci. 2012;13:38–50.
77. Vanden Broeck L, Callaerts P, Dermaut B. TDP-43-mediated neurodegeneration: towards a loss-of-function hypothesis? Trends Mol Med. 2014;20:66–71.
78. Li YR, King OD, Shorter J, Gitler AD. Stress granules as crucibles of ALS pathogenesis. J Cell Biol. 2013;201:361–72.
79. Weishaupt JH, Hyman T, Dikic I. Common molecular pathways in amyotrophic lateral sclerosis and frontotemporal dementia. Trends Mol Med. 2016;22:769–83.
80. Coyne AN, Zaepfel BL, Zarnescu DC. Failure to deliver and translate—new insights into RNA dysregulation in ALS. Front Cell Neurosci. 2017;11:243.
81. Ramaswami M, Taylor JP, Parker R. Altered ribostasis: RNA-protein granules in degenerative disorders. Cell. 2013;154:727–36.
82. Aulas A, Vande Velde C. Alterations in stress granule dynamics driven by TDP-43 and FUS: a link to pathological inclusions in ALS? Front Cell Neurosci. 2015;9:423.
83. Dewey CM, Cenik B, Sephton CF, Johnson BA, Herz J, Yu G. TDP-43 aggregation in neurodegeneration: are stress granules the key? Brain Res. 2012;1462:16–25.
84. Wolozin B. Regulated protein aggregation: stress granules and neurodegeneration. Mol Neurodegener. 2012;7:56.
85. Takanashi K, Yamaguchi A. Aggregation of ALS-linked FUS mutant sequesters RNA binding proteins and impairs RNA granules formation. Biochem Biophys Res Commun. 2014;452:600–7.
86. Che M-X, Jiang L-L, Li H-Y, Jiang Y-J, Hu H-Y. TDP-35 sequesters TDP-43 into cytoplasmic inclusions through binding with RNA. FEBS Lett. 2015;589:1920–8.
87. Elden AC, Kim H-J, Hart MP, Chen-Plotkin AS, Johnson BS, Fang X, et al. Ataxin-2 intermediate-length polyglutamine expansions are associated with increased risk for ALS. Nature. 2010;466:1069–75.
88. Voigt A, Herholz D, Fiesel FC, Kaur K, Müller D, Karsten P, et al. TDP-43-mediated neuron loss in vivo requires RNA-binding activity. PLoS One. 2010;5:e12247.
89. Daigle JG, Lanson NA, Smith RB, Casci I, Maltare A, Monaghan J, et al. RNA-binding ability of FUS regulates neurodegeneration, cytoplasmic mislocalization and incorporation

into stress granules associated with FUS carrying ALS-linked mutations. Hum Mol Genet. 2013;22:1193–205.

90. Sun Z, Diaz Z, Fang X, Hart MP, Chesi A, Shorter J, et al. Molecular determinants and genetic modifiers of aggregation and toxicity for the ALS disease protein FUS/TLS. PLoS Biol. 2011;9:e1000614.

91. Buchan JR. mRNP granules: assembly, function, and connections with disease. RNA Biol. 2014;11:1019–30.

92. Kedersha NL, Gupta M, Li W, Miller I, Anderson P. RNA-binding proteins TIA-1 and TIAR link the phosphorylation of eIF-2 alpha to the assembly of mammalian stress granules. J Cell Biol. 1999;147:1431–42.

93. Anderson P, Kedersha N. Stress granules: the Tao of RNA triage. Trends Biochem Sci. 2008;33:141–50.

94. Buchan JR, Parker R. Eukaryotic stress granules: the ins and outs of translation. Mol Cell. 2009;36:932–41.

95. Protter DSW, Parker R. Principles and properties of stress granules. Trends Cell Biol. 2016;26:668–79.

96. Kedersha N, Stoecklin G, Ayodele M, Yacono P, Lykke-Andersen J, Fritzler MJ, et al. Stress granules and processing bodies are dynamically linked sites of mRNP remodeling. J Cell Biol. 2005;169:871–84.

97. Buchan JR, Muhlrad D, Parker R. P bodies promote stress granule assembly in Saccharomyces cerevisiae. J Cell Biol. 2008;183:441–55.

98. Kedersha N, Anderson P. Stress granules: sites of mRNA triage that regulate mRNA stability and translatability. Biochem Soc Trans. 2002;30:963–9.

99. Lloyd RE. Regulation of stress granules and P-bodies during RNA virus infection. Wiley Interdiscip Rev RNA. 2013;4:317–31.

100. Buchan JR, Yoon J-H, Parker R. Stress-specific composition, assembly and kinetics of stress granules in Saccharomyces cerevisiae. J Cell Sci. 2011;124:228–39.

101. Wheeler JR, Matheny T, Jain S, Abrisch R, Parker R. Distinct stages in stress granule assembly and disassembly. elife. 2016;5. https://doi.org/10.7554/eLife.18413.

102. Yang X, Shen Y, Garre E, Hao X, Krumlinde D, Cvijović M, et al. Stress granule-defective mutants deregulate stress responsive transcripts. PLoS Genet. 2014;10:e1004763.

103. Kedersha N, Cho MR, Li W, Yacono PW, Chen S, Gilks N, et al. Dynamic shuttling of TIA-1 accompanies the recruitment of mRNA to mammalian stress granules. J Cell Biol. 2000;151:1257–68.

104. Kedersha N, Chen S, Gilks N, Li W, Miller IJ, Stahl J, et al. Evidence that ternary complex (eIF2-GTP-tRNAiMet)-deficient preinitiation complexes are core constituents of mammalian stress granules. Mol Biol Cell. 2002;13:195–210.

105. Mollet S, Cougot N, Wilczynska A, Dautry F, Kress M, Bertrand E, et al. Translationally repressed mRNA transiently cycles through stress granules during stress. Mol Biol Cell. 2008;19:4469–79.

106. Bounedjah O, Desforges B, Wu T-D, Pioche-Durieu C, Marco S, Hamon L, et al. Free mRNA in excess upon polysome dissociation is a scaffold for protein multimerization to form stress granules. Nucleic Acids Res. 2014;42:8678–91.

107. Kimball SR, Horetsky RL, Ron D, Jefferson LS, Harding HP. Mammalian stress granules represent sites of accumulation of stalled translation initiation complexes. Am J Physiol Cell Physiol. 2003;284:C273–84.

108. Souquere S, Mollet S, Kress M, Dautry F, Pierron G, Weil D. Unravelling the ultrastructure of stress granules and associated P-bodies in human cells. J Cell Sci. 2009;122:3619–26.

109. Emara MM, Ivanov P, Hickman T, Dawra N, Tisdale S, Kedersha N, et al. Angiogenin-induced tRNA-derived stress-induced RNAs promote stress-induced stress granule assembly. J Biol Chem. 2010;285:10959–68.

110. Leung AKL, Calabrese JM, Sharp PA. Quantitative analysis of Argonaute protein reveals microRNA-dependent localization to stress granules. Proc Natl Acad Sci U S A. 2006;103:18125–30.

111. Royo H, Basyuk E, Marty V, Marques M, Bertrand E, Cavaillé J. Bsr, a nuclear-retained RNA with monoallelic expression. Mol Biol Cell. 2007;18:2817–27.

112. Hubstenberger A, Courel M, Bénard M, Souquere S, Ernoult-Lange M, Chouaib R, et al. P-body purification reveals the condensation of repressed mRNA regulons. Mol Cell. 2017;68:144–157.e5.

113. Clemson CM, Hutchinson JN, Sara SA, Ensminger AW, Fox AH, Chess A, et al. An architectural role for a nuclear noncoding RNA: NEAT1 RNA is essential for the structure of paraspeckles. Mol Cell. 2009;33:717–26.

114. Tourrière H, Chebli K, Zekri L, Courselaud B, Blanchard JM, Bertrand E, et al. The RasGAP-associated endoribonuclease G3BP assembles stress granules. J Cell Biol. 2003;160:823–31.

115. Gilks N, Kedersha N, Ayodele M, Shen L, Stoecklin G, Dember LM, et al. Stress granule assembly is mediated by prion-like aggregation of TIA-1. Mol Biol Cell. 2004;15:5383–98.

116. Bentmann E, Neumann M, Tahirovic S, Rodde R, Dormann D, Haass C. Requirements for stress granule recruitment of fused in sarcoma (FUS) and TAR DNA-binding protein of 43 kDa (TDP-43). J Biol Chem. 2012;287:23079–94.

117. Kato M, Han TW, Xie S, Shi K, Du X, Wu LC, et al. Cell-free formation of RNA granules: low complexity sequence domains form dynamic fibers within hydrogels. Cell. 2012;149:753–67.

118. Molliex A, Temirov J, Lee J, Coughlin M, Kanagaraj AP, Kim HJ, et al. Phase separation by low complexity domains promotes stress granule assembly and drives pathological fibrillization. Cell. 2015;163:123–33.

119. Lin Y, Protter DSW, Rosen MK, Parker R. Formation and maturation of phase-separated liquid droplets by RNA-binding proteins. Mol Cell. 2015;60:208–19.

120. Jain S, Wheeler JR, Walters RW, Agrawal A, Barsic A, Parker R. ATPase-modulated stress granules contain a diverse proteome and substructure. Cell. 2016;164:487–98.

121. Chalupníková K, Lattmann S, Selak N, Iwamoto F, Fujiki Y, Nagamine Y. Recruitment of the RNA helicase RHAU to stress granules via a unique RNA-binding domain. J Biol Chem. 2008;283:35186–98.

122. Banani SF, Lee HO, Hyman AA, Rosen MK. Biomolecular condensates: organizers of cellular biochemistry. Nat Rev Mol Cell Biol. 2017;18:285–98.

123. Brangwynne CP, Eckmann CR, Courson DS, Rybarska A, Hoege C, Gharakhani J, et al. Germline P granules are liquid droplets that localize by controlled dissolution/condensation. Science. 2009;324:1729–32.

124. Patel A, Lee HO, Jawerth L, Maharana S, Jahnel M, Hein MY, et al. A liquid-to-solid phase transition of the ALS Protein FUS accelerated by disease mutation. Cell. 2015;162:1066–77.

125. Fromm SA, Kamenz J, Nöldeke ER, Neu A, Zocher G, Sprangers R. In vitro reconstitution of a cellular phase-transition process that involves the mRNA decapping machinery. Angew Chem Int Ed. 2014;53:7354–9.

126. Nott TJ, Petsalaki E, Farber P, Jervis D, Fussner E, Plochowietz A, et al. Phase transition of a disordered nuage protein generates environmentally responsive membraneless organelles. Mol Cell. 2015;57:936–47.

127. Jain A, Vale RD. RNA phase transitions in repeat expansion disorders. Nature. 2017;546:243–7.

128. Conicella AE, Zerze GH, Mittal J, Fawzi NL. ALS mutations disrupt phase separation mediated by α-helical structure in the TDP-43 low-complexity C-terminal domain. Structure. 2016;24:1537–49.

129. Abrakhi S, Kretov DA, Desforges B, Dobra I, Bouhss A, Pastré D, et al. Nanoscale analysis reveals the maturation of neurodegeneration-associated protein aggregates: grown in mRNA granules then released by stress granule proteins. ACS Nano. 2017;11:7189–200.

130. Murakami T, Qamar S, Lin JQ, Schierle GSK, Rees E, Miyashita A, et al. ALS/FTD mutation-induced phase transition of FUS liquid droplets and reversible hydrogels into irreversible hydrogels impairs RNP granule function. Neuron. 2015;88:678–90.
131. Cherkasov V, Hofmann S, Druffel-Augustin S, Mogk A, Tyedmers J, Stoecklin G, et al. Coordination of translational control and protein homeostasis during severe heat stress. Curr Biol. 2013;23:2452–62.
132. Hilliker A, Gao Z, Jankowsky E, Parker R. The DEAD-box protein Ded1 modulates translation by the formation and resolution of an eIF4F-mRNA complex. Mol Cell. 2011;43:962–72.
133. Mazroui R, Di Marco S, Kaufman RJ, Gallouzi I-E. Inhibition of the ubiquitin-proteasome system induces stress granule formation. Mol Biol Cell. 2007;18:2603–18.
134. Walters RW, Muhlrad D, Garcia J, Parker R. Differential effects of Ydj1 and Sis1 on Hsp70-mediated clearance of stress granules in Saccharomyces cerevisiae. RNA. 2015;21:1660–71.
135. Walters RW, Parker R. Coupling of ribostasis and proteostasis: Hsp70 proteins in mRNA metabolism. Trends Biochem Sci. 2015;40:552–9.
136. Ganassi M, Mateju D, Bigi I, Mediani L, Poser I, Lee HO, et al. A surveillance function of the HSPB8-BAG3-HSP70 chaperone complex ensures stress granule integrity and dynamism. Mol Cell. 2016;63:796–810.
137. Meyer H, Bug M, Bremer S. Emerging functions of the VCP/p97 AAA-ATPase in the ubiquitin system. Nat Cell Biol. 2012;14:117–23.
138. Ohn T, Anderson P. The role of posttranslational modifications in the assembly of stress granules. Wiley Interdiscip Rev RNA. 2010;1:486–93.
139. Kedersha N, Panas MD, Achorn CA, Lyons S, Tisdale S, Hickman T, et al. G3BP-Caprin1-USP10 complexes mediate stress granule condensation and associate with 40S subunits. J Cell Biol. 2016;212:845–60.
140. Tsai N-P, Ho P-C, Wei L-N. Regulation of stress granule dynamics by Grb7 and FAK signalling pathway. EMBO J. 2008;27:715–26.
141. De Leeuw F, Zhang T, Wauquier C, Huez G, Kruys V, Gueydan C. The cold-inducible RNA-binding protein migrates from the nucleus to cytoplasmic stress granules by a methylation-dependent mechanism and acts as a translational repressor. Exp Cell Res. 2007;313:4130–44.
142. Carpio MA, López Sambrooks C, Durand ES, Hallak ME. The arginylation-dependent association of calreticulin with stress granules is regulated by calcium. Biochem J. 2010;429:63–72.
143. Ohn T, Kedersha N, Hickman T, Tisdale S, Anderson P. A functional RNAi screen links O-GlcNAc modification of ribosomal proteins to stress granule and processing body assembly. Nat Cell Biol. 2008;10:1224–31.
144. Kwon S, Zhang Y, Matthias P. The deacetylase HDAC6 is a novel critical component of stress granules involved in the stress response. Genes Dev. 2007;21:3381–94.
145. Jayabalan AK, Sanchez A, Park RY, Yoon SP, Kang G-Y, Baek J-H, et al. NEDDylation promotes stress granule assembly. Nat Commun. 2016;7:12125.
146. Leung AKL, Vyas S, Rood JE, Bhutkar A, Sharp PA, Chang P. Poly(ADP-ribose) regulates stress responses and microRNA activity in the cytoplasm. Mol Cell. 2011;42:489–99.
147. Han TW, Kato M, Xie S, Wu LC, Mirzaei H, Pei J, et al. Cell-free formation of RNA granules: bound RNAs identify features and components of cellular assemblies. Cell. 2012;149:768–79.
148. Ivanov PA, Chudinova EM, Nadezhdina ES. Disruption of microtubules inhibits cytoplasmic ribonucleoprotein stress granule formation. Exp Cell Res. 2003;290:227–33.
149. Loschi M, Leishman CC, Berardone N, Boccaccio GL. Dynein and kinesin regulate stress-granule and P-body dynamics. J Cell Sci. 2009;122:3973–82.
150. Fujimura K, Katahira J, Kano F, Yoneda Y, Murata M. Microscopic dissection of the process of stress granule assembly. Biochim Biophys Acta. 2009;1793:1728–37.
151. Chernov KG, Barbet A, Hamon L, Ovchinnikov LP, Curmi PA, Pastré D. Role of microtubules in stress granule assembly. J Biol Chem. 2009;284:36569–80.
152. Zurla C, Lifland AW, Santangelo PJ. Characterizing mRNA interactions with RNA granules during translation initiation inhibition. PLoS One. 2011;6:e19727.

153. Nadezhdina ES, Lomakin AJ, Shpilman AA, Chudinova EM, Ivanov PA. Microtubules govern stress granule mobility and dynamics. Biochim Biophys Acta Mol Cell Res. 2010;1803:361–71.
154. Krisenko MO, Higgins RL, Ghosh S, Zhou Q, Trybula JS, Wang W-H, et al. Syk is recruited to stress granules and promotes their clearance through autophagy. J Biol Chem. 2015;290:27803–15.
155. Alberti S, Mateju D, Mediani L, Carra S. Granulostasis: protein quality control of RNP granules. Front Mol Neurosci. 2017;10:84.
156. Colombrita C, Zennaro E, Fallini C, Weber M, Sommacal A, Buratti E, et al. TDP-43 is recruited to stress granules in conditions of oxidative insult. J Neurochem. 2009;111:1051–61.
157. Freibaum BD, Chitta RK, High AA, Taylor JP. Global analysis of TDP-43 interacting proteins reveals strong association with RNA splicing and translation machinery. J Proteome Res. 2010;9:1104–20.
158. Parker SJ, Meyerowitz J, James JL, Liddell JR, Crouch PJ, Kanninen KM, et al. Endogenous TDP-43 localized to stress granules can subsequently form protein aggregates. Neurochem Int. 2012;60:415–24.
159. Ayala V, Granado-Serrano AB, Cacabelos D, Naudí A, Ilieva EV, Boada J, et al. Cell stress induces TDP-43 pathological changes associated with ERK1/2 dysfunction: implications in ALS. Acta Neuropathol. 2011;122:259–70.
160. Meyerowitz J, Parker SJ, Vella LJ, Ng DC, Price KA, Liddell JR, et al. C-Jun N-terminal kinase controls TDP-43 accumulation in stress granules induced by oxidative stress. Mol Neurodegener. 2011;6:57.
161. Volkening K, Leystra-Lantz C, Yang W, Jaffee H, Strong MJ. Tar DNA binding protein of 43 kDa (TDP-43), 14-3-3 proteins and copper/zinc superoxide dismutase (SOD1) interact to modulate NFL mRNA stability. Implications for altered RNA processing in amyotrophic lateral sclerosis (ALS). Brain Res. 2009;1305:168–82.
162. Andersson MK, Ståhlberg A, Arvidsson Y, Olofsson A, Semb H, Stenman G, et al. The multifunctional FUS, EWS and TAF15 proto-oncoproteins show cell type-specific expression patterns and involvement in cell spreading and stress response. BMC Cell Biol. 2008;9:37.
163. Sama RRK, Ward CL, Kaushansky LJ, Lemay N, Ishigaki S, Urano F, et al. FUS/TLS assembles into stress granules and is a prosurvival factor during hyperosmolar stress. J Cell Physiol. 2013;228:2222–31.
164. Vance C, Scotter EL, Nishimura AL, Troakes C, Mitchell JC, Kathe C, et al. ALS mutant FUS disrupts nuclear localization and sequesters wild-type FUS within cytoplasmic stress granules. Hum Mol Genet. 2013;22:2676–88.
165. Lenzi J, De Santis R, de Turris V, Morlando M, Laneve P, Calvo A, et al. ALS mutant FUS proteins are recruited into stress granules in induced pluripotent stem cell-derived motoneurons. Dis Model Mech. 2015;8:755–66.
166. Dormann D, Rodde R, Edbauer D, Bentmann E, Fischer I, Hruscha A, et al. ALS-associated fused in sarcoma (FUS) mutations disrupt Transportin-mediated nuclear import. EMBO J. 2010;29:2841–57.
167. Liu G, Coyne AN, Pei F, Vaughan S, Chaung M, Daniela C, Zarnescu JRB. Endocytosis regulates TDP-43 toxicity and turnover. Nat Commun. 2017;8(1):2092.
168. Aulas A, Stabile S, Vande Velde C. Endogenous TDP-43, but not FUS, contributes to stress granule assembly via G3BP. Mol Neurodegener. 2012;7:54.
169. Johnson BS, Snead D, Lee JJ, McCaffery JM, Shorter J, Gitler AD. TDP-43 is intrinsically aggregation-prone, and amyotrophic lateral sclerosis-linked mutations accelerate aggregation and increase toxicity. J Biol Chem. 2009;284:20329–39.
170. Guo W, Chen Y, Zhou X, Kar A, Ray P, Chen X, et al. An ALS-associated mutation affecting TDP-43 enhances protein aggregation, fibril formation and neurotoxicity. Nat Struct Mol Biol. 2011;18:822–30.
171. Orrù S, Coni P, Floris A, Littera R, Carcassi C, Sogos V, et al. Reduced stress granule formation and cell death in fibroblasts with the A382T mutation of TARDBP gene: evidence for loss of TDP-43 nuclear function. Hum Mol Genet. 2016;25:4473–83.

172. Guil S, Long JC, Caceres JF. hnRNP A1 relocalization to the stress granules reflects a role in the stress response. Mol Cell Biol. 2006;26:5744–58.
173. Nonhoff U, Ralser M, Welzel F, Piccini I, Balzereit D, Yaspo M-L, et al. Ataxin-2 interacts with the DEAD/H-box RNA helicase DDX6 and interferes with P-bodies and stress granules. Mol Biol Cell. 2007;18:1385–96.
174. Seguin SJ, Morelli FF, Vinet J, Amore D, De Biasi S, Poletti A, et al. Inhibition of autophagy, lysosome and VCP function impairs stress granule assembly. Cell Death Differ. 2014;21(12):1838–51.
175. Wang T, Xu W, Qin M, Yang Y, Bao P, Shen F, et al. Pathogenic mutations in the valosin-containing protein/p97(VCP) N-domain inhibit the SUMOylation of VCP and lead to impaired stress response. J Biol Chem. 2016;291:14373–84.
176. Monahan Z, Shewmaker F, Pandey UB. Stress granules at the intersection of autophagy and ALS. Brain Res. 2016;1649(Pt B):189–200.
177. Al-Saif A, Al-Mohanna F, Bohlega S. A mutation in sigma-1 receptor causes juvenile amyotrophic lateral sclerosis. Ann Neurol. 2011;70:913–9.
178. Yu A, Shibata Y, Shah B, Calamini B, Lo DC, Morimoto RI. Protein aggregation can inhibit clathrin-mediated endocytosis by chaperone competition. Proc Natl Acad Sci U S A. 2014;111:E1481–90.
179. Scotter EL, Vance C, Nishimura AL, Lee Y-B, Chen H-J, Urwin H, et al. Differential roles of the ubiquitin proteasome system and autophagy in the clearance of soluble and aggregated TDP-43 species. J Cell Sci. 2014;127:1263–78.
180. Wang X, Fan H, Ying Z, Li B, Wang H, Wang G. Degradation of TDP-43 and its pathogenic form by autophagy and the ubiquitin-proteasome system. Neurosci Lett. 2010;469:112–6.
181. Witke W, Podtelejnikov AV, Di Nardo A, Sutherland JD, Gurniak CB, Dotti C, et al. In mouse brain profilin I and profilin II associate with regulators of the endocytic pathway and actin assembly. EMBO J. 1998;17:967–76.
182. Tanaka Y, Nonaka T, Suzuki G, Kametani F, Hasegawa M. Gain-of-function profilin 1 mutations linked to familial amyotrophic lateral sclerosis cause seed-dependent intracellular TDP-43 aggregation. Hum Mol Genet. 2016;25:1420–33.
183. Daoud H, Zhou S, Noreau A, Sabbagh M, Belzil V, Dionne-Laporte A, et al. Exome sequencing reveals SPG11 mutations causing juvenile ALS. Neurobiol Aging. 2012;33:839.e5–9.
184. Belzil VV, Bauer PO, Prudencio M, Gendron TF, Stetler CT, Yan IK, et al. Reduced C9orf72 gene expression in c9FTD/ALS is caused by histone trimethylation, an epigenetic event detectable in blood. Acta Neuropathol. 2013;126:895–905.
185. Fratta P, Mizielinska S, Nicoll AJ, Zloh M, Fisher EMC, Parkinson G, et al. C9orf72 hexanucleotide repeat associated with amyotrophic lateral sclerosis and frontotemporal dementia forms RNA G-quadruplexes. Sci Rep. 2012;2:1016.
186. Waite AJ, Bäumer D, East S, Neal J, Morris HR, Ansorge O, et al. Reduced C9orf72 protein levels in frontal cortex of amyotrophic lateral sclerosis and frontotemporal degeneration brain with the C9ORF72 hexanucleotide repeat expansion. Neurobiol Aging. 2014;35:1779. e5–1779.e13.
187. Xiao S, MacNair L, McLean J, McGoldrick P, McKeever P, Soleimani S, et al. C9orf72 isoforms in amyotrophic lateral sclerosis and frontotemporal lobar degeneration. Brain Res. 2016;1647:43–9.
188. Gitler AD, Tsuiji H. There has been an awakening: emerging mechanisms of C9orf72 mutations in FTD/ALS. Brain Res. 2016;1647:19–29.
189. Tang BL. C9orf72's interaction with Rab GTPases—modulation of membrane traffic and autophagy. Front Cell Neurosci. 2016;10:228.
190. Freibaum BD, Taylor JP. The role of dipeptide repeats in C9ORF72-related ALS-FTD. Front Mol Neurosci. 2017;10:35.
191. Nassif M, Woehlbier U, Manque PA. The enigmatic role of C9ORF72 in autophagy. Front Neurosci. 2017;11:442.
192. Zhang D, Iyer LM, He F, Aravind L. Discovery of novel DENN proteins: implications for the evolution of eukaryotic intracellular membrane structures and human disease. Front Genet. 2012;3:1–10.

193. Levine TP, Daniels RD, Gatta AT, Wong LH, Hayes MJ. The product of C9orf72, a gene strongly implicated in neurodegeneration, is structurally related to DENN Rab-GEFs. Bioinformatics. 2013;29:499–503.
194. Farg MA, Sundaramoorthy V, Sultana JM, Yang S, Atkinson RAK, Levina V, et al. C9ORF72, implicated in amytrophic lateral sclerosis and frontotemporal dementia, regulates endosomal trafficking. Hum Mol Genet. 2014;23:3579–95.
195. Webster CP, Smith EF, Bauer CS, Moller A, Hautbergue GM, Ferraiuolo L, et al. The C9orf72 protein interacts with Rab1a and the ULK1 complex to regulate initiation of autophagy. EMBO J. 2016;35:1656–76.
196. Sellier C, Campanari M-L, Julie Corbier C, Gaucherot A, Kolb-Cheynel I, Oulad-Abdelghani M, et al. Loss of C9ORF72 impairs autophagy and synergizes with polyQ Ataxin-2 to induce motor neuron dysfunction and cell death. EMBO J. 2016;35:1276–97.
197. Mori K, Weng S-M, Arzberger T, May S, Rentzsch K, Kremmer E, et al. The C9orf72 GGGGCC repeat is translated into aggregating dipeptide-repeat proteins in FTLD/ALS. Science. 2013;339:1335–8.
198. Zu T, Liu Y, Bañez-Coronel M, Reid T, Pletnikova O, Lewis J, et al. RAN proteins and RNA foci from antisense transcripts in C9ORF72 ALS and frontotemporal dementia. Proc Natl Acad Sci U S A. 2013;110:E4968–77.
199. Kwon I, Xiang S, Kato M, Wu L, Theodoropoulos P, Wang T, et al. Poly-dipeptides encoded by the C9orf72 repeats bind nucleoli, impede RNA biogenesis, and kill cells. Science. 2014;345:1139–45.
200. Wen X, Tan W, Westergard T, Krishnamurthy K, Markandaiah SS, Shi Y, et al. Antisense proline-arginine RAN dipeptides linked to C9ORF72-ALS/FTD form toxic nuclear aggregates that initiate in vitro and in vivo neuronal death. Neuron. 2014;84:1213–25.
201. Gendron TF, Bieniek KF, Zhang Y-J, Jansen-West K, Ash PEA, Caulfield T, et al. Antisense transcripts of the expanded C9ORF72 hexanucleotide repeat form nuclear RNA foci and undergo repeat-associated non-ATG translation in c9FTD/ALS. Acta Neuropathol. 2013;126:829–44.
202. Haeusler AR, Donnelly CJ, Rothstein JD. The expanding biology of the C9orf72 nucleotide repeat expansion in neurodegenerative disease. Nat Rev Neurosci. 2016;17:383–95.
203. Rossi S, Serrano A, Gerbino V, Giorgi A, Di Francesco L, Nencini M, et al. Nuclear accumulation of mRNAs underlies G4C2-repeat-induced translational repression in a cellular model of C9orf72 ALS. J Cell Sci. 2015;128:1787–99.
204. Bunton-Stasyshyn RKA, Saccon RA, Fratta P, Fisher EMC. SOD1 function and its implications for amyotrophic lateral sclerosis pathology. Neuroscientist. 2015;21:519–29.
205. Borchelt DR, Lee MK, Slunt HS, Guarnieri M, Xu ZS, Wong PC, et al. Superoxide dismutase 1 with mutations linked to familial amyotrophic lateral sclerosis possesses significant activity. Proc Natl Acad Sci U S A. 1994;91:8292–6.
206. Renton AE, Chiò A, Traynor BJ. State of play in amyotrophic lateral sclerosis genetics. Nat Neurosci. 2014;17:17–23.
207. Geser F, Robinson JL, Malunda JA, Xie SX, Clark CM, Kwong LK, et al. Pathological 43-kDa transactivation response DNA-binding protein in older adults with and without severe mental illness. Arch Neurol. 2010;67:1238–50.
208. Uchino A, Takao M, Hatsuta H, Sumikura H, Nakano Y, Nogami A, et al. Incidence and extent of TDP-43 accumulation in aging human brain. Acta Neuropathol Commun. 2015;3:35.
209. Baloh RH. TDP-43: the relationship between protein aggregation and neurodegeneration in amyotrophic lateral sclerosis and frontotemporal lobar degeneration. FEBS J. 2011;278:3539–49.
210. Hara T, Nakamura K, Matsui M, Yamamoto A, Nakahara Y, Suzuki-Migishima R, et al. Suppression of basal autophagy in neural cells causes neurodegenerative disease in mice. Nature. 2006;441:885–9.
211. Komatsu M, Waguri S, Chiba T, Murata S, Iwata J, Tanida I, et al. Loss of autophagy in the central nervous system causes neurodegeneration in mice. Nature. 2006;441:880–4.

212. Rubinsztein DC, Mariño G, Kroemer G. Autophagy and aging. Cell. 2011;146:682–95.
213. Haigis MC, Yankner BA. The aging stress response. Mol Cell. 2010;40:333–44.
214. Lipinski MM, Zheng B, Lu T, Yan Z, Py BF, Ng A, et al. Genome-wide analysis reveals mechanisms modulating autophagy in normal brain aging and in Alzheimer's disease. Proc Natl Acad Sci. 2010;107:14164–9.
215. Shibata M, Lu T, Furuya T, Degterev A, Mizushima N, Yoshimori T, et al. Regulation of intra-cellular accumulation of mutant Huntingtin by Beclin 1. J Biol Chem. 2006;281:14474–85.
216. Frake RA, Ricketts T, Menzies FM, Rubinsztein DC. Autophagy and neurodegeneration. J Clin Invest. 2015;125:65–74.
217. Barmada SJ, Serio A, Arjun A, Bilican B, Daub A, Ando DM, et al. Autophagy induction enhances TDP43 turnover and survival in neuronal ALS models. Nat Chem Biol. 2014;10:677–85.
218. Wang I-F, Guo B-S, Liu Y-C, Wu C-C, Yang C-H, Tsai K-J, et al. Autophagy activators rescue and alleviate pathogenesis of a mouse model with proteinopathies of the TAR DNA-binding protein 43. Proc Natl Acad Sci U S A. 2012;109:15024–9.
219. Koga H, Kaushik S, Cuervo AM. Protein homeostasis and aging: the importance of exquisite quality control. Ageing Res Rev. 2011;10:205–15.
220. Brehme M, Voisine C, Rolland T, Wachi S, Soper JH, Zhu Y, et al. A chaperome subnetwork safeguards proteostasis in aging and neurodegenerative disease. Cell Rep. 2014;9:1135–50.
221. Kaarniranta K, Oksala N, Karjalainen HM, Suuronen T, Sistonen L, Helminen HJ, et al. Neuronal cells show regulatory differences in the hsp70 gene response. Brain Res Mol Brain Res. 2002;101:136–40.
222. Batulan Z, Shinder GA, Minotti S, He BP, Doroudchi MM, Nalbantoglu J, et al. High threshold for induction of the stress response in motor neurons is associated with failure to activate HSF1. J Neurosci. 2003;23:5789–98.
223. Rolfe DF, Brown GC. Cellular energy utilization and molecular origin of standard metabolic rate in mammals. Physiol Rev. 1997;77:731–58.
224. Stranahan AM, Mattson MP. Recruiting adaptive cellular stress responses for successful brain ageing. Nat Rev Neurosci. 2012;13:209–16.
225. Shaw PJ, Ince PG, Falkous G, Mantle D. Oxidative damage to protein in sporadic motor neuron disease spinal cord. Ann Neurol. 1995;38:691–5.
226. Ferrante RJ, Browne SE, Shinobu LA, Bowling AC, Baik MJ, MacGarvey U, et al. Evidence of increased oxidative damage in both sporadic and familial amyotrophic lateral sclerosis. J Neurochem. 1997;69:2064–74.
227. Abe K, Pan LH, Watanabe M, Kato T, Itoyama Y. Induction of nitrotyrosine-like immunoreactivity in the lower motor neuron of amyotrophic lateral sclerosis. Neurosci Lett. 1995;199:152–4.
228. Barber SC, Shaw PJ. Oxidative stress in ALS: key role in motor neuron injury and therapeutic target. Free Radic Biol Med. 2010;48:629–41.
229. Kagias K, Nehammer C, Pocock R. Neuronal responses to physiological stress. Front Genet. 2012;3:222.
230. Lechler MC, Crawford ED, Groh N, Widmaier K, Jung R, Kirstein J, et al. Reduced insulin/IGF-1 signaling restores the dynamic properties of key stress granule proteins during aging. Cell Rep. 2017;18:454–67.
231. Kenyon CJ. The genetics of ageing. Nature. 2010;464:504–12.
232. Stout GJ, Stigter ECA, Essers PB, Mulder KW, Kolkman A, Snijders DS, et al. Insulin/IGF-1-mediated longevity is marked by reduced protein metabolism. Mol Syst Biol. 2013;9:679.
233. Kanai Y, Dohmae N, Hirokawa N. Kinesin transports RNA. Neuron. 2004;43:513–25.
234. Fallini C, Bassell GJ, Rossoll W. The ALS disease protein TDP-43 is actively transported in motor neuron axons and regulates axon outgrowth. Hum Mol Genet. 2012;21:3703–18.
235. Gopal PP, Nirschl JJ, Klinman E, Holzbaur ELF. Amyotrophic lateral sclerosis-linked mutations increase the viscosity of liquid-like TDP-43 RNP granules in neurons. Proc Natl Acad Sci U S A. 2017;114:E2466–75.

236. Alami NH, Smith RB, Carrasco MA, Williams LA, Winborn CS, Han SSW, et al. Axonal transport of TDP-43 mRNA granules is impaired by ALS-causing mutations. Neuron. 2014;81:536–43.

237. Feiler MS, Strobel B, Freischmidt A, Helferich AM, Kappel J, Brewer BM, et al. TDP-43 is intercellularly transmitted across axon terminals. J Cell Biol. 2015;211:897–911.

238. Iguchi Y, Eid L, Parent M, Soucy G, Bareil C, Riku Y, et al. Exosome secretion is a key pathway for clearance of pathological TDP-43. Brain. 2016;139:3187–201.

239. Grad LI, Yerbury JJ, Turner BJ, Guest WC, Pokrishevsky E, O'Neill MA, et al. Intercellular propagated misfolding of wild-type Cu/Zn superoxide dismutase occurs via exosome-dependent and -independent mechanisms. Proc Natl Acad Sci U S A. 2014;111:3620–5.

240. Petrov D, Mansfield C, Moussy A, Hermine O. ALS clinical trials review: 20 years of failure. Are we any closer to registering a new treatment? Front Aging Neurosci. 2017;9:68.

241. Lagier-Tourenne C, Baughn M, Rigo F, Sun S, Liu P, Li H-R, et al. Targeted degradation of sense and antisense C9orf72 RNA foci as therapy for ALS and frontotemporal degeneration. Proc Natl Acad Sci U S A. 2013;110:E4530–9.

242. Donnelly CJ, Zhang P-W, Pham JT, Haeusler AR, Mistry NA, Vidensky S, et al. RNA toxicity from the ALS/FTD C9ORF72 expansion is mitigated by antisense intervention. Neuron. 2013;80:415–28.

243. van Zundert B, Brown RH. Silencing strategies for therapy of SOD1-mediated ALS. Neurosci Lett. 2017;636:32–9.

244. Forostyak S, Sykova E. Neuroprotective potential of cell-based therapies in ALS: from bench to bedside. Front Neurosci. 2017;11:591.

245. Sonenberg N, Hinnebusch AG. Regulation of translation initiation in eukaryotes: mechanisms and biological targets. Cell. 2009;136:731–45.

246. Jennings MD, Pavitt GD. A new function and complexity for protein translation initiation factor eIF2B. Cell Cycle. 2014;13:2660–5.

247. Kim H-J, Raphael AR, LaDow ES, McGurk L, Weber RA, Trojanowski JQ, et al. Therapeutic modulation of eIF2α phosphorylation rescues TDP-43 toxicity in amyotrophic lateral sclerosis disease models. Nat Genet. 2014;46:152–60.

248. Axten JM, Medina JR, Feng Y, Shu A, Romeril SP, Grant SW, et al. Discovery of 7-Methyl-5-(1-{[3-(trifluoromethyl)phenyl]acetyl}-2,3-dihydro-1 H -indol-5-yl)-7 H-pyrrolo[2,3- d] pyrimidin-4-amine (GSK2606414), a potent and selective first-in-class inhibitor of protein kinase R (PKR)-like endoplasmic reticulum kinase (PERK). J Med Chem. 2012;55:7193–207.

249. Sidrauski C, McGeachy AM, Ingolia NT, Walter P. The small molecule ISRIB reverses the effects of eIF2α phosphorylation on translation and stress granule assembly. elife. 2015;4

250. Sidrauski C, Tsai JC, Kampmann M, Hearn BR, Vedantham P, Jaishankar P, et al. Pharmacological dimerization and activation of the exchange factor eIF2B antagonizes the integrated stress response. elife. 2015;4:e07314.

251. Halliday M, Radford H, Sekine Y, Moreno J, Verity N, le Quesne J, et al. Partial restoration of protein synthesis rates by the small molecule ISRIB prevents neurodegeneration without pancreatic toxicity. Cell Death Dis. 2015;6:e1672.

252. Castillo K, Nassif M, Valenzuela V, Rojas F, Matus S, Mercado G, et al. Trehalose delays the progression of amyotrophic lateral sclerosis by enhancing autophagy in motoneurons. Autophagy. 2013;9:1308–20.

253. Zhang X, Chen S, Song L, Tang Y, Shen Y, Jia L, et al. MTOR-independent, autophagic enhancer trehalose prolongs motor neuron survival and ameliorates the autophagic flux defect in a mouse model of amyotrophic lateral sclerosis. Autophagy. 2014;10:588–602.

254. Hetz C, Thielen P, Matus S, Nassif M, Court F, Kiffin R, et al. XBP-1 deficiency in the nervous system protects against amyotrophic lateral sclerosis by increasing autophagy. Genes Dev. 2009;23:2294–306.

255. Zhang X, Li L, Chen S, Yang D, Wang Y, Zhang X, et al. Rapamycin treatment augments motor neuron degeneration in SOD1(G93A) mouse model of amyotrophic lateral sclerosis. Autophagy. 2011;7:412–25.

Chapter 8
Deregulation of RNA Metabolism in Microsatellite Expansion Diseases

Chaitali Misra, Feikai Lin, and Auinash Kalsotra

Abstract RNA metabolism impacts different steps of mRNA life cycle including splicing, polyadenylation, nucleo-cytoplasmic export, translation, and decay. Growing evidence indicates that defects in any of these steps lead to devastating diseases in humans. This chapter reviews the various RNA metabolic mechanisms that are disrupted in Myotonic Dystrophy—a trinucleotide repeat expansion disease—due to dysregulation of RNA-Binding Proteins. We also compare Myotonic Dystrophy to other microsatellite expansion disorders and describe how some of these mechanisms commonly exert direct versus indirect effects toward disease pathologies.

Keywords Microsatellite repeat expansions · Post-transcriptional gene regulation · RNA toxicity · Alternative splicing and polyadenylation · RNA-binding proteins

8.1 Introduction

Gene expression is a highly coordinated multistep process, which allows organisms to integrate intrinsic and environmental information to exert appropriate cellular functions. The expression of most genes can be regulated at distinct stages of RNA metabolism including synthesis or transcription, post-transcriptional processing or maturation, nucleo-cytoplasmic export, translation, as well as degradation at a rate that is often dictated by transcript- and cell-type-specific cues. Although transcription is a general point of control, many co- and post-transcriptional pre-mRNA processing events add substantial capacity to tune overall gene expression [1]. The

C. Misra · F. Lin
Department of Biochemistry, University of Illinois, Urbana-Champaign, IL, USA

A. Kalsotra (✉)
Department of Biochemistry, University of Illinois, Urbana-Champaign, IL, USA

Carl R. Woese Institute of Genomic Biology, University of Illinois,
Urbana-Champaign, IL, USA
e-mail: kalsotra@illinois.edu

© Springer International Publishing AG, part of Springer Nature 2018
R. Sattler, C. J. Donnelly (eds.), *RNA Metabolism in Neurodegenerative Diseases*, Advances in Neurobiology 20,
https://doi.org/10.1007/978-3-319-89689-2_8

typical pre-mRNA processing events comprise 5′ capping, splicing, and 3′ polyadenylation, which are directly linked to the nucleo-cytoplasmic export and eventual fate of mRNAs. RNA-Binding Proteins (RBPs) are essential in carrying out these processing events in both the nucleus and cytoplasm by interacting with RNA sequence or structural elements and forming distinct mRNA-protein (mRNP) complexes [2]. Disruption of RBP function(s), therefore, frequently results in deleterious RNA metabolism defects that in some cases become pathogenic [3, 4].

Neurodegenerative diseases are a heterogeneous group of neurological disorders characterized by progressive degeneration of structure and function of the central or peripheral nervous systems. Aberrant RNA metabolism is increasingly implicated in neurodegenerative diseases, a subset of which are caused by the expansion of short repetitive elements (microsatellites) within particular genes [5]. The causative repeat expansion mutation for this group of disorders is unstable because the repeat size changes through generations and even within an individual, as different tissues have cell populations with variable repeat length and in some cases the repeat length varies within the same tissue [6]. The severity of a repeat expansion disease is dependent on numerous variables, including the length of the repeat, its sequence context, and the native function of the protein-coding gene with which the repeat is associated. A typical pathogenic feature of these diseases is the accumulation of repeat-containing transcripts into aberrant RNA foci, which can sequester RBPs and prevent them from performing their normal functions [7–9]. Interestingly, once the repeat length cross a critical number, the repeat-containing RNAs can undergo phase separation—partitioning into granules due to multivalent base-pairing between repeat RNAs—or spontaneous gelation to form RNA foci, explaining why disease symptoms appear to be triggered after the expansions have reached a particular threshold number [10].

8.2 Toxicity of Coding and Noncoding Microsatellite Repeat Expansions

Over 25 human genes with tandem repeat expansions have been identified to date, and these disease-causing repeats can occur in the coding or noncoding regions [6] (Fig. 8.1 and Table 8.1). Majority of the microsatellites arise due to the expansion of trinucleotide repeats. However, expanded tetranucleotide, pentanucleotide, and hexanucleotide repeats are also detected. In the early 1990s, two microsatellites were discovered providing the first evidence that simple repeat expansions are linked to human disease. Fragile X Syndrome (FXS)—an X-linked disorder caused by CGG repeat expansions in the 5′ untranslated region (UTR) of the *FMR1* gene— is the most prevalent form of inherited cognitive impairment and mental retardation [11–16]. The repeat expansion in FXS causes loss of *FMR1* gene product FMRP, a polyribosome-associated RBP that binds ~4% of brain mRNAs and regulates their expression—either enhancing or suppressing translation through unknown mechanisms [17–20].

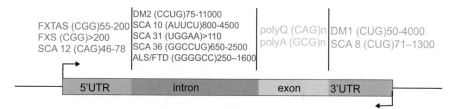

Fig. 8.1 Origin and expansion of microsatellite repeats in human disease. Schematic of the gene location for various disease-associated repeat expansions. Types of repeat expansions are indicated within the parentheses along with the range of expanded repeat numbers (UTR: untranslated region)

Table 8.1 Summary of the tissue-specific symptoms of the repeat expansion diseases with the disease-associated gene

Defected mRNA region	Disease	Defected gene	Tissue-specific clinical symptoms	
			Neuronal tissues	Other tissues
5'UTR	FXTAS	*FMR1* [158]	Ataxia [159], brain atrophy, white matter lesions [160, 161], cognitive decline, parkinsonism [160], peripheral neuropathy, autonomic dysfunction and short-term memory loss [162]	Premature ovarian failure, hypothyroidism in female [159], limb proximal muscle weakness [160]
	FXS	*FMR1* [14]	Autism [163], mental retardation, developmental delay and increased susceptibility to seizures [15]	Macroorchidism [15], cardiac murmur [164], hyperflexible joints, hernias, flat feet [165]
	SCA12	*PPP2R2B* [166]	Ataxia, cerebral and/or cerebellar atrophy [167], seizures [22]	Dysarthria, action tremors in upper limbs [167]
Intron	DM2	*ZNF9* [168]	Cognitive impairment [169], intellectual disability, sleepiness and fatigue [170], brain atrophy, white and grey matter abnormalities [171, 172]	Myotonia, muscle dysfunction, cardiac arrhythmia [40, 173], hypertrophy calf muscles [174]
	ALS	*C9ORF72* [28, 29]	Motor neuron degeneration, frontotemporal lobar dysfunction, dementia and cognitive impairment [175]	Progressive spasticity, muscle wasting, weakness and muscle atrophy [28]
	FTD	*C9ORF72* [28, 29]	Frontotemporal lobar dysfunction, motor neuron dysfunction [176], changes in personality, behavior, and language ability, dementia [175]	Fasciculation, muscle atrophy, weakness [177].

(continued)

Table 8.1 (continued)

Defected mRNA region	Disease	Defected gene	Tissue-specific clinical symptoms	
			Neuronal tissues	Other tissues
Coding region	Polyglutamine (PolyQ) diseases			
	SBMA	*AR* [21]	Lower motor neuron degeneration [178], androgen insensitivity [22]	Muscle weakness, gynecomastia and reduced fertility [22, 178]
	HD	*HTT* [179]	Cognitive decline and dementia [22], dystonia [180]	Chorea [22], movement disorder [181]
	DRPLA	*ATN1* [182–184]	White matter lesion, neural loss, ataxia, seizures, choreoathetosis, dementia [22, 185], myoclonus, epilepsy [184]	Chorea, incoordination [185]
	SCA 1, 2, 3, 6, 7, 17	*ATXN1* [186, 187], *ATXN2* [188–190], *ATXN3* [191], *CACNA1A* [192], *ATXN7* [193], *TBP* [194]	Ataxia, tremor, and dysarthria, parkinsonism (SCA3), retinal dystrophy (SCA7), seizures (SCA17) [22].	Slurred speech (SCA1); hyporeflexia (SCA2); cardiac dysfunction (SCA7) [22]
	Poly Alanine (Poly A) diseases			
	OPMD (OPMD)	*PABPN1* [195]	– (no data)	Eyelid ptosis and dysphagia [195], involuntary muscle weakness [196].
	XLMR	*ARX* [197]	Cognitive impairment [198], mental retardation [199], dysarthria [200]	Involuntary hand movements (MRXS), growth abnormality [200]
3′UTR	DM1	*DMPK* [37]	Neuropsychiatric disturbances, cognitive defeats, sleepiness and fatigue; brain atrophy [169], white and grey matter abnormalities [201], mood disorder, emotion problem and memory problem	Myotonia, muscle wasting, cardiac arrhythmias, insulin resistance, gastrointestinal dysfunctions, posterior iridescent cataracts [54]
	SCA8	*ATXN8* [202]	Cerebellar atrophy [203], progressive ataxia [204]	Limb ataxia, dysarthria, nystagmus, spasticity [22]

Abbreviations: *FXTAS* fragile X-associated tremor/ataxia syndrome, *FXS* fragile X Syndrome, *SCA12* spinocerebellar ataxia type 12, *DM2* myotonic dystrophy type 2, *ALS* amyotrophic lateral sclerosis, *FTD* frontotemporal degeneration, *SBMA* spinal and bulbar muscular atrophy, *HD* Huntington disease, *DRPLA* dentatorubral pallidoluysian atrophy, *SCA* spinocerebellar ataxias, *PolyA* polyalanine diseases, *OPMD* oculopharyngeal muscular dystrophy, *XLMR* syndromic and non-syndromic X-linked mental retardation, *DM1* myotonic dystrophy type 1

Spinal and bulbar muscular atrophy (SBMA)—the other microsatellite disease discovered along with FXS—arises due to a CAG repeat expansion in the coding region of the X chromosome-linked androgen receptor (*AR*) gene [21]. The discovery of SBMA was soon followed by the elucidation of a similar mutation as the basis for a group of disorders now known as the polyglutamine (polyQ) neurodegenerative diseases (Table 8.1). Along with SBMA, the polyQ diseases include Huntington disease (HD), dentatorubral-pallidoluysian atrophy, and six spinocerebellar ataxias (SCA) 1, 2, 3, 6, 7, and 17 [22]. As a group, these nine diseases are among the more common forms of inherited neurodegeneration. The translation of exons containing CAG repeats gives rise to elongated stretches of polyQs in mutant proteins, which aggregate into nuclear or cytoplasmic inclusions in the diseased brain [23–25]. Several observations indicate that the CAG repeat-containing RNAs, in the absence of coding for a protein, may also be a source of toxicity in polyQ diseases [26, 27]. GGGGCC hexanucleotide repeat expansion in the *C9ORF72* gene has gained much attention in the past few years and is now considered the most frequent inherited cause of Amyotrophic lateral sclerosis (ALS) and Frontotemporal dementia (FTD) [28, 29]. Pathology occurs due to the toxicity of expanded repeats, which are transcribed in both the sense and antisense directions and give rise to distinct sets of intracellular RNA and protein aggregates [30–33].

Myotonic Dystrophy (DM) is part of a group of diseases characterized by repeat expansions in noncoding regions of genes. DM is defined in two clinical and molecular forms: myotonic dystrophy type 1 (DM1), and type 2 (DM2), both of which are inherited in an autosomal dominant fashion. The combined worldwide incidence of DM is approximately 1 in 8000 [34, 35]. DM1 is the most prevalent form of adult onset muscular dystrophy [36] and is caused by a CTG repeat expansion in the 3′ UTR of *Dystrophia Myotonica Protein Kinase* (*DMPK*) gene [37, 38]. DM2, on the other hand, is caused by a CCTG repeat expansion in an intron of *Zinc Finger Protein 9* (*ZNF9*) gene [39]. While 5–37 repeats are considered normal, DM1 patients can have up to several thousand CTG repeats, which can reduce expression of DMPK [40] (Fig. 8.2a). *DMPK* is expressed in multiple tissues, and the major symptoms of the disease include muscle hyperexcitability (myotonia), progressive muscle wasting, cardiac defects, insulin resistance, and neuropsychiatric disturbances [41–44]. Table 8.1 provides further description of tissue-specific symptoms observed in DM and other microsatellite expansion disorders.

8.3 RNA Metabolism Defects in Myotonic Dystrophy

Closely after the discovery of repeats, the *DMPK* haploinsufficiency model was put forward to explain the DM1 pathology. However, the removal of *DMPK* gene in mice failed to recapitulate the major neuromuscular symptoms of DM1 [45, 46]. A separate hypothesis proposed that expanded CTG repeats might affect the expression of nearby genes. Although the adjacent gene, *SIX5*, exhibits reduced expression in DM1 patients [47], *Six5* knockout mice also do not reproduce DM1 muscle

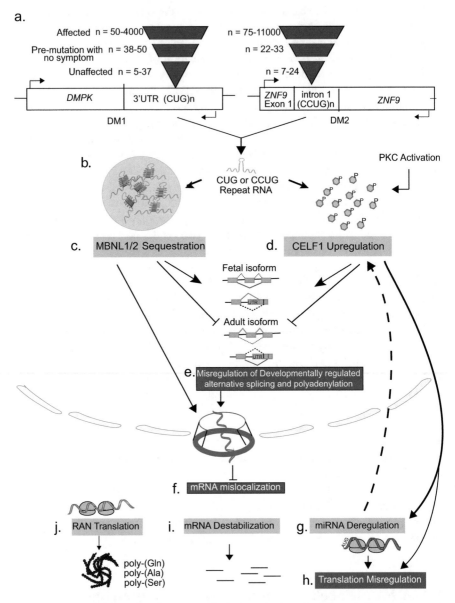

Fig. 8.2 Schematic showing different pathological mechanisms for Myotonic Dystrophy type 1 (DM1) and 2 (DM2). (a) Causative mutation for DM1 is CUG repeat expansion in 3′UTR of *DMPK* gene and for DM2 is CCUG repeat expansion in intron 1 of *ZNF9* gene. The severity of the disease is dependent on the number of repeats. Although these mutations are in two different genes, the disease mechanisms for both diseases are surprisingly similar. Most of the pathology is consistent with the toxic RNA gain-of-function mechanism and affects general RNA metabolism in both the nucleus and cytoplasm. (b) After transcription, the repeat-containing transcripts form stable hairpin loop comprising secondary structures (pink), which aggregate to form ribonuclear foci. (c) Members of the Mbnl family of RNA-binding proteins (RBPs) MBNL1/2 (purple) bind the CUG

pathology [48]. Instead, the CTG repeats alone, regardless of the gene context, are sufficient to induce pathogenic features of DM1 [49, 50]. The predominant pathology of DM1 actually stems from the toxic effects of expanded CUG RNA, which disrupts the normal activity of certain RBPs. Further support for the RNA toxicity model comes from the finding that although the repeat expansion in DM2 is on an entirely different gene, both diseases exhibit similar symptoms.

In both DM1 and DM2, the RNAs with expanded repeats (CUG in DM1; and CCUG in DM2) fold into stable hairpin loops that accumulate as ribonuclear foci in the nuclei of affected tissues [9] (Fig. 8.2b). These expanded RNA transcripts directly trap RBPs such as muscleblind-like proteins (MBNLs) and cause upregulation of CUG-binding protein 1 (CELF1) family of alternative splicing factors [51–54], which results in aberrant splicing of many transcripts and a broad, multi-systemic phenotype (Fig. 8.2c, d). Alternative pre-mRNA splicing generates much of the transcriptome diversity in higher eukaryotes as it enables the production of multiple transcripts with potentially different functions from each individual gene [55]. Alternative splicing decisions are generally influenced by *cis*-acting regulatory elements within pre-mRNAs that promote or inhibit exon recognition, as well as expression/activity of *trans*-acting factors (e.g., MBNL and CELF proteins) that bind to these *cis* elements and regulate the accessibility of the spliceosome to splice sites [56]. The misregulated splicing events in DM are usually developmentally regulated and exhibit an adult-to-embryonic switch in splicing patterns (Fig. 8.2e). Some of these embryonic isoforms fail to meet the adult tissue requirements and thus directly contribute to the overall disease pathology [54].

8.3.1   Misregulation of mRNA Processing

MBNL loss-of-function in DM1 and DM2 is a prominent example of RBP sequestration by disease-associated microsatellite expansion RNAs. The MBNL proteins were initially identified in *Drosophila melanogaster* for their requirement in muscle development and eye differentiation [57], and they were later shown as direct regulators of alternative splicing [58]. There are three MBNL paralogues in mammals, named MBNL1–3. MBNL1 and MBNL2 are widely expressed across many tissues,

---

Fig. 8.2 (continued) or CCUG repeats and are sequestered in the ribonuclear foci. (d) Hyperphosphorylation by PKC stabilizes another RBP, CELF1, resulting in its gain-of-function. (e) Both MBNL and CELF proteins regulate various aspects of RNA metabolism during normal development. Alterations in their functional levels due to toxic repeat RNA cause adult-to-fetal reversion of splicing and polyadenylation for many pre-mRNAs in the nucleus. (f) MBNL depletion also leads to cellular mislocalization of many mRNAs. CELF1 gain-of-function further affects (g) miRNA metabolism and (h) mRNA translation. (i) Dysregulation of MBNL and CELF activity in the cytoplasm also affects mRNA stability through various mechanisms. (j) Both sense and antisense CUG/CCUG-containing transcripts are subject to RAN translation in all three frames giving rise to homopolymeric polypeptides that accumulate in the cytoplasm and form pathological intracellular aggregates

including brain, heart, muscle, and liver, whereas MBNL3 expression is restricted to the placenta [59]. In a majority of tissues, *MBNL1* and *MBNL2* mRNA levels rise during differentiation [60, 61]. Besides their roles in pre-mRNA processing, MBNLs also influence gene expression by regulating cellular mRNA transport, stability as well as microRNA biogenesis [62–67]. The high expression of MBNL1 in the heart and skeletal muscle is consistent with the most severe DM phenotypes in these tissues. For instance, independent of the repeat expansion, *Mbnl1* deletion in mice reproduces many of the cardinal symptoms of DM1 such as myotonia, myopathy, cataracts, and misregulation of developmentally regulated RNA processing [63, 68].

The expanded repeat-containing RNAs in DM sequester MBNL1, 2, and 3 in nuclear RNA foci [69–71], and this protein redistribution explains the inhibition of their normal functions predominantly in alternative splicing and polyadenylation, microRNA processing, and mRNA localization [58, 62, 67, 72–75]. The MBNL loss-of-function hypothesis is further supported by studies on *Mbnl* single- and compound-knockout mice, which recapitulate many of the DM phenotypes [68, 76–78]. The extent of symptoms, however, varies depending on the tissue context, relative concentrations of MBNL paralogues, and the degree to which they are sequestered [78]. For instance, compared to skeletal muscle, only few splicing defects are observed in the brains of *Mbnl1* knockout mice [63, 79]. Alternatively, *Mbnl2* knockout mice exhibit a number of DM-related central nervous system abnormalities including irregular REM sleep propensity and deficits in spatial memory [76], which is consistent with the observation that MBNL2 expression in the brain is higher than MBNL1 [59]. MBNL2 is directly sequestered by repeat expansions in the brain tissue of human DM patients resulting in misregulation of alternative splicing and polyadenylation of its normal RNA targets [80]. One of the most misspliced mRNA due to loss of MBNL2 is human microtubule-associated protein tau (MAPT) in the DM1 frontal cortex [80]. RNA toxicity mediated through MBNL2 sequestration leads to abnormal expression of tau isoforms and the progressive appearance of neurofibrillary tangles composed of intraneuronal aggregates of hyper-phosphorylated tau protein [81].

More recently, MBNL proteins were found to serve essential roles in poly(A) site selection for many transcripts (Fig. 8.2e). By integrating HITS-CLIP and RNA-seq from MBNL knockout cells and transgenic DM1 mouse model, along with minigene reporter studies, Swanson and colleagues demonstrated that MBNL proteins directly suppress or activate polyadenylation for thousands of pre-mRNAs [75, 80]. Thus, MBNL proteins coordinate multiple pre-mRNA processing steps and their sequestration in DM depletes them from their normal RNA targets.

Besides MBNL loss-of-function, there is accumulation and aberrant sub-cellular distribution of another splicing factor CELF1 in DM. CELF proteins are normally downregulated during postnatal striated muscle development, which facilitates fetal-to-adult splicing transitions in hundreds of muscle transcripts [61, 82]. CELF1 actually does not colocalize with RNA foci [83], and its upregulation in DM1 occurs through two separate mechanisms. First, CELF1 protein is stabilized through its hyper-phosphorylation [84]; and second, reduced levels of microRNAs in DM1 derepress CELF1 protein translation [85, 86] (Fig. 8.2d, g, h). The situation is less clear in DM2, with conflicting reports of normal [87, 88] and increasing CELF1 protein levels [89] in patient tissues and cells. It is interesting to note that for many pre-mRNAs whose splicing is disrupted in DM1, CELF1 and MBNL1 regulate them in an antagonistic manner [58, 61, 90–92]. The antagonism, however, is not due to direct competition for the binding site as both CELF1 and MBNL1 bind and regulate splicing independently via distinct *cis*-acting RNA motifs.

In addition to MBNL and CELF proteins, other RNA splicing factors are implicated in DM. For instance, hnRNP H binds to *DMPK*-derived CUG-expanded RNAs in vitro and increased hnRNP H levels may also contribute toward DM pathogenesis [93]. hnRNP H forms a repressor complex with MBNL1 and nine other proteins (hnRNP H2, H3, F, A2/B1, K, L, DDX5, DDX17, and DHX9) in normal myoblast extracts but elevated hnRNP H levels in DM1 disrupt the stoichiometry of these complexes which affects splicing of specific pre-mRNAs [94, 95]. Since expanded CUG repeat RNAs fold into hairpin structures [96], the partial recruitment and colocalization of the RNA helicase p68/DDX5 with RNA foci may also have a contributing role in splicing dysregulation. Moreover, p68/DDX5 can modulate MBNL1-binding activity, and its colocalization with nuclear RNA foci can further stimulate MBNL1 binding to repeat RNAs [97].

8.3.2 Misregulation of mRNA Localization and Stability

Following transcription, newly synthesized and fully processed mRNAs are bound by specific RBPs to form export-competent mRNPs, which help their transport through the nuclear pore complex (NPC). Some pre-mRNAs are processed at the speckle periphery before being exported and repeat-containing nuclear foci can colocalize at the periphery of nuclear speckles, a non-membrane bound nuclear assembly of macromolecules including splicing factors. The presence of expanded CUG repeats may, therefore, prevent entry of other RNAs into the nuclear speckle [98, 99]. However, in DM2, the mutant *ZNF9* mRNA is exported normally as the expanded CCUG repeats are removed during splicing. The nuclear foci formed by DM2 intronic repeats are widely dispersed in the nucleoplasm and not associated with nuclear speckles. Also, it is not yet clear whether the DM1 and/or DM2 nuclear foci contain partially degraded fragments of CUG or CCUG repeats or larger intact RNAs respectively.

As discussed above, CELF1 upregulation and MBNL sequestration by the CUG repeats in DM1 cause misprocessing of hundreds of transcripts. Aberrant processing results in nucleocytoplasmic export defects for many of these transcripts. Furthermore, MBNL proteins are localized both in the nucleus and cytoplasm and several studies have demonstrated their direct roles in mRNA localization [62, 100] (Fig. 8.2f). For instance, by interacting with the 3'-UTR of Integrin α3, MBNL2 moves it to the plasma membrane for its local translation [64]. Similarly, MBNL1 also plays major roles in mRNA localization and membrane-associated translation. Transcriptome-wide analyses of subcellular compartments from mouse myoblasts showed widespread defects in mRNA localization upon combined depletion of MBNL1 and MBNL2 [62]. Many of the mislocalized mRNAs encode for secreted proteins, extracellular matrix components, and proteins involved in cell–cell communication. MBNL depletion in DM can thus have a significant impact on mRNA localization potentially affecting proper neuromuscular junction formation.

In the cytoplasm, MBNLs also regulate mRNA stability [101] (Fig. 8.2i). MBNL1 specifically recognizes YGCY-containing motifs within the 3′-UTR regions and destabilizes the target mRNAs through unknown mechanisms [65, 92]. CELF1, on the other hand, induces mRNA decay of short-lived transcripts through interactions with GU-rich elements (GREs) in their 3′-UTR and possibly recruitment of poly(A)-specific ribonuclease, which promotes deadenylation of target transcripts [102–104]. Many of the CELF mRNA targets with GREs encode proteins essential for muscle cell development and function [105–108]. Interestingly, CELF1 binds to the mRNAs coding for SRP protein subunits and promotes their decay [109]. Signal recognition particle (SRP) is a cytoplasmic ribonucleoprotein complex, which regulates the translation of secreted and membrane-associated proteins. It is likely that the CELF1 overexpression contributes to the faster turnover of SRP mRNAs and the reduced SRP levels thereby attenuate the protein secretory pathway in DM1 [109].

8.3.3 Misregulation of mRNA Translation

CELF1 is additionally involved in the regulation of mRNA translation [106, 110–112] (Fig. 8.2h). The affinity of CELF1 toward its mRNA targets can be modulated through phosphorylation [113]. For instance, phosphorylated CELF1 interacts with a subunit of initiation factor eIF2, leading to the recruitment of translational machinery to target mRNAs [106]. In myoblasts, AKT phosphorylates CELF1 and increases its affinity for *CCND1* mRNA. During myoblast-to-myotube differentiation, cyclinD3-cdk4/6 phosphorylates CELF1, which increases CELF1 interaction with 5′-UTR of *p21* mRNA (a cell cycle inhibitor) and enhances its translation. Myoblasts from DM1 patients show an increased interaction between CELF1 and AKT and have reduced cyclinD3-CDK4/6 levels during differentiation [105]. Moreover, DM1 myoblasts during differentiation show a reduced ability to withdraw from cell cycle, which may be due to the altered translation of P21 or myogenic transcription factor MEF2A by CELF1 [111, 112].

mRNA translation in DM1 is also affected due to microRNA deregulation (Fig. 8.2g). A subset of developmentally regulated microRNAs associated with cardiac arrhythmias is downregulated in the hearts of DM1 patients and mice [67, 86]. Downregulation of these microRNAs recapitulates particular gene expression deficits seen in DM1 hearts including enhanced protein levels of miR-1 targets CX43 and Cav1.2 as well as miR-23a/b target CELF1 [67, 86]. In DM1 and DM2 skeletal muscle biopsies, both the levels and cellular distribution of several evolutionarily conserved microRNAs are altered affecting their downstream targets [114–117]. Furthermore, specific microRNAs are differentially detected in peripheral blood plasma of DM1 patients, which inversely correlate with skeletal muscle strength and may serve as noninvasive biomarkers [118]. More recently, reduced expression of miR-200c/141 tumor suppressor family was shown to correlate with increased oncologic risk in women with DM1 especially for gynecologic, brain, and thyroid cancer [119].

Besides altering cellular translation through misregulation of RBPs and microR-NAs, the microsatellite expansions also promote unconventional translation of repeats in multiple reading frames producing homopolymeric peptides that aggregate in both the nucleus and the cytoplasm [120] (Fig. 8.2j). Designated as Repeat Associated Non-AUG Translation (RAN translation), it was first described for the expanded CAG and CTG repeats that cause spinocerebellar ataxia 8 (SCA8) and DM1, respectively [120]. Interestingly, the efficiency of RAN translation increases with the size of repeats and when RNA forms hairpin-like structures [121]. Additionally, the cells making the toxic RAN protein products are prone to apoptosis as detected in tissues of affected patients, indicating a potential contribution of RAN to pathogenesis. In addition to DM1, Zu et al. recently demonstrated that in DM2 the tetranucleotide expansion repeats are bidirectionally transcribed, and the resulting transcripts are RAN translated, producing tetrapeptide expansion proteins with Leu-Pro-Ala-Cys (LPAC) from the sense strand or Gln-Ala-Gly-Arg (QAGR) repeats from the antisense strand [122]. These RAN proteins were readily detected in the DM2 patient brains; however, the specific roles of these RAN proteins regarding toxicity, mechanism of action, and their regulation are yet to be determined.

Since their original discovery, RAN translation has now been observed in many other repeat-expansion diseases, including ALS/FTD, FXTAS, and Huntington's disease [52, 123]. However, the exact mechanisms initiating translation from these repeats likely differ across diverse sequence contexts [124]. For instance, in case of *FMR1*, expanded CGG repeats in the 5′-UTR initiate CAP-dependent RAN translation upstream of the canonical AUG start codon, producing FMRpolyGlycine and FMRpolyAlanine in FXTAS [123, 125]. In contrast to FXTAS, the expanded repeats in DM1 exist within the 3′ UTR of *DMPK* mRNA, which is not in the normal path of ribosome scanning; thus, unconventional ribosome interactions must contribute in their translation. For *HTT* in Huntington's disease, the CAG repeats are in the ORF, and canonical translation starts at the native AUG codon upstream of the repeats. But in some instances, HTTpolySerine and HTTpolyAlanine proteins are also produced due to RAN-translation and frame shifting from the normal HTTpolyGlutamine frame of the repeats [126]. Finally, in case of ALS/FTD, the GGGGCC repeats are within *C9ORF72* intron, and the RAN-translation generates polyGlycine-Alanine, polyGlycine-Arginine, and polyGlycine-Proline dipeptide products [31, 127]. The RAN translation in this case, however, may occur from the intron retained transcript, spliced lariat, or a 3′ truncated RNA generated due to stalled transcription [124, 128].

8.4 Disrupted Function of RBPs in Other Microsatellite Expansion Disorders

Recent paradigm-shifting advances have established that defective RNA processing through disrupted function of RBPs is central to many other repeat expansion diseases (Table 8.2). For instance, RBP defects occur in both familial and sporadic cases of ALS/FTD [129, 130]. Mutations in *TARDBP* and *FUS* genes respectively encoding TDP-43 and FUS/TLS proteins result in abnormal aggregation of these proteins in neurons and are considered pathogenic for ALS/FTD. TDP-43 and FUS/TLS are RNA/DNA-binding proteins, with noticeable structural and functional similarities.

TDP-43 functions in multiple RNA processing steps including pre-mRNA splicing [131–134], RNA stability [135–137], and transport [138]. Similar to TDP-43, FUS interacts with serine-arginine (SR) proteins that serve diverse roles in splicing [139] and regulates transcription by recruiting other RBPs through noncoding RNAs [140]. Hence, the association of TDP43 and FUS/TLS with ALS and FTD is

Table 8.2 Common postulated pathological mechanisms and associated RNA-Binding Proteins (RBPs) for disease-associated microsatellite repeat expansions

	Diseases name	RBPs	Pathological mechanism
(a)	FXTAS, FXS	FMRP [11–14] Pur α and hnRNP A2/B1 [205, 206], CELF1 [207], Sam68 [208]	mRNP transport and translation [209–213] Nuclear Foci and RBP Sequestration leads to changes in expression and cellular distribution of several proteins [214, 215], RNA Splicing [208, 216].
(b)	DM1/2	MBNL1/2/3 [49, 68, 72–74]; and CELF [39, 83, 84, 90, 91, 111, 217], HnRNP H [93], p68/DDX5 [97]	Nuclear Foci and RBP Sequestration, RNA splicing [58, 61, 62, 82, 218, 219] and polyadenylation misregulation [75, 80], miRNA biogenesis [67, 86, 115], Translation and cellular localization disruption [62, 99], Intracellular aggregation by non-canonical RAN translation [122]
(c)	ALS/ FTD	TDP-43 [220, 221] FUS [222, 223], TAF15 [141, 142], EWSR1 [143, 144] hnRNPA1 and hnRNPA2B1 [146], Ataxin 2 [145], TIA1 [224]	Nuclear foci [225], Splicing misregulation [132, 134, 226], translation, and RNA transport [227], impaired cytoplasmic localization [154, 228, 229], mutated LCD domain mediated cytoplasmic inclusions [146, 230–233]
(d)	SCA 8	MBNL/CELF [234], Staufen [235]	RNA Splicing [234], RAN Translation [120]

Abbreviations: *FXTAS* fragile X-associated tremor/ataxia syndrome, *FXS* fragile X syndrome, *DM1/2* myotonic dystrophy type 1/2, *ALS* amyotrophic lateral sclerosis, *FTD* frontotemporal degeneration, *SCA* spinocerebellar ataxias, *FMRP* fragile X mental retardation protein, *CELF* CUGBP Elav-like family member, *mbnl* Muscleblind like splicing regulator, *hnRNPs* heterogeneous nuclear ribonucleoprotein, *TAF15* TATA box-binding protein-associated factor 15, *EWSR1* Ewing sarcoma breakpoint region 1, *TIA1* T cell intracytoplasmic antigen

redirecting research efforts toward identifying additional RBPs that are mutated in neurological diseases, defining their normal RNA substrates and determining the misprocessed RNAs that underlie particular disease symptoms. In fact, mutations in several other RBPs that are functionally and structurally similar to FUS/TLS such as TAF15 [141, 142] and EWSR1 [143, 144], as well as the less closely related RBPs—Ataxin 2 [145], hnRNPA2B1 [146], hnRNPA1 [146], and Matrin3 [147] were recently identified. Among these RBPs, TDP-43, FUS, and hnRNPA1 harbor low complexity domains (LCDs), which can polymerize and drive phase separation to form dynamic membrane-less organelles or liquid droplets. For instance, a 57-residue segment within the FUS-LCD was recently shown to assemble into a fibril core that promotes phase-separation and hydrogel formation. Interestingly, phosphorylation of the core-forming residues by DNA-dependent protein kinase dissolves the FUS-LCD liquid droplets providing a molecular basis for the dynamics of LCD polymerization and phase separation [148].

Disease-associated mutations within LCDs of RBPs also enhance prion-like properties and accelerate the shift from liquid to solid phase disturbing proper ribonucleoprotein (RNP) formation [127, 149, 150]. These mutations likely trigger protein aggregation due to aberrant self-assembly of LCDs. The cytoplasmic aggregation of RBPs not only affects their typical functions in RNA metabolism but also diminishes general nucleocytoplasmic trafficking, a common consequence of ALS-initiating mutations [151–153]. While the exact reasons impeding nuclear/cytoplasmic transport in ALS are not yet fully established, multiple independent mechanisms have been proposed. For example, nucleocytoplasmic trafficking defects can arise due to proteotoxicity caused by cytoplasmic β-sheet containing protein aggregations [154], direct interactions between repeat RNAs and nuclear import factors [153], or inhibition by RAN translation-products of repeat RNAs [151]. Interestingly, arginine-containing dipeptide repeats produced from RAN translation of hexanucleotide GGGGCC expansions in ALS interact with LCDs of RBPs, which disrupts the dynamics and functions of membrane-less organelle formation by LCDs [155, 156]. Furthermore, subsets of these arginine-containing dipeptides frequently bind to the LCDs encoded by the nuclear pore proteins blocking the transport of macromolecules into and out of the nucleus [157]. Thus, interaction of RAN translation products with LCDs is a yet another pathogenic mechanism that interferes with the normal function of RBPs in microsatellite expansion disorders.

8.5 Conclusions

The past decade has seen remarkable progress in our understanding of the molecular pathogenesis of microsatellite repeat expansion disorders. Although the repeats may vary in terms of their length and location within a gene or the multiple ways through which they cause disease, one commonality of microsatellite expansions is the production of toxic RNA species containing repeats. Mechanistically, the pathology arises either due to loss-of-function of the affected gene, or gain-of-function of the repeat-containing RNAs. Regarding loss-of-function, the repeats can induce transcriptional silencing of the affected gene through epigenetic modifications or produce a non-functional protein that contains a long stretch of homopolymeric amino acids. In case of gain-of-function, the RNAs with expanded repeats often sequester RBPs and thus disrupt their normal activities. Alternatively, the translated protein with a repetitive stretch of homopolymeric peptide sequence can misfold, aggregate, and trap critical cellular proteins causing nucleo-cytoplasmic export defects and further proteotoxicity. For a number of repeat expansion disorders, there is an intricate overlap of such loss- and gain-of-function mechanisms resulting in complex molecular pathologies. We envision that for many repeat expansions, the future investigations will be geared toward determining the unique versus overlapping disease mechanisms, dissecting direct versus indirect RNA metabolism defects, and finally, understanding whether alterations in RNA metabolism occur early or during late stages of the disease.

Acknowledgments A.K. is supported by grants from the US National Institute of Health (R01HL126845), Muscular Dystrophy Association (MDA514335), and the Center for Advanced Study at the University of Illinois. C.M. is supported by the American Heart Association postdoctoral fellowship (16POST29950018).

References

1. Arif W, Datar G, Kalsotra A. Intersections of post-transcriptional gene regulatory mechanisms with intermediary metabolism. Biochim Biophys Acta. 2017;1860(3):349–62.
2. Lewis CJ, Pan T, Kalsotra A. RNA modifications and structures cooperate to guide RNA-protein interactions. Nat Rev Mol Cell Biol. 2017;18(3):202–10.
3. Scotti MM, Swanson MS. RNA mis-splicing in disease. Nat Rev Genet. 2016;17(1):19–32.
4. Brinegar AE, Cooper TA. Roles for RNA-binding proteins in development and disease. Brain Res. 2016;1647:1–8.
5. Mirkin SM. Expandable DNA repeats and human disease. Nature. 2007;447(7147):932–40.
6. La Spada AR, Taylor JP. Repeat expansion disease: progress and puzzles in disease pathogenesis. Nat Rev Genet. 2010;11(4):247–58.
7. Renoux AJ, Todd PK. Neurodegeneration the RNA way. Prog Neurobiol. 2012;97(2):173–89.
8. O'Rourke JR, Swanson MS. Mechanisms of RNA-mediated disease. J Biol Chem. 2009;284(12):7419–23.
9. Zhang N, Ashizawa T. RNA toxicity and foci formation in microsatellite expansion diseases. Curr Opin Genet Dev. 2017;44:17–29.
10. Jain A, Vale RD. RNA phase transitions in repeat expansion disorders. Nature. 2017;546(7657):243–7.
11. Heitz D, Rousseau F, Devys D, Saccone S, Abderrahim H, Le Paslier D, Cohen D, Vincent A, Toniolo D, Della Valle G, et al. Isolation of sequences that span the fragile X and identification of a fragile X-related CpG island. Science. 1991;251(4998):1236–9.
12. Kremer EJ, Pritchard M, Lynch M, Yu S, Holman K, Baker E, Warren ST, Schlessinger D, Sutherland GR, Richards RI. Mapping of DNA instability at the fragile X to a trinucleotide repeat sequence p(CCG)n. Science. 1991;252(5013):1711–4.

13. Oberle I, Rousseau F, Heitz D, Kretz C, Devys D, Hanauer A, Boue J, Bertheas MF, Mandel JL. Instability of a 550-base pair DNA segment and abnormal methylation in fragile X syndrome. Science. 1991;252(5009):1097–102.

14. Verkerk AJ, Pieretti M, Sutcliffe JS, Fu YH, Kuhl DP, Pizzuti A, Reiner O, Richards S, Victoria MF, Zhang FP, et al. Identification of a gene (FMR-1) containing a CGG repeat coincident with a breakpoint cluster region exhibiting length variation in fragile X syndrome. Cell. 1991;65(5):905–14.

15. Bhakar AL, Dolen G, Bear MF. The pathophysiology of fragile X (and what it teaches us about synapses). Annu Rev Neurosci. 2012;35:417–43.

16. Santoro MR, Bray SM, Warren ST. Molecular mechanisms of fragile X syndrome: a twenty-year perspective. Annu Rev Pathol. 2012;7:219–45.

17. Ashley CT Jr, Wilkinson KD, Reines D, Warren ST. FMR1 protein: conserved RNP family domains and selective RNA binding. Science. 1993;262(5133):563–6.

18. Brown V, Small K, Lakkis L, Feng Y, Gunter C, Wilkinson KD, Warren ST. Purified recombinant Fmrp exhibits selective RNA binding as an intrinsic property of the fragile X mental retardation protein. J Biol Chem. 1998;273(25):15521–7.

19. Siomi H, Choi M, Siomi MC, Nussbaum RL, Dreyfuss G. Essential role for KH domains in RNA binding: impaired RNA binding by a mutation in the KH domain of FMR1 that causes fragile X syndrome. Cell. 1994;77(1):33–9.

20. Siomi H, Siomi MC, Nussbaum RL, Dreyfuss G. The protein product of the fragile X gene, FMR1, has characteristics of an RNA-binding protein. Cell. 1993;74(2):291–8.

21. La Spada AR, Wilson EM, Lubahn DB, Harding AE, Fischbeck KH. Androgen receptor gene mutations in X-linked spinal and bulbar muscular atrophy. Nature. 1991;352(6330):77–9.

22. Orr HT, Zoghbi HY. Trinucleotide repeat disorders. Annu Rev Neurosci. 2007;30:575–621.

23. Davies SW, Turmaine M, Cozens BA, DiFiglia M, Sharp AH, Ross CA, Scherzinger E, Wanker EE, Mangiarini L, Bates GP. Formation of neuronal intranuclear inclusions underlies the neurological dysfunction in mice transgenic for the HD mutation. Cell. 1997;90(3):537–48.

24. DiFiglia M, Sapp E, Chase KO, Davies SW, Bates GP, Vonsattel JP, Aronin N. Aggregation of huntingtin in neuronal intranuclear inclusions and dystrophic neurites in brain. Science. 1997;277(5334):1990–3.

25. Paulson HL, Perez MK, Trottier Y, Trojanowski JQ, Subramony SH, Das SS, Vig P, Mandel JL, Fischbeck KH, Pittman RN. Intranuclear inclusions of expanded polyglutamine protein in spinocerebellar ataxia type 3. Neuron. 1997;19(2):333–44.

26. Li LB, Bonini NM. Roles of trinucleotide-repeat RNA in neurological disease and degeneration. Trends Neurosci. 2010;33(6):292–8.

27. Gatchel JR, Zoghbi HY. Diseases of unstable repeat expansion: mechanisms and common principles. Nat Rev Genet. 2005;6(10):743–55.

28. DeJesus-Hernandez M, Mackenzie IR, Boeve BF, Boxer AL, Baker M, Rutherford NJ, Nicholson AM, Finch NA, Flynn H, Adamson J, et al. Expanded GGGGCC hexanucleotide repeat in noncoding region of C9ORF72 causes chromosome 9p-linked FTD and ALS. Neuron. 2011;72(2):245–56.

29. Renton AE, Majounie E, Waite A, Simon-Sanchez J, Rollinson S, Gibbs JR, Schymick JC, Laaksovirta H, van Swieten JC, Myllykangas L, et al. A hexanucleotide repeat expansion in C9ORF72 is the cause of chromosome 9p21-linked ALS-FTD. Neuron. 2011;72(2):257–68.

30. Ash PE, Bieniek KF, Gendron TF, Caulfield T, Lin WL, Dejesus-Hernandez M, van Blitterswijk MM, Jansen-West K, Paul JW III, Rademakers R, et al. Unconventional translation of C9ORF72 GGGGCC expansion generates insoluble polypeptides specific to c9FTD/ALS. Neuron. 2013;77(4):639–46.

31. Mori K, Weng SM, Arzberger T, May S, Rentzsch K, Kremmer E, Schmid B, Kretzschmar HA, Cruts M, Van Broeckhoven C, et al. The C9orf72 GGGGCC repeat is translated into aggregating dipeptide-repeat proteins in FTLD/ALS. Science. 2013;339(6125):1335–8.

32. Xu Z, Poidevin M, Li X, Li Y, Shu L, Nelson DL, Li H, Hales CM, Gearing M, Wingo TS, et al. Expanded GGGGCC repeat RNA associated with amyotrophic lateral scle-

rosis and frontotemporal dementia causes neurodegeneration. Proc Natl Acad Sci U S A. 2013;110(19):7778–83.

33. Zu T, Liu Y, Banez-Coronel M, Reid T, Pletnikova O, Lewis J, Miller TM, Harms MB, Falchook AE, Subramony SH, et al. RAN proteins and RNA foci from antisense transcripts in C9ORF72 ALS and frontotemporal dementia. Proc Natl Acad Sci U S A. 2013;110(51):E4968–77.

34. Faustino NA, Cooper TA. Pre-mRNA splicing and human disease. Genes Dev. 2003;17(4):419–37.

35. Wheeler TM. Myotonic dystrophy: therapeutic strategies for the future. Neurotherapeutics. 2008;5(4):592–600.

36. Harper P. Myotonic dystrophy. London: W.B. Saunders; 2001.

37. Brook JD, McCurrach ME, Harley HG, Buckler AJ, Church D, Aburatani H, Hunter K, Stanton VP, Thirion JP, Hudson T, et al. Molecular basis of myotonic dystrophy: expansion of a trinucleotide (CTG) repeat at the 3′ end of a transcript encoding a protein kinase family member. Cell. 1992;68(4):799–808.

38. Mahadevan M, Tsilfidis C, Sabourin L, Shutler G, Amemiya C, Jansen G, Neville C, Narang M, Barcelo J, O'Hoy K, et al. Myotonic dystrophy mutation: an unstable CTG repeat in the 3′ untranslated region of the gene. Science. 1992;255(5049):1253–5.

39. Liquori CL, Ricker K, Moseley ML, Jacobsen JF, Kress W, Naylor SL, Day JW, Ranum LP. Myotonic dystrophy type 2 caused by a CCTG expansion in intron 1 of ZNF9. Science. 2001;293(5531):864–7.

40. Yum K, Wang ET, Kalsotra A. Myotonic dystrophy: disease repeat range, penetrance, age of onset, and relationship between repeat size and phenotypes. Curr Opin Genet Dev. 2017;44:30–7.

41. Groh WJ, Groh MR, Saha C, Kincaid JC, Simmons Z, Ciafaloni E, Pourmand R, Otten RF, Bhakta D, Nair GV, et al. Electrocardiographic abnormalities and sudden death in myotonic dystrophy type 1. N Engl J Med. 2008;358(25):2688–97.

42. Heatwole C, Bode R, Johnson N, Quinn C, Martens W, McDermott MP, Rothrock N, Thornton C, Vickrey B, Victorson D, et al. Patient-reported impact of symptoms in myotonic dystrophy type 1 (PRISM-1). Neurology. 2012;79(4):348–57.

43. Phillips MF, Harper PS. Cardiac disease in myotonic dystrophy. Cardiovasc Res. 1997;33(1):13–22.

44. Salehi LB, Bonifazi E, Stasio ED, Gennarelli M, Botta A, Vallo L, Iraci R, Massa R, Antonini G, Angelini C, et al. Risk prediction for clinical phenotype in myotonic dystrophy type 1: data from 2,650 patients. Genet Test. 2007;11(1):84–90.

45. Jansen G, Groenen PJ, Bachner D, Jap PH, Coerwinkel M, Oerlemans F, van den Broek W, Gohlsch B, Pette D, Plomp JJ, et al. Abnormal myotonic dystrophy protein kinase levels produce only mild myopathy in mice. Nat Genet. 1996;13(3):316–24.

46. Berul CI, Maguire CT, Aronovitz MJ, Greenwood J, Miller C, Gehrmann J, Housman D, Mendelsohn ME, Reddy S. DMPK dosage alterations result in atrioventricular conduction abnormalities in a mouse myotonic dystrophy model. J Clin Invest. 1999;103(4):R1–7.

47. Thornton CA, Wymer JP, Simmons Z, McClain C, Moxley RT III. Expansion of the myotonic dystrophy CTG repeat reduces expression of the flanking DMAHP gene. Nat Genet. 1997;16(4):407–9.

48. Klesert TR, Cho DH, Clark JI, Maylie J, Adelman J, Snider L, Yuen EC, Soriano P, Tapscott SJ. Mice deficient in Six5 develop cataracts: implications for myotonic dystrophy. Nat Genet. 2000;25(1):105–9.

49. Mankodi A, Logigian E, Callahan L, McClain C, White R, Henderson D, Krym M, Thornton CA. Myotonic dystrophy in transgenic mice expressing an expanded CUG repeat. Science. 2000;289(5485):1769–73.

50. Gomes-Pereira M, Cooper TA, Gourdon G. Myotonic dystrophy mouse models: towards rational therapy development. Trends Mol Med. 2011;17(9):506–17.

51. Echeverria GV, Cooper TA. RNA-binding proteins in microsatellite expansion disorders: mediators of RNA toxicity. Brain Res. 2012;1462:100–11.

52. Cleary JD, Ranum LP. Repeat associated non-ATG (RAN) translation: new starts in micro-satellite expansion disorders. Curr Opin Genet Dev. 2014;26:6–15.
53. Mohan A, Goodwin M, Swanson MS. RNA-protein interactions in unstable microsatellite diseases. Brain Res. 2014;1584:3–14.
54. Chau A, Kalsotra A. Developmental insights into the pathology of and therapeutic strategies for DM1: back to the basics. Dev Dyn. 2015;244(3):377–90.
55. Nilsen TW, Graveley BR. Expansion of the eukaryotic proteome by alternative splicing. Nature. 2010;463(7280):457–63.
56. Kalsotra A, Cooper TA. Functional consequences of developmentally regulated alternative splicing. Nat Rev Genet. 2011;12(10):715–29.
57. Begemann G, Paricio N, Artero R, Kiss I, Perez-Alonso M, Mlodzik M. muscleblind, a gene required for photoreceptor differentiation in Drosophila, encodes novel nuclear Cys3His-type zinc-finger-containing proteins. Development. 1997;124(21):4321–31.
58. Ho TH, Charlet BN, Poulos MG, Singh G, Swanson MS, Cooper TA. Muscleblind proteins regulate alternative splicing. EMBO J. 2004;23(15):3103–12.
59. Kanadia RN, Urbinati CR, Crusselle VJ, Luo D, Lee YJ, Harrison JK, Oh SP, Swanson MS. Developmental expression of mouse muscleblind genes Mbnl1, Mbnl2 and Mbnl3. Gene Expr Patterns. 2003;3(4):459–62.
60. Konieczny P, Stepniak-Konieczna E, Sobczak K. MBNL proteins and their target RNAs, interaction and splicing regulation. Nucleic Acids Res. 2014;42(17):10873–87.
61. Kalsotra A, Xiao X, Ward AJ, Castle JC, Johnson JM, Burge CB, Cooper TA. A postnatal switch of CELF and MBNL proteins reprograms alternative splicing in the developing heart. Proc Natl Acad Sci U S A. 2008;105(51):20333–8.
62. Wang ET, Cody NA, Jog S, Biancolella M, Wang TT, Treacy DJ, Luo S, Schroth GP, Housman DE, Reddy S, et al. Transcriptome-wide regulation of pre-mRNA splicing and mRNA local-ization by muscleblind proteins. Cell. 2012;150(4):710–24.
63. Du H, Cline MS, Osborne RJ, Tuttle DL, Clark TA, Donohue JP, Hall MP, Shiue L, Swanson MS, Thornton CA, et al. Aberrant alternative splicing and extracellular matrix gene expres-sion in mouse models of myotonic dystrophy. Nat Struct Mol Biol. 2010;17(2):187–93.
64. Adereth Y, Dammai V, Kose N, Li R, Hsu T. RNA-dependent integrin alpha3 protein localiza-tion regulated by the Muscleblind-like protein MLP1. Nat Cell Biol. 2005;7(12):1240–7.
65. Masuda A, Andersen HS, Doktor TK, Okamoto T, Ito M, Andresen BS, Ohno K. CUGBP1 and MBNL1 preferentially bind to 3′ UTRs and facilitate mRNA decay. Sci Rep. 2012;2:209.
66. Osborne RJ, Lin X, Welle S, Sobczak K, O'Rourke JR, Swanson MS, Thornton CA. Transcriptional and post-transcriptional impact of toxic RNA in myotonic dystrophy. Hum Mol Genet. 2009;18(8):1471–81.
67. Rau F, Freyermuth F, Fugier C, Villemin JP, Fischer MC, Jost B, Dembele D, Gourdon G, Nicole A, Duboc D, et al. Misregulation of miR-1 processing is associated with heart defects in myotonic dystrophy. Nat Struct Mol Biol. 2011;18(7):840–5.
68. Kanadia RN, Johnstone KA, Mankodi A, Lungu C, Thornton CA, Esson D, Timmers AM, Hauswirth WW, Swanson MS. A muscleblind knockout model for myotonic dystrophy. Science. 2003;302(5652):1978–80.
69. Kino Y, Mori D, Oma Y, Takeshita Y, Sasagawa N, Ishiura S. Muscleblind protein, MBNL1/EXP, binds specifically to CHHG repeats. Hum Mol Genet. 2004;13(5):495–507.
70. Warf MB, Berglund JA. MBNL binds similar RNA structures in the CUG repeats of myo-tonic dystrophy and its pre-mRNA substrate cardiac troponin T. RNA. 2007;13(12):2238–51.
71. Yuan Y, Compton SA, Sobczak K, Stenberg MG, Thornton CA, Griffith JD, Swanson MS. Muscleblind-like 1 interacts with RNA hairpins in splicing target and pathogenic RNAs. Nucleic Acids Res. 2007;35(16):5474–86.
72. Fardaei M, Rogers MT, Thorpe HM, Larkin K, Hamshere MG, Harper PS, Brook JD. Three proteins, MBNL, MBLL and MBXL, co-localize in vivo with nuclear foci of expanded-repeat transcripts in DM1 and DM2 cells. Hum Mol Genet. 2002;11(7):805–14.

73. Jiang H, Mankodi A, Swanson MS, Moxley RT, Thornton CA. Myotonic dystrophy type 1 is associated with nuclear foci of mutant RNA, sequestration of muscleblind proteins and deregulated alternative splicing in neurons. Hum Mol Genet. 2004;13(24):3079–88.

74. Miller JW, Urbinati CR, Teng-Umnuay P, Stenberg MG, Byrne BJ, Thornton CA, Swanson MS. Recruitment of human muscleblind proteins to (CUG)(n) expansions associated with myotonic dystrophy. EMBO J. 2000;19(17):4439–48.

75. Batra R, Charizanis K, Manchanda M, Mohan A, Li M, Finn DJ, Goodwin M, Zhang C, Sobczak K, Thornton CA, et al. Loss of MBNL leads to disruption of developmentally regulated alternative polyadenylation in RNA-mediated disease. Mol Cell. 2014;56(2):311–22.

76. Charizanis K, Lee KY, Batra R, Goodwin M, Zhang C, Yuan Y, Shiue L, Cline M, Scotti MM, Xia G, et al. Muscleblind-like 2-mediated alternative splicing in the developing brain and dysregulation in myotonic dystrophy. Neuron. 2012;75(3):437–50.

77. Poulos MG, Batra R, Li M, Yuan Y, Zhang C, Darnell RB, Swanson MS. Progressive impairment of muscle regeneration in muscleblind-like 3 isoform knockout mice. Hum Mol Genet. 2013;22(17):3547–58.

78. Lee KY, Li M, Manchanda M, Batra R, Charizanis K, Mohan A, Warren SA, Chamberlain CM, Finn D, Hong H, et al. Compound loss of muscleblind-like function in myotonic dystrophy. EMBO Mol Med. 2013;5(12):1887–900.

79. Suenaga K, Lee KY, Nakamori M, Tatsumi Y, Takahashi MP, Fujimura H, Jinnai K, Yoshikawa H, Du H, Ares M Jr, et al. Muscleblind-like 1 knockout mice reveal novel splicing defects in the myotonic dystrophy brain. PLoS One. 2012;7(3):e33218.

80. Goodwin M, Mohan A, Batra R, Lee KY, Charizanis K, Fernandez Gomez FJ, Eddarkaoui S, Sergeant N, Buee L, Kimura T, et al. MBNL sequestration by toxic RNAs and RNA misprocessing in the myotonic dystrophy brain. Cell Rep. 2015;12(7):1159–68.

81. Sergeant N, Sablonniere B, Schraen-Maschke S, Ghestem A, Maurage CA, Wattez A, Vermersch P, Delacourte A. Dysregulation of human brain microtubule-associated tau mRNA maturation in myotonic dystrophy type 1. Hum Mol Genet. 2001;10(19):2143–55.

82. Ladd AN, Charlet N, Cooper TA. The CELF family of RNA binding proteins is implicated in cell-specific and developmentally regulated alternative splicing. Mol Cell Biol. 2001;21(4):1285–96.

83. Fardaei M, Larkin K, Brook JD, Hamshere MG. In vivo co-localisation of MBNL protein with DMPK expanded-repeat transcripts. Nucleic Acids Res. 2001;29(13):2766–71.

84. Kuyumcu-Martinez NM, Wang GS, Cooper TA. Increased steady-state levels of CUGBP1 in myotonic dystrophy 1 are due to PKC-mediated hyperphosphorylation. Mol Cell. 2007;28(1):68–78.

85. Kalsotra A, Wang K, Li PF, Cooper TA. MicroRNAs coordinate an alternative splicing network during mouse postnatal heart development. Genes Dev. 2010;24(7):653–8.

86. Kalsotra A, Singh RK, Gurha P, Ward AJ, Creighton CJ, Cooper TA. The Mef2 transcription network is disrupted in myotonic dystrophy heart tissue, dramatically altering miRNA and mRNA expression. Cell Rep. 2014;6(2):336–45.

87. Lin X, Miller JW, Mankodi A, Kanadia RN, Yuan Y, Moxley RT, Swanson MS, Thornton CA. Failure of MBNL1-dependent post-natal splicing transitions in myotonic dystrophy. Hum Mol Genet. 2006;15(13):2087–97.

88. Pelletier R, Hamel F, Beaulieu D, Patry L, Haineault C, Tarnopolsky M, Schoser B, Puymirat J. Absence of a differentiation defect in muscle satellite cells from DM2 patients. Neurobiol Dis. 2009;36(1):181–90.

89. Salisbury E, Schoser B, Schneider-Gold C, Wang GL, Huichalaf C, Jin B, Sirito M, Sarkar P, Krahe R, Timchenko NA, et al. Expression of RNA CCUG repeats dysregulates translation and degradation of proteins in myotonic dystrophy 2 patients. Am J Pathol. 2009;175(2):748–62.

90. Savkur RS, Philips AV, Cooper TA. Aberrant regulation of insulin receptor alternative splicing is associated with insulin resistance in myotonic dystrophy. Nat Genet. 2001;29(1):40–7.

91. Charlet BN, Savkur RS, Singh G, Philips AV, Grice EA, Cooper TA. Loss of the muscle-specific chloride channel in type 1 myotonic dystrophy due to misregulated alternative splicing. Mol Cell. 2002;10(1):45–53.

92. Wang ET, Ward AJ, Cherone JM, Giudice J, Wang TT, Treacy DJ, Lambert NJ, Freese P, Saxena T, Cooper TA, et al. Antagonistic regulation of mRNA expression and splicing by CELF and MBNL proteins. Genome Res. 2015;25(6):858–71.
93. Kim DH, Langlois MA, Lee KB, Riggs AD, Puymirat J, Rossi JJ. HnRNP H inhibits nuclear export of mRNA containing expanded CUG repeats and a distal branch point sequence. Nucleic Acids Res. 2005;33(12):3866–74.
94. Paul S, Dansithong W, Kim D, Rossi J, Webster NJ, Comai L, Reddy S. Interaction of muscleblind, CUG-BP1 and hnRNP H proteins in DM1-associated aberrant IR splicing. EMBO J. 2006;25(18):4271–83.
95. Paul S, Dansithong W, Jog SP, Holt I, Mittal S, Brook JD, Morris GE, Comai L, Reddy S. Expanded CUG repeats dysregulate RNA splicing by altering the stoichiometry of the muscleblind 1 complex. J Biol Chem. 2011;286(44):38427–38.
96. Krzyzosiak WJ, Sobczak K, Wojciechowska M, Fiszer A, Mykowska A, Kozlowski P. Triplet repeat RNA structure and its role as pathogenic agent and therapeutic target. Nucleic Acids Res. 2012;40(1):11–26.
97. Laurent FX, Sureau A, Klein AF, Trouslard F, Gasnier E, Furling D, Marie J. New function for the RNA helicase p68/DDX5 as a modifier of MBNL1 activity on expanded CUG repeats. Nucleic Acids Res. 2012;40(7):3159–71.
98. Holt I, Mittal S, Furling D, Butler-Browne GS, Brook JD, Morris GE. Defective mRNA in myotonic dystrophy accumulates at the periphery of nuclear splicing speckles. Genes Cells. 2007;12(9):1035–48.
99. Smith KP, Byron M, Johnson C, Xing Y, Lawrence JB. Defining early steps in mRNA transport: mutant mRNA in myotonic dystrophy type I is blocked at entry into SC-35 domains. J Cell Biol. 2007;178(6):951–64.
100. Taliaferro JM, Vidaki M, Oliveira R, Olson S, Zhan L, Saxena T, Wang ET, Graveley BR, Gertler FB, Swanson MS, et al. Distal alternative last exons localize mRNAs to neural projections. Mol Cell. 2016;61(6):821–33.
101. Wang ET, Taliaferro JM, Lee JA, Sudhakaran IP, Rossoll W, Gross C, Moss KR, Bassell GJ. Dysregulation of mRNA localization and translation in genetic disease. J Neurosci. 2016;36(45):11418–26.
102. Vlasova IA, Tahoe NM, Fan D, Larsson O, Rattenbacher B, Sternjohn JR, Vasdewani J, Karypis G, Reilly CS, Bitterman PB, et al. Conserved GU-rich elements mediate mRNA decay by binding to CUG-binding protein 1. Mol Cell. 2008;29(2):263–70.
103. Lee JE, Lee JY, Wilusz J, Tian B, Wilusz CJ. Systematic analysis of cis-elements in unstable mRNAs demonstrates that CUGBP1 is a key regulator of mRNA decay in muscle cells. PLoS One. 2010;5(6):e11201.
104. Moraes KC, Wilusz CJ, Wilusz J. CUG-BP binds to RNA substrates and recruits PARN deadenylase. RNA. 2006;12(6):1084–91.
105. Timchenko L. Molecular mechanisms of muscle atrophy in myotonic dystrophies. Int J Biochem Cell Biol. 2013;45(10):2280–7.
106. Dasgupta T, Ladd AN. The importance of CELF control: molecular and biological roles of the CUG-BP, Elav-like family of RNA-binding proteins. Wiley Interdiscip Rev RNA. 2012;3(1):104–21.
107. Rattenbacher B, Beisang D, Wiesner DL, Jeschke JC, von Hohenberg M, St Louis-Vlasova IA, Bohjanen PR. Analysis of CUGBP1 targets identifies GU-repeat sequences that mediate rapid mRNA decay. Mol Cell Biol. 2010;30(16):3970–80.
108. Zhang L, Lee JE, Wilusz J, Wilusz CJ. The RNA-binding protein CUGBP1 regulates stability of tumor necrosis factor mRNA in muscle cells: implications for myotonic dystrophy. J Biol Chem. 2008;283(33):22457–63.
109. Russo J, Lee JE, Lopez CM, Anderson J, Nguyen TP, Heck AM, Wilusz J, Wilusz CJ. The CELF1 RNA-binding protein regulates decay of signal recognition particle mRNAs and limits its secretion in mouse myoblasts. PLoS One. 2017;12(1):e0170680.
110. Vlasova-St Louis I, Dickson AM, Bohjanen PR, Wilusz CJ. CELFish ways to modulate mRNA decay. Biochim Biophys Acta. 2013;1829(6–7):695–707.

111. Timchenko NA, Iakova P, Cai ZJ, Smith JR, Timchenko LT. Molecular basis for impaired muscle differentiation in myotonic dystrophy. Mol Cell Biol. 2001;21(20):6927–38.
112. Timchenko NA, Patel R, Iakova P, Cai ZJ, Quan L, Timchenko LT. Overexpression of CUG triplet repeat-binding protein, CUGBP1, in mice inhibits myogenesis. J Biol Chem. 2004;279(13):13129–39.
113. Salisbury E, Sakai K, Schoser B, Huichalaf C, Schneider-Gold C, Nguyen H, Wang GL, Albrecht JH, Timchenko LT. Ectopic expression of cyclin D3 corrects differentiation of DM1 myoblasts through activation of RNA CUG-binding protein, CUGBP1. Exp Cell Res. 2008;314(11–12):2266–78.
114. Gambardella S, Rinaldi F, Lepore SM, Viola A, Loro E, Angelini C, Vergani L, Novelli G, Botta A. Overexpression of microRNA-206 in the skeletal muscle from myotonic dystrophy type 1 patients. J Transl Med. 2010;8:48.
115. Perbellini R, Greco S, Sarra-Ferraris G, Cardani R, Capogrossi MC, Meola G, Martelli F. Dysregulation and cellular mislocalization of specific miRNAs in myotonic dystrophy type 1. Neuromuscul Disord. 2011;21(2):81–8.
116. Fernandez-Costa JM, Garcia-Lopez A, Zuniga S, Fernandez-Pedrosa V, Felipo-Benavent A, Mata M, Jaka O, Aiastui A, Hernandez-Torres F, Aguado B, et al. Expanded CTG repeats trigger miRNA alterations in Drosophila that are conserved in myotonic dystrophy type 1 patients. Hum Mol Genet. 2013;22(4):704–16.
117. Greco S, Perfetti A, Fasanaro P, Cardani R, Capogrossi MC, Meola G, Martelli F. Deregulated microRNAs in myotonic dystrophy type 2. PLoS One. 2012;7(6):e39732.
118. Perfetti A, Greco S, Bugiardini E, Cardani R, Gaia P, Gaetano C, Meola G, Martelli F. Plasma microRNAs as biomarkers for myotonic dystrophy type 1. Neuromuscul Disord. 2014;24(6):509–15.
119. Fernandez-Torron R, Garcia-Puga M, Emparanza JI, Maneiro M, Cobo AM, Poza JJ, Espinal JB, Zulaica M, Ruiz I, Martorell L, et al. Cancer risk in DM1 is sex-related and linked to miRNA-200/141 downregulation. Neurology. 2016;87(12):1250–7.
120. Zu T, Gibbens B, Doty NS, Gomes-Pereira M, Huguet A, Stone MD, Margolis J, Peterson M, Markowski TW, Ingram MA, et al. Non-ATG-initiated translation directed by microsatellite expansions. Proc Natl Acad Sci U S A. 2011;108(1):260–5.
121. Cleary JD, Ranum LP. Repeat-associated non-ATG (RAN) translation in neurological disease. Hum Mol Genet. 2013;22(R1):R45–51.
122. Zu T, Cleary JD, Liu Y, Banez-Coronel M, Bubenik JL, Ayhan F, Ashizawa T, Xia G, Clark HB, Yachnis AT, et al. RAN translation regulated by Muscleblind proteins in myotonic dystrophy type 2. Neuron. 2017;95(6):1292–1305 e1295.
123. Kearse MG, Todd PK. Repeat-associated non-AUG translation and its impact in neurodegenerative disease. Neurotherapeutics. 2014;11(4):721–31.
124. Green KM, Linsalata AE, Todd PK. RAN translation-what makes it run? Brain Res. 2016;1647:30–42.
125. Todd PK, Oh SY, Krans A, He F, Sellier C, Frazer M, Renoux AJ, Chen KC, Scaglione KM, Basrur V, et al. CGG repeat-associated translation mediates neurodegeneration in fragile X tremor ataxia syndrome. Neuron. 2013;78(3):440–55.
126. Banez-Coronel M, Ayhan F, Tarabochia AD, Zu T, Perez BA, Tusi SK, Pletnikova O, Borchelt DR, Ross CA, Margolis RL, et al. RAN translation in Huntington disease. Neuron. 2015;88(4):667–77.
127. Harrison AF, Shorter J. RNA-binding proteins with prion-like domains in health and disease. Biochem J. 2017;474(8):1417–38.
128. Tran H, Almeida S, Moore J, Gendron TF, Chalasani U, Lu Y, Du X, Nickerson JA, Petrucelli L, Weng Z, et al. Differential toxicity of nuclear RNA foci versus dipeptide repeat proteins in a Drosophila model of C9ORF72 FTD/ALS. Neuron. 2015;87(6):1207–14.
129. Liu EY, Cali CP, Lee EB. RNA metabolism in neurodegenerative disease. Dis Models Mech. 2017;10(5):509–18.
130. Ling SC, Polymenidou M, Cleveland DW. Converging mechanisms in ALS and FTD: disrupted RNA and protein homeostasis. Neuron. 2013;79(3):416–38.

131. Buratti E, Dork T, Zuccato E, Pagani F, Romano M, Baralle FE. Nuclear factor TDP-43 and SR proteins promote in vitro and in vivo CFTR exon 9 skipping. EMBO J. 2001;20(7):1774–84.
132. Ling JP, Pletnikova O, Troncoso JC, Wong PC. TDP-43 repression of nonconserved cryptic exons is compromised in ALS-FTD. Science. 2015;349(6248):650–5.
133. Shiga A, Ishihara T, Miyashita A, Kuwabara M, Kato T, Watanabe N, Yamahira A, Kondo C, Yokoseki A, Takahashi M, et al. Alteration of POLDIP3 splicing associated with loss of function of TDP-43 in tissues affected with ALS. PLoS One. 2012;7(8):e43120.
134. Tollervey JR, Curk T, Rogelj B, Briese M, Cereda M, Kayikci M, Konig J, Hortobagyi T, Nishimura AL, Zupunski V, et al. Characterizing the RNA targets and position-dependent splicing regulation by TDP-43. Nat Neurosci. 2011;14(4):452–8.
135. Costessi L, Porro F, Iaconcig A, Muro AF. TDP-43 regulates beta-adducin (Add2) transcript stability. RNA Biol. 2014;11(10):1280–90.
136. Liu X, Li D, Zhang W, Guo M, Zhan Q. Long non-coding RNA gadd7 interacts with TDP-43 and regulates Cdk6 mRNA decay. EMBO J. 2012;31(23):4415–27.
137. Strong MJ, Volkening K, Hammond R, Yang W, Strong W, Leystra-Lantz C, Shoesmith C. TDP43 is a human low molecular weight neurofilament (hNFL) mRNA-binding protein. Mol Cell Neurosci. 2007;35(2):320–7.
138. Alami NH, Smith RB, Carrasco MA, Williams LA, Winborn CS, Han SSW, Kiskinis E, Winborn B, Freibaum BD, Kanagaraj A, et al. Axonal transport of TDP-43 mRNA granules is impaired by ALS-causing mutations. Neuron. 2014;81(3):536–43.
139. Yang L, Embree LJ, Tsai S, Hickstein DD. Oncoprotein TLS interacts with serine-arginine proteins involved in RNA splicing. J Biol Chem. 1998;273(43):27761–4.
140. Wang X, Arai S, Song X, Reichart D, Du K, Pascual G, Tempst P, Rosenfeld MG, Glass CK, Kurokawa R. Induced ncRNAs allosterically modify RNA-binding proteins in cis to inhibit transcription. Nature. 2008;454(7200):126–30.
141. Couthouis J, Hart MP, Shorter J, DeJesus-Hernandez M, Erion R, Oristano R, Liu AX, Ramos D, Jethava N, Hosangadi D, et al. A yeast functional screen predicts new candidate ALS disease genes. Proc Natl Acad Sci U S A. 2011;108(52):20881–90.
142. Ticozzi N, Vance C, Leclerc AL, Keagle P, Glass JD, McKenna-Yasek D, Sapp PC, Silani V, Bosco DA, Shaw CE, et al. Mutational analysis reveals the FUS homolog TAF15 as a candidate gene for familial amyotrophic lateral sclerosis. Am J Med Genet B Neuropsychiatric Genet. 2011;156B(3):285–90.
143. Neumann M, Bentmann E, Dormann D, Jawaid A, DeJesus-Hernandez M, Ansorge O, Roeber S, Kretzschmar HA, Munoz DG, Kusaka H, et al. FET proteins TAF15 and EWS are selective markers that distinguish FTLD with FUS pathology from amyotrophic lateral sclerosis with FUS mutations. Brain. 2011;134(Pt 9):2595–609.
144. Couthouis J, Hart MP, Erion R, King OD, Diaz Z, Nakaya T, Ibrahim F, Kim HJ, Mojsilovic-Petrovic J, Panossian S, et al. Evaluating the role of the FUS/TLS-related gene EWSR1 in amyotrophic lateral sclerosis. Hum Mol Genet. 2012;21(13):2899–911.
145. Elden AC, Kim HJ, Hart MP, Chen-Plotkin AS, Johnson BS, Fang X, Armakola M, Geser F, Greene R, Lu MM, et al. Ataxin-2 intermediate-length polyglutamine expansions are associated with increased risk for ALS. Nature. 2010;466(7310):1069–75.
146. Kim HJ, Kim NC, Wang YD, Scarborough EA, Moore J, Diaz Z, MacLea KS, Freibaum B, Li S, Molliex A, et al. Mutations in prion-like domains in hnRNPA2B1 and hnRNPA1 cause multisystem proteinopathy and ALS. Nature. 2013;495(7442):467–73.
147. Johnson JO, Pioro EP, Boehringer A, Chia R, Feit H, Renton AE, Pliner HA, Abramzon Y, Marangi G, Winborn BJ, et al. Mutations in the Matrin 3 gene cause familial amyotrophic lateral sclerosis. Nat Neurosci. 2014;17(5):664–6.
148. Murray DT, Kato M, Lin Y, Thurber KR, Hung I, McKnight SL, Tycko R. Structure of FUS protein fibrils and its relevance to self-assembly and phase separation of low-complexity domains. Cell. 2017;171(3):615–627 e616.
149. Murakami T, Qamar S, Lin JQ, Schierle GS, Rees E, Miyashita A, Costa AR, Dodd RB, Chan FT, Michel CH, et al. ALS/FTD mutation-induced phase transition of FUS liquid droplets

and reversible hydrogels into irreversible hydrogels impairs RNP granule function. Neuron. 2015;88(4):678–90.

150. Taylor JP, Brown RH Jr, Cleveland DW. Decoding ALS: from genes to mechanism. Nature. 2016;539(7628):197–206.

151. Freibaum BD, Lu Y, Lopez-Gonzalez R, Kim NC, Almeida S, Lee KH, Badders N, Valentine M, Miller BL, Wong PC, et al. GGGGCC repeat expansion in C9orf72 compromises nucleo-cytoplasmic transport. Nature. 2015;525(7567):129–33.

152. Jovicic A, Mertens J, Boeynaems S, Bogaert E, Chai N, Yamada SB, Paul JW III, Sun S, Herdy JR, Bieri G, et al. Modifiers of C9orf72 dipeptide repeat toxicity connect nucleocyto-plasmic transport defects to FTD/ALS. Nat Neurosci. 2015;18(9):1226–9.

153. Zhang K, Donnelly CJ, Haeusler AR, Grima JC, Machamer JB, Steinwald P, Daley EL, Miller SJ, Cunningham KM, Vidensky S, et al. The C9orf72 repeat expansion disrupts nucleocyto-plasmic transport. Nature. 2015;525(7567):56–61.

154. Woerner AC, Frottin F, Hornburg D, Feng LR, Meissner F, Patra M, Tatzelt J, Mann M, Winklhofer KF, Hartl FU, et al. Cytoplasmic protein aggregates interfere with nucleocyto-plasmic transport of protein and RNA. Science. 2016;351(6269):173–6.

155. Lee KH, Zhang P, Kim HJ, Mitrea DM, Sarkar M, Freibaum BD, Cika J, Coughlin M, Messing J, Molliex A, et al. C9orf72 dipeptide repeats impair the assembly, dynamics, and function of membrane-less organelles. Cell. 2016;167(3):774–788 e717.

156. Lin Y, Mori E, Kato M, Xiang S, Wu L, Kwon I, McKnight SL. Toxic PR poly-dipeptides encoded by the C9orf72 repeat expansion target LC domain polymers. Cell. 2016;167(3):789–802 e712.

157. Shi KY, Mori E, Nizami ZF, Lin Y, Kato M, Xiang S, Wu LC, Ding M, Yu Y, Gall JG, et al. Toxic PRn poly-dipeptides encoded by the C9orf72 repeat expansion block nuclear import and export. Proc Natl Acad Sci U S A. 2017;114(7):E1111–7.

158. Hagerman RJ, Hull CE, Safanda JF, Carpenter I, Staley LW, O'Connor RA, Seydel C, Mazzocco MM, Snow K, Thibodeau SN, et al. High functioning fragile X males: demonstra-tion of an unmethylated fully expanded FMR-1 mutation associated with protein expression. Am J Med Genet. 1994;51(4):298–308.

159. Coffey SM, Cook K, Tartaglia N, Tassone F, Nguyen DV, Pan R, Bronsky HE, Yuhas J, Borodyanskaya M, Grigsby J, et al. Expanded clinical phenotype of women with the FMR1 premutation. Am J Med Genet A. 2008;146A(8):1009–16.

160. Jacquemont S, Hagerman RJ, Leehey M, Grigsby J, Zhang L, Brunberg JA, Greco C, Des Portes V, Jardini T, Levine R, et al. Fragile X premutation tremor/ataxia syndrome: molecu-lar, clinical, and neuroimaging correlates. Am J Hum Genet. 2003;72(4):869–78.

161. Greco CM, Hagerman RJ, Tassone F, Chudley AE, Del Bigio MR, Jacquemont S, Leehey M, Hagerman PJ. Neuronal intranuclear inclusions in a new cerebellar tremor/ataxia syndrome among fragile X carriers. Brain. 2002;125(Pt 8):1760–71.

162. Leehey MA. Fragile X-associated tremor/ataxia syndrome: clinical phenotype, diagnosis, and treatment. J Investig Med. 2009;57(8):830–6.

163. Penagarikano O, Mulle JG, Warren ST. The pathophysiology of fragile x syndrome. Annu Rev Genomics Hum Genet. 2007;8:109–29.

164. Sreeram N, Wren C, Bhate M, Robertson P, Hunter S. Cardiac abnormalities in the fragile X syndrome. Br Heart J. 1989;61(3):289–91.

165. Lozano R, Azarang A, Wilaisakditipakorn T, Hagerman RJ. Fragile X syndrome: a review of clinical management. Intractable Rare Dis Res. 2016;5(3):145–57.

166. Holmes SE, O'Hearn EE, McInnis MG, Gorelick-Feldman DA, Kleiderlein JJ, Callahan C, Kwak NG, Ingersoll-Ashworth RG, Sherr M, Sumner AJ, et al. Expansion of a novel CAG trinucleotide repeat in the 5′ region of PPP2R2B is associated with SCA12. Nat Genet. 1999;23(4):391–2.

167. O'Hearn E, Holmes SE, Calvert PC, Ross CA, Margolis RL. SCA-12: tremor with cer-ebellar and cortical atrophy is associated with a CAG repeat expansion. Neurology. 2001;56(3):299–303.

168. Ranum LP, Rasmussen PF, Benzow KA, Koob MD, Day JW. Genetic mapping of a second myotonic dystrophy locus. Nat Genet. 1998;19(2):196–8.
169. Peric S, Rakocevic Stojanovic V, Mandic Stojmenovic G, Ilic V, Kovacevic M, Parojcic A, Pesovic J, Mijajlovic M, Savic-Pavicevic D, Meola G. Clusters of cognitive impairment among different phenotypes of myotonic dystrophy type 1 and type 2. Neurol Sci. 2017;38(3):415–23.
170. Tieleman AA, Knoop H, van de Logt AE, Bleijenberg G, van Engelen BG, Overeem S. Poor sleep quality and fatigue but no excessive daytime sleepiness in myotonic dystrophy type 2. J Neurol Neurosurg Psychiatry. 2010;81(9):963–7.
171. Hund E, Jansen O, Koch MC, Ricker K, Fogel W, Niedermaier N, Otto M, Kuhn E, Meinck HM. Proximal myotonic myopathy with MRI white matter abnormalities of the brain. Neurology. 1997;48(1):33–7.
172. Schneider-Gold C, Bellenberg B, Prehn C, Krogias C, Schneider R, Klein J, Gold R, Lukas C. Cortical and subcortical grey and white matter atrophy in myotonic dystrophies Type 1 and 2 is associated with cognitive impairment, depression and daytime sleepiness. PLoS One. 2015;10(6):e0130352.
173. Meola G, Sansone V, Perani D, Scarone S, Cappa S, Dragoni C, Cattaneo E, Cotelli M, Gobbo C, Fazio F, et al. Executive dysfunction and avoidant personality trait in myotonic dystrophy type 1 (DM-1) and in proximal myotonic myopathy (PROMM/DM-2). Neuromuscul Disord. 2003;13(10):813–21.
174. Meola G, Cardani R. Myotonic dystrophy type 2: an update on clinical aspects, genetic and pathomolecular mechanism. J Neuromuscul Dis. 2015;2(s2):S59–71.
175. Giordana MT, Ferrero P, Grifoni S, Pellerino A, Naldi A, Montuschi A. Dementia and cognitive impairment in amyotrophic lateral sclerosis: a review. Neurol Sci. 2011;32(1):9–16.
176. Graff-Radford NR, Woodruff BK. Frontotemporal dementia. Semin Neurol. 2007;27(1):48–57.
177. Kirshner HS. Frontotemporal dementia and primary progressive aphasia, a review. Neuropsychiatr Dis Treat. 2014;10:1045–55.
178. Thomas PS Jr, Fraley GS, Damian V, Woodke LB, Zapata F, Sopher BL, Plymate SR, La Spada AR. Loss of endogenous androgen receptor protein accelerates motor neuron degeneration and accentuates androgen insensitivity in a mouse model of X-linked spinal and bulbar muscular atrophy. Hum Mol Genet. 2006;15(14):2225–38.
179. The Huntington's Disease Collaborative Research Group. A novel gene containing a trinucleotide repeat that is expanded and unstable on Huntington's disease chromosomes. Cell. 1993;72(6):971–83.
180. Louis ED, Lee P, Quinn L, Marder K. Dystonia in Huntington's disease: prevalence and clinical characteristics. Mov Disord. 1999;14(1):95–101.
181. Kim SD, Fung VS. An update on Huntington's disease: from the gene to the clinic. Curr Opin Neurol. 2014;27(4):477–83.
182. Koide R, Ikeuchi T, Onodera O, Tanaka H, Igarashi S, Endo K, Takahashi H, Kondo R, Ishikawa A, Hayashi T, et al. Unstable expansion of CAG repeat in hereditary dentatorubral-pallidoluysian atrophy (DRPLA). Nat Genet. 1994;6(1):9–13.
183. Nagafuchi S, Yanagisawa H, Sato K, Shirayama T, Ohsaki E, Bundo M, Takeda T, Tadokoro K, Kondo I, Murayama N, et al. Dentatorubral and pallidoluysian atrophy expansion of an unstable CAG trinucleotide on chromosome 12p. Nat Genet. 1994;6(1):14–8.
184. Ikeuchi T, Koide R, Tanaka H, Onodera O, Igarashi S, Takahashi H, Kondo R, Ishikawa A, Tomoda A, Miike T, et al. Dentatorubral-pallidoluysian atrophy: clinical features are closely related to unstable expansions of trinucleotide (CAG) repeat. Ann Neurol. 1995;37(6):769–75.
185. Nucifora FC Jr, Ellerby LM, Wellington CL, Wood JD, Herring WJ, Sawa A, Hayden MR, Dawson VL, Dawson TM, Ross CA. Nuclear localization of a non-caspase truncation product of atrophin-1, with an expanded polyglutamine repeat, increases cellular toxicity. J Biol Chem. 2003;278(15):13047–55.

186. Klement IA, Skinner PJ, Kaytor MD, Yi H, Hersch SM, Clark HB, Zoghbi HY, Orr HT. Ataxin-1 nuclear localization and aggregation: role in polyglutamine-induced disease in SCA1 transgenic mice. Cell. 1998;95(1):41–53.
187. Skinner PJ, Vierra-Green CA, Emamian E, Zoghbi HY, Orr HT. Amino acids in a region of ataxin-1 outside of the polyglutamine tract influence the course of disease in SCA1 transgenic mice. NeuroMolecular Med. 2002;1(1):33–42.
188. Sanpei K, Takano H, Igarashi S, Sato T, Oyake M, Sasaki H, Wakisaka A, Tashiro K, Ishida Y, Ikeuchi T, et al. Identification of the spinocerebellar ataxia type 2 gene using a direct identification of repeat expansion and cloning technique, DIRECT. Nat Genet. 1996;14(3):277–84.
189. Pulst SM, Nechiporuk A, Nechiporuk T, Gispert S, Chen XN, Lopes-Cendes I, Pearlman S, Starkman S, Orozco-Diaz G, Lunkes A, et al. Moderate expansion of a normally biallelic trinucleotide repeat in spinocerebellar ataxia type 2. Nat Genet. 1996;14(3):269–76.
190. Imbert G, Saudou F, Yvert G, Devys D, Trottier Y, Garnier JM, Weber C, Mandel JL, Cancel G, Abbas N, et al. Cloning of the gene for spinocerebellar ataxia 2 reveals a locus with high sensitivity to expanded CAG/glutamine repeats. Nat Genet. 1996;14(3):285–91.
191. Kawaguchi Y, Okamoto T, Taniwaki M, Aizawa M, Inoue M, Katayama S, Kawakami H, Nakamura S, Nishimura M, Akiguchi I, et al. CAG expansions in a novel gene for Machado-Joseph disease at chromosome 14q32.1. Nat Genet. 1994;8(3):221–8.
192. Zhuchenko O, Bailey J, Bonnen P, Ashizawa T, Stockton DW, Amos C, Dobyns WB, Subramony SH, Zoghbi HY, Lee CC. Autosomal dominant cerebellar ataxia (SCA6) associated with small polyglutamine expansions in the alpha 1A-voltage-dependent calcium channel. Nat Genet. 1997;15(1):62–9.
193. David G, Abbas N, Stevanin G, Durr A, Yvert G, Cancel G, Weber C, Imbert G, Saudou F, Antoniou E, et al. Cloning of the SCA7 gene reveals a highly unstable CAG repeat expansion. Nat Genet. 1997;17(1):65–70.
194. Koide R, Kobayashi S, Shimohata T, Ikeuchi T, Maruyama M, Saito M, Yamada M, Takahashi H, Tsuji S. A neurological disease caused by an expanded CAG trinucleotide repeat in the TATA-binding protein gene: a new polyglutamine disease? Hum Mol Genet. 1999;8(11):2047–53.
195. Brais B, Bouchard JP, Xie YG, Rochefort DL, Chretien N, Tome FM, Lafreniere RG, Rommens JM, Uyama E, Nohira O, et al. Short GCG expansions in the PABP2 gene cause oculopharyngeal muscular dystrophy. Nat Genet. 1998;18(2):164–7.
196. Anvar SY, Raz Y, Verway N, van der Sluijs B, Venema A, Goeman JJ, Vissing J, van der Maarel SM, t Hoen PA, van Engelen BG, et al. A decline in PABPN1 induces progressive muscle weakness in oculopharyngeal muscle dystrophy and in muscle aging. Aging. 2013;5(6):412–26.
197. Stromme P, Mangelsdorf ME, Shaw MA, Lower KM, Lewis SM, Bruyere H, Lutcherath V, Gedeon AK, Wallace RH, Scheffer IE, et al. Mutations in the human ortholog of Aristaless cause X-linked mental retardation and epilepsy. Nat Genet. 2002;30(4):441–5.
198. Ropers HH, Hamel BC. X-linked mental retardation. Nat Rev Genet. 2005;6(1):46–57.
199. Nawara M, Szczaluba K, Poirier K, Chrzanowska K, Pilch J, Bal J, Chelly J, Mazurczak T. The ARX mutations: a frequent cause of X-linked mental retardation. Am J Med Genet A. 2006;140(7):727–32.
200. Messaed C, Rouleau GA. Molecular mechanisms underlying polyalanine diseases. Neurobiol Dis. 2009;34(3):397–405.
201. Winblad S, Lindberg C, Hansen S. Cognitive deficits and CTG repeat expansion size in classical myotonic dystrophy type 1 (DM1). Behav Brain Funct. 2006;2:16.
202. Koob MD, Moseley ML, Schut LJ, Benzow KA, Bird TD, Day JW, Ranum LP. An untranslated CTG expansion causes a novel form of spinocerebellar ataxia (SCA8). Nat Genet. 1999;21(4):379–84.
203. Ikeda Y, Dalton JC, Moseley ML, Gardner KL, Bird TD, Ashizawa T, Seltzer WK, Pandolfo M, Milunsky A, Potter NT, et al. Spinocerebellar ataxia type 8: molecular genetic comparisons and haplotype analysis of 37 families with ataxia. Am J Hum Genet. 2004;75(1):3–16.
204. Paulson HL. The spinocerebellar ataxias. J Neuroophthalmol. 2009;29(3):227–37.

205. Jin P, Duan R, Qurashi A, Qin Y, Tian D, Rosser TC, Liu H, Feng Y, Warren ST. Pur alpha binds to rCGG repeats and modulates repeat-mediated neurodegeneration in a Drosophila model of fragile X tremor/ataxia syndrome. Neuron. 2007;55(4):556–64.

206. Sofola OA, Jin P, Qin Y, Duan R, Liu H, de Haro M, Nelson DL, Botas J. RNA-binding proteins hnRNP A2/B1 and CUGBP1 suppress fragile X CGG premutation repeat-induced neurodegeneration in a Drosophila model of FXTAS. Neuron. 2007;55(4):565–71.

207. Timchenko LT, Miller JW, Timchenko NA, DeVore DR, Datar KV, Lin L, Roberts R, Caskey CT, Swanson MS. Identification of a (CUG)n triplet repeat RNA-binding protein and its expression in myotonic dystrophy. Nucleic Acids Res. 1996;24(22):4407–14.

208. Sellier C, Rau F, Liu Y, Tassone F, Hukema RK, Gattoni R, Schneider A, Richard S, Willemsen R, Elliott DJ, et al. Sam68 sequestration and partial loss of function are associated with splicing alterations in FXTAS patients. EMBO J. 2010;29(7):1248–61.

209. Pieretti M, Zhang FP, Fu YH, Warren ST, Oostra BA, Caskey CT, Nelson DL. Absence of expression of the FMR-1 gene in fragile X syndrome. Cell. 1991;66(4):817–22.

210. Feng Y, Gutekunst CA, Eberhart DE, Yi H, Warren ST, Hersch SM, Fragile X. mental retardation protein: nucleocytoplasmic shuttling and association with somatodendritic ribosomes. J Neurosci. 1997;17(5):1539–47.

211. Khandjian EW, Huot ME, Tremblay S, Davidovic L, Mazroui R, Bardoni B. Biochemical evidence for the association of fragile X mental retardation protein with brain polyribosomal ribonucleoparticles. Proc Natl Acad Sci U S A. 2004;101(36):13357–62.

212. Stefani G, Fraser CE, Darnell JC, Darnell RB. Fragile X mental retardation protein is associated with translating polyribosomes in neuronal cells. J Neurosci. 2004;24(33):7272–6.

213. Greenough WT, Klintsova AY, Irwin SA, Galvez R, Bates KE, Weiler IJ. Synaptic regulation of protein synthesis and the fragile X protein. Proc Natl Acad Sci U S A. 2001;98(13):7101–6.

214. Hokkanen S, Feldmann HM, Ding H, Jung CK, Bojarski L, Renner-Muller I, Schuller U, Kretzschmar H, Wolf E, Herms J. Lack of Pur-alpha alters postnatal brain development and causes megalencephaly. Hum Mol Genet. 2012;21(3):473–84.

215. Qurashi A, Li W, Zhou JY, Peng J, Jin P. Nuclear accumulation of stress response mRNAs contributes to the neurodegeneration caused by Fragile X premutation rCGG repeats. PLoS Genet. 2011;7(6):e1002102.

216. Li Y, Jin P. RNA-mediated neurodegeneration in fragile X-associated tremor/ataxia syndrome. Brain Res. 2012;1462:112–7.

217. Wang GS, Kearney DL, De Biasi M, Taffet G, Cooper TA. Elevation of RNA-binding protein CUGBP1 is an early event in an inducible heart-specific mouse model of myotonic dystrophy. J Clin Invest. 2007;117(10):2802–11.

218. Ho TH, Savkur RS, Poulos MG, Mancini MA, Swanson MS, Cooper TA. Colocalization of muscleblind with RNA foci is separable from mis-regulation of alternative splicing in myotonic dystrophy. J Cell Sci. 2005;118(Pt 13):2923–33.

219. Ladd AN, Stenberg MG, Swanson MS, Cooper TA. Dynamic balance between activation and repression regulates pre-mRNA alternative splicing during heart development. Dev Dyn. 2005;233(3):783–93.

220. Neumann M, Sampathu DM, Kwong LK, Truax AC, Micsenyi MC, Chou TT, Bruce J, Schuck T, Grossman M, Clark CM, et al. Ubiquitinated TDP-43 in frontotemporal lobar degeneration and amyotrophic lateral sclerosis. Science. 2006;314(5796):130–3.

221. Arai T, Hasegawa M, Akiyama H, Ikeda K, Nonaka T, Mori H, Mann D, Tsuchiya K, Yoshida M, Hashizume Y, et al. TDP-43 is a component of ubiquitin-positive tau-negative inclusions in frontotemporal lobar degeneration and amyotrophic lateral sclerosis. Biochem Biophys Res Commun. 2006;351(3):602–11.

222. Kwiatkowski TJ Jr, Bosco DA, Leclerc AL, Tamrazian E, Vanderburg CR, Russ C, Davis A, Gilchrist J, Kasarskis EJ, Munsat T, et al. Mutations in the FUS/TLS gene on chromosome 16 cause familial amyotrophic lateral sclerosis. Science. 2009;323(5918):1205–8.

223. Vance C, Rogelj B, Hortobagyi T, De Vos KJ, Nishimura AL, Sreedharan J, Hu X, Smith B, Ruddy D, Wright P, et al. Mutations in FUS, an RNA processing protein, cause familial amyotrophic lateral sclerosis type 6. Science. 2009;323(5918):1208–11.

224. Mackenzie IR, Nicholson AM, Sarkar M, Messing J, Purice MD, Pottier C, Annu K, Baker M, Perkerson RB, Kurti A, et al. TIA1 mutations in amyotrophic lateral sclerosis and frontotemporal dementia promote phase separation and alter stress granule dynamics. Neuron. 2017;95(4):808–16. e809

225. Todd PK. Making sense of the antisense transcripts in C9FTD/ALS. Acta Neuropathol. 2013;126(6):785–7.

226. Polymenidou M, Lagier-Tourenne C, Hutt KR, Huelga SC, Moran J, Liang TY, Ling SC, Sun E, Wancewicz E, Mazur C, et al. Long pre-mRNA depletion and RNA missplicing contribute to neuronal vulnerability from loss of TDP-43. Nat Neurosci. 2011;14(4):459–68.

227. Ito D, Suzuki N. Conjoint pathologic cascades mediated by ALS/FTLD-U linked RNA-binding proteins TDP-43 and FUS. Neurology. 2011;77(17):1636–43.

228. Ito D, Seki M, Tsunoda Y, Uchiyama H, Suzuki N. Nuclear transport impairment of amyotrophic lateral sclerosis-linked mutations in FUS/TLS. Ann Neurol. 2011;69(1):152–62.

229. Dormann D, Rodde R, Edbauer D, Bentmann E, Fischer I, Hruscha A, Than ME, Mackenzie IR, Capell A, Schmid B, et al. ALS-associated fused in sarcoma (FUS) mutations disrupt Transportin-mediated nuclear import. EMBO J. 2010;29(16):2841–57.

230. King OD, Gitler AD, Shorter J. The tip of the iceberg: RNA-binding proteins with prion-like domains in neurodegenerative disease. Brain Res. 2012;1462:61–80.

231. Kato M, Han TW, Xie S, Shi K, Du X, Wu LC, Mirzaei H, Goldsmith EJ, Longgood J, Pei J, et al. Cell-free formation of RNA granules: low complexity sequence domains form dynamic fibers within hydrogels. Cell. 2012;149(4):753–67.

232. Lim L, Wei Y, Lu Y, Song J. ALS-causing mutations significantly perturb the self-assembly and interaction with nucleic acid of the intrinsically disordered prion-like domain of TDP-43. PLoS Biol. 2016;14(1):e1002338.

233. Molliex A, Temirov J, Lee J, Coughlin M, Kanagaraj AP, Kim HJ, Mittag T, Taylor JP. Phase separation by low complexity domains promotes stress granule assembly and drives pathological fibrillization. Cell. 2015;163(1):123–33.

234. Daughters RS, Tuttle DL, Gao W, Ikeda Y, Moseley ML, Ebner TJ, Swanson MS, Ranum LP. RNA gain-of-function in spinocerebellar ataxia type 8. PLoS Genet. 2009;5(8):e1000600.

235. Mutsuddi M, Marshall CM, Benzow KA, Koob MD, Rebay I. The spinocerebellar ataxia 8 noncoding RNA causes neurodegeneration and associates with staufen in Drosophila. Curr Biol. 2004;14(4):302–8.

Chapter 9
Mechanisms Associated with TDP-43 Neurotoxicity in ALS/FTLD

Marc Shenouda, Ashley B. Zhang, Anna Weichert, and Janice Robertson

Abstract The discovery of TDP-43 as a major disease protein in amyotrophic lateral sclerosis (ALS) and frontotemporal lobar degeneration (FTLD) was first made in 2006. Prior to 2006 there were only 11 publications related to TDP-43, now there are over 2000, indicating the importance of TDP-43 to unraveling the complex molecular mechanisms that underpin the pathogenesis of ALS/FTLD. Subsequent to this discovery, TDP-43 pathology was also found in other neurodegenerative diseases, including Alzheimer's disease, the significance of which is still in the early stages of exploration. TDP-43 is a predominantly nuclear DNA/RNA-binding protein, one of a number of RNA-binding proteins that are now known to be linked with ALS/FTLD, including Fused in Sarcoma (FUS), heterogeneous nuclear ribonucleoprotein A1 (hnRNP A1), and heterogeneous nuclear ribonucleoprotein A2/B1 (hnRNP A2/B1). However, what sets TDP-43 apart is the vast number of cases in which TDP-43 pathology is present, providing a point of convergence, the understanding of which could lead to broadly applicable therapeutics. Here we will focus on TDP-43 in ALS/FTLD, its nuclear and cytoplasmic functions, and consequences should these functions go awry.

Keywords TDP-43 · ALS · FTLD · RNA · Granules

9.1 Discovery of TDP-43 as a Disease Relevant Protein in Amyotrophic Lateral Sclerosis and Frontotemporal Lobar Degeneration

Amyotrophic lateral sclerosis (ALS), also known as Lou Gehrig's disease, is an adult-onset neurodegenerative disease characterized by loss of motor neurons from the motor cortex, brain stem, and spinal cord, causing progressive paralysis and

M. Shenouda · A. B. Zhang · A. Weichert · J. Robertson (✉)
Tanz Centre for Research in Neurodegenerative Diseases and Department of Laboratory Medicine and Pathobiology, University of Toronto, Toronto, ON M5T 2S8, Canada
e-mail: jan.robertson@utoronto.ca

© Springer International Publishing AG, part of Springer Nature 2018
R. Sattler, C. J. Donnelly (eds.), *RNA Metabolism in Neurodegenerative Diseases*, Advances in Neurobiology 20,
https://doi.org/10.1007/978-3-319-89689-2_9

239

death due to respiratory failure 2–5 years from diagnosis [1]. In populations of European origin, the median incidence of ALS is 2.08 per 100,000 with a median prevalence of 5.4 per 100,000 [2]. Approximately 5–10% of ALS cases exhibit Mendelian inheritance, primarily autosomal dominant, known as familial ALS (fALS), with the remaining 90–95% of sporadic ALS (sALS) cases showing no apparent family history of disease. To date, the genes associated with about 70% of fALS cases have been identified, and a genetic component has also been found in approximately 15% of sALS cases [3]. The major genes accounting for the greatest number of ALS cases are *superoxide dismutase* 1 (*SOD1*; 15–20% of fALS, 1% sALS), *TAR DNA-Binding Protein-43* (*TARDBP*; 4% of fALS, <1% sALS), *Fused in Sarcoma* (*FUS*; 4% of fALS, <1% sALS), and *C9orf72* (30–40% of fALS, 7% sALS) [3]. Clinically, ALS requires involvement of both upper and lower motor neurons with three sites of focal onset, limb (70%), bulbar (25%), and diaphragm (5%), spreading to other regions as the disease progresses [1]. Although manifesting as a motor neuron disease, neuropsychological, neuroimaging, and histological studies have indicated extramotor involvement, and as such ALS has been proposed as a multisystems disorder, affecting other neuronal subtypes [4]. ALS has clear clinical, neuropathological and genetic overlap with frontotemporal dementia (FTD) [5], an umbrella term for a group of disorders affecting the frontal and temporal lobes causing impairments in behavior, language, or executive function [6]. FTD is the second most common form of dementia below the age of 65. There are three major clinical syndromes of FTD: (1) behavioral variant FTD (bvFTD), characterized by changes in personality and social behaviors; and two forms of primary progressive aphasia, (2) semantic dementia (SD) where there is impaired word comprehension and decline in semantic memory; and (3) progressive non-fluent aphasia (PNFA), characterized by impaired speech production [7, 8]. Up to 50% of patients with ALS exhibit clinical signs of frontotemporal dysfunction with 10–15% fulfilling the diagnostic criteria of FTD, usually bvFTD [9]. This is evident at autopsy with neuropathological markers showing degeneration of the frontal and temporal lobes, known as frontotemporal lobar degeneration (FTLD), the term that will be used herein.

In 2006, the clinical overlap between ALS and FTLD was substantiated when two groups independently identified the TAR DNA-Binding Protein-43 (TDP-43) encoded by *TARDBP*, as a core component of the ubiquitinated inclusions pathognomonic of disease [10, 11]. Soon thereafter, primarily autosomal dominant mutations in *TARDBP* were found associated with ~4% of fALS and <1% of sALS cases, indicating a direct role for TDP-43 in disease causation [3, 12, 13]. TDP-43 is a predominantly nuclear DNA/RNA-binding protein that has numerous functions related to RNA metabolism [14, 15]. In ALS, TDP-43 is mislocalized from the nucleus to the cytoplasm of neurons and glia of the primary motor cortex, brainstem motor nuclei, and spinal cord. In the neuronal cytoplasm, pathological TDP-43 appears as non-ubiquitinated diffuse/granules, termed pre-inclusions [16], or as ubiquitinated skein-like inclusions, or compact Lewy body-like inclusions [10, 11, 17]. These structures appear to be a continuum of the same pathology [18]. TDP-43 in pathological inclusions (but not normal nuclear TDP-43) is phosphorylated on

a

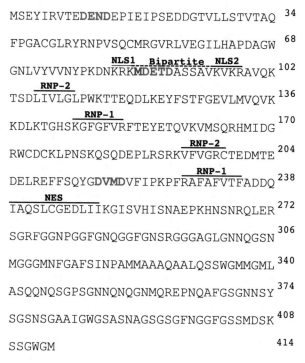

```
MSEYIRVTEDENDEPIEIPSEDDGTVLLSTVTAQ 34

FPGACGLRYRNPVSQCMRGVRLVEGILHAPDAGW 68
              NLS1  Bipartite  NLS2
GNLVYVVNYPKDNKRKMDETDASSAVKVKRAVQK 102
     RNP-2
TSDLIVLGLPWKTTEQDLKEYFSTFGEVLMVQVK 136
            RNP-1
KDLKTGHSKGFGFVRFTEYETQVKVMSQRHMIDG 170
                RNP-2
RWCDCKLPNSKQSQDEPLRSRKVFVGRCTEDMTE 204
              RNP-1
DELREFFSQYGDVMDVFIPKPFRAFAFVTFADDQ 238
        NES
IAQSLCGEDLIIKGISVHISNAEPKHNSNRQLER 272

SGRFGGNPGGFGNQGGFGNSRGGGAGLGNNQGSN 306

MGGGMNFGAFSINPAMMAAAQAALQSSWGMMGML 340

ASQQNQSGPSGNNQNQGNMQREPNQAFGSGNNSY 374

SGSNSGAAIGWGSASNAGSGSGFNGGFGSSMDSK 408

SSGWGM                             414
```

b

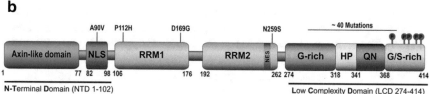

Fig. 9.1 Sequence and domain structure of TDP-43. (**a**) Amino acid sequence of TDP-43 showing location of the NLS (residues 82–98), the NES (residues 239–250), RNP1 and RNP2 motifs, the caspase 3 cleavage sites (DEND, DETD, DVMD in red), alternative translation start site Met 85 (blue). (**b**) Domain structure of TDP-43 with location of pathological phosphorylation sites (serines 379, 403, 404, 409, and 410) shown in red

at least five sites, Ser 379, Ser 403/404, Ser 409/410, all present within the Glycine-Serine-rich domain, and can be detected using specific antibodies (Fig. 9.1) [19]. Evidence suggests that these sites are phosphorylated by casein kinase 1 [19]. TDP-43 pathology is present in over 95% of ALS cases, irrespective of the causal genetic mutation [20], with the key exception of cases caused by mutations in FUS [20, 21] and SOD1, where only rare TDP-43 inclusions are observed [22–24]. Interestingly, a recent study described nuclear clearing of TDP-43 in pyramidal Betz cells in the absence of cytoplasmic inclusions [25]. These cells occurred alongside Betz cells with traces of granular or skein-like TDP-43 pathology as well as Betz cells with robust pathology [25, 26]. TDP-43 pathology is present in about 50% of FTLD cases (FTLD-TDP), with the other 45% of cases exhibiting tau pathology (FTLD-tau), and the remaining 5% characterized by inclusion bodies comprised of FUS (FTLD-FUS) or other FET (*F*US-*E*wings Sarcoma Protein-*T*AF15) proteins (FTLD-FET) [26].

In FTLD-TDP, there are four main subtypes of TDP-43 pathology that are based on anatomical distribution and morphology, and exhibit clinical and genetic correlations [26–28]. Type A is characterized by compact neuronal cytoplasmic inclusions (NCI), short and thick dystrophic neurites (DN), and occasional lentiform neuronal intranuclear inclusions (NII) primarily in layer II of the neocortex; and is associated with bvFTD or PNFA, and cases caused by mutations in progranulin. Type B exhibits diffuse, granular, and compact NCI as well as abundant "wispy dot-like profiles" throughout all cortical layers, with few DN and NII, and is associated with bvFTD and FTLD-ALS. Type C is characterized by long DN in upper cortical layers with few NCI, and this pathology is associated with SD. Type D is a rare pathology characterized by lentiform NII, short DN, and rare NCI throughout all cortical layers, and is specifically associated with inclusion body myopathy with Paget's disease of bone and frontotemporal dementia caused by mutations in vasolin-containing protein [26, 29]. These distinctive patterns of TDP-43 neuropathology may be reflective of differing disease mechanisms, and this is supported by the association of each subtype with different clinical presentations [27, 28]. An exception are cases caused by G_4C_2 hexanucleotide repeat expansions in C9orf72, where features of Type A and Type B TDP-43 pathology have been observed [29, 30]. The reason for this heterogeneity is uncertain. It is worth noting here that dipeptide repeat (DPR) protein pathology generated through repeat-associated non-ATG (RAN) translation of the G_4C_2 repeats is present in ALS-FTLD cases caused by C9orf72 mutations [31–35]. The anatomical distribution of DPR (poly-GA, poly-GR, poly-PR, poly-PA, poly-GP) pathology differs from TDP-43, with TDP-43 reported as being the better correlate of areas of neurodegeneration [35, 36]. However, poly-GR was recently shown to colocalize with phosphorylated TDP-43 in dendrites of the motor cortex in C9orf72 ALS cases [37].

TDP-43 pathology has also been associated with a number of other neurodegenerative diseases, including Alzheimer's disease and hippocampal sclerosis [39–41], Huntingtons's disease [42], Lewy body-related diseases (Parkinson's Disease, Parkinson-Dementia, Dementia with Lewy bodies) [39, 43–45], argyrophilic grain disease [46], and late stage chronic traumatic encephalopathy [47]. These differing TDP-43 pathologies across numerous diseases are subsumed by the collective term TDP-43 proteinopathy.

9.2   TDP-43 Structure and Function

Ou et al. [48] first identified a protein binding to the pyrimidine-rich motif within the LTR region of HIV-1 by screening a HeLa cell library using a TAR DNA probe. Subsequent analysis using Northern and Western blot identified the protein as TAR

DNA-binding protein of 43 kDa transcribed from a 2.8 kb transcript. Thus, the protein was named TAR-DNA-binding protein of 43 kDa (TDP-43) [48]. Later, TDP-43 was found to function as a splicing regulator by binding to the (TG)m polymorphic repeat region near the 3′-splice site of exon 9 in the cystic fibrosis transmembrane conductance regulator (*CFTR*) pre-mRNA causing exon skipping [49, 50]. Since that time, TDP-43 has been associated with numerous aspects of RNA processing, including transcriptional regulation, pre-mRNA splicing, miRNA biogenesis, lncRNA/ncRNA expression/regulation, as well as mRNA stability, transport and translation (reviewed in [15].

The human TAR DNA-binding protein gene (*TARDBP*) is located on chromosome 1p36, and has six exons with exons 2–6 encoding the 414-aa TDP-43 protein. TDP-43 is evolutionarily conserved in mouse, *Drosophila melanogaster*, and *C. elegans* [51]. TDP-43 is ubiquitously expressed and is essential for embryological development [52–54]. Structurally, TDP-43 contains all the elements of a heterogenous ribonuclear protein (hnRNP) [48–50], comprising of two RNA Recognition Motifs, RRM1 (residues 106–176) and RRM2 (residues 191–262), an N-terminal domain (residues 1–102), and an intrinsically disordered C-terminal domain (residues 274–414) [55] (Fig. 9.1). The RRM domains each contain conserved octamer and hexamer consensus sequences known as Ribonucleoprotein 1 (RNP1) and Ribonucleoprotein 2 (RNP2) [59]. The N-terminal domain (specifically residues 1–77) adopts a Ubiquitin-like [56] or DIshevelled and aXin (DIX)-domain-like fold [57] in solution. The C-terminal domain shares homology with prion-like domains and mediates interactions with other proteins such as hnRNP A/B and hnRNP A2, necessary for the splicing functions of TDP-43 [58–61]. The C-terminal domain is subdivided into four regions: a glycine-rich motif (267–317); a hydrophobic segment (318–340), which adopts a marginally stable α-helical conformation in aqueous solution [62]; an aggregation prone glutamine/asparagine (Q/N)-rich region (341–367), which binds hnRNP A2 [63–66]; and a glycine-serine-rich region (368–414), where the majority of pathological phosphorylation sites are located [19] (Fig. 9.1). The majority of TDP-43 mutations associated with ALS/FTLD are located in the C-terminal domain [20]. TDP-43 is a predominantly nuclear protein, with low levels present in the cytoplasm [67]. The nucleocytoplasmic shuttling of TDP-43 is regulated by a classical bipartite nuclear localization sequence (NLS) at the N-terminus composed of K82RK84 (NLS1) and K95VKR98 (NLS2), and a nuclear export sequence (NES) in RRM2, I239AQSLCGEDLII250 [68, 69].

TDP-43 regulates the expression and splicing of multiple RNA targets, including processing/expression of its own transcript [49, 70–72]. TDP-43 binds with high affinity to UG repeats in target RNAs through the RRM1 and RRM2 domains, with RRM1 indispensable for RNA binding [49, 72]. Native TDP-43 forms functional dimers and oligomers under physiological conditions, interacting through the N-terminal domains, spatially separating the aggregation prone C-terminal domains [73–77]. This dimerization/oligomerization is required for the nuclear splicing activity of TDP-43 [73, 74]. Oligomeric TDP-43 is predicted to provide higher binding affinities for longer contiguous UG-repeats [73]. Furthermore, oligomeric TDP-43 could also bring distal sites into close proximity either within the same

RNA or multiple RNAs, creating loops that potentiate splicing [73, 78]. Three disease mutations have so far been identified in the RNA-binding domain of TDP-43: P112H [79], D169G [12], and N259S [80]. While the D169G mutation does not affect RNA-binding affinity [81], it was found to increase the thermal stability of TDP-43 promoting cleavage by caspase-3 both in vitro and in culture. This produced increased levels of a 35 kDa C-terminal fragment, which enhanced cellular toxicity [82, 83].

RNA targets of TDP-43 have been identified using RNA precipitation techniques in cell culture, mouse brain, and human tissue [71, 84–86]. Using conventional cloning, over 100 TDP-43 RNA targets were identified in SHSY5Y cells, binding predominantly to UG-rich motifs [86]. Binding was mainly to intronic regions (82%), but also to 3′-UTRs and noncoding RNA. Pertinent TDP-43-binding targets included the transmembrane synaptic protein neurexin-1 (*NXRN1*) and the RNA editing enzyme *ADARB2* [86]. In rat primary cortical neurons, 4352 TDP-43 RNA targets were identified, 1971 mapping to introns, 910 to exons, and 1471 to both introns and exons [84]. Gene ontology (GO) analysis of the intronic reads showed enrichment for synaptic formation and function, and regulation of neurotransmitter processes, with *Nrxn1–3* and *Nlgn1–3* (neuroligin) identified as top hits. GO terms for exonic targets were related to splicing, RNA processing, and maturation. Notable targets included *Tardbp, Fus, Grn, Mapt, Atxn1 and 2*, and *Adarb1* [84]. In mouse brain, over 6000 TDP-43 RNA targets were identified, with TDP-43 binding preferentially to pre-mRNAs with long introns [71]. Depletion of TDP-43 using antisense oligonucleotides led to expression changes in 601 mRNAs and 965 splicing changes. Most of the target genes were downregulated, and GO analysis revealed an enrichment for synaptic activity and function, several genes of which were validated by independent qRT-PCR, including *Nrxn 1 and 3*, and *Nlgn*. Interestingly, pertinent to cholinergic neurons, there was a significant reduction in *Chat* levels [71]. Of the splicing changes, *sortilin 1*, which encodes SORT1 a neuronal progranulin receptor, had the highest splicing score [71]. Subsequent studies have shown that abnormal splicing of human *sortilin 1* caused by TDP-43 depletion generates a non-functional progranulin receptor that acts as a decoy to antagonize progranulin uptake [87]. This is an important link since mutations in progranulin cause haploinsufficiency in FTLD-TDP, and the splice isoform of SORT1 is elevated in FTLD-TDP tissues [87, 88]. Studies in healthy human and FTLD-TDP cortical brain tissue revealed binding of TDP-43 to UG-rich regions in noncoding RNAs and 3′UTRs of mRNAs again with greatest binding to intronic sequences [85]. GO terms of exons regulated by TDP-43 were diverse and included organ morphogenesis, neural tube closure, mitotic cycle, and cell surface receptor linked signaling pathway. Splicing transcripts included those involved in neuronal survival or development, and seven were relevant to neurodegenerative diseases, including *CNTFR* and *KIF1B* [85]. The most significant changes in FTLD-TDP versus healthy controls were increased TDP-43 binding to nuclear paraspeckle assembly transcript 1 (*NEAT1*) and *NEAT2* (*MALAT1*) lncRNAs, and correspondingly these transcripts were increased in FTLD-TDP tissue [85]. The genes with decreased TDP-43 binding were *neurexin 3*

(*NRXN3*) and glial excitatory amino acid transporter-2 (*EAAT2*). EAAT2 levels are decreased in ALS which would cause reduced glutamate clearance at the synaptic cleft and contribute to glutamate excitotoxicty [89, 90]. It is interesting that *NRXN* and/or *NLGN* were consistently identified as TDP-43-binding targets in all four studies, and more generally, genes associated with synaptic function were highly represented [71, 84–86]. Similarly, studies in NSC-34 cells identified TDP-43 mRNA targets enriched for GO terms related to neuron differentiation and dendrite development [91], and *syntaxin 1A* is upregulated in TDP-43 silenced primary neurons [92]. These findings implicate a role for TDP-43 in regulating synaptic integrity/function. TDP-43 has been identified as a neuronal activity responsive factor colocalizing with FMRP and staufen-1 in neuronal transport RNA granules [93]. The number of TDP-43 granules in the somatodendritic compartment of rat primary hippocampal neurons or axons of mouse primary motor neurons increased with neuronal stimulation caused by depolarization or in response to BDNF, respectively [93, 94]. TDP-43 co-associates with RNA and other RBPs in transport granules to deliver translationally dormant mRNAs to synaptic sites, where synaptic activity releases the mRNAs and promotes their localized translation [95–97]. Transport of TDP-43 RNA granules is microtubule-dependent and bidirectional, and involves several motor proteins [98]. Disease-associated mutations in TDP-43 impair this transport of TDP-43 granules and this could lead to synaptic deficits [98]. Indeed, loss or gain of the TDP-43 homolog in Drosophila caused synaptic dysfunction [99–101], and early synaptic loss is a feature of various lines of TDP-43 transgenic mice [102–105]. Collectively, these findings demonstrate the importance of TDP-43 in maintaining synaptic integrity.

Recently, it was demonstrated that TDP-43 represses the splicing of non-conserved cryptic exons which, if expressed, introduce frameshifts and/or premature stop codons causing nonsense-mediated decay [106]. A series of cryptic exons, found to be expressed after knockdown of TDP-43 in HeLa cells, were also found expressed in ALS-FTLD brain tissue [106]. This indicates that aberrant proteins could be expressed or lost as a consequence of TDP-43 depletion, and this could contribute to the disease mechanism. Moreover, the cryptic exons repressed by TDP-43 were highly variable between cell types, and this may give clues to selective vulnerability in disease [107]. To uncover the motor neuron-specific changes caused by abnormal TDP-43, a recent study using translational affinity purification, which allows for the isolation of polysomes from specific cell types, was used to identify the transcripts being actively translated in spinal motor neurons of an A315T mutant TDP-43 transgenic mouse model [108, 109]. Methenyltetrahydrofolate synthetase domain containing (*Mthfsd*) and DEAD (Asp-Glu-Ala-Asp) box polypeptide 58 (*Ddx58*, also known as RIG-1) were identified and found to have altered expression at the protein level in spinal motor neurons of both the A315T TDP-43 mutant mouse and ALS cases [108]. *Mthfsd* is a novel stress granule protein, and was recently ranked in the top ten by IBM® Watson as an ALS-relevant RNA-binding protein [108, 110].

9.3   TDP-43 and Mitochondria

A recent study has shown that TDP-43 is localized to mitochondria, repressing expression of mitochondrial RNAs, and mutations in TDP-43 enhance this localization [111]. Structural damage and fragmentation of mitochondria is thought to be an early event in disease, representing a potential upstream source of motor neuron degeneration [112–114]. The expression of wild type and ALS mutant TDP-43 (M337V, Q331K and A315T) results in vacuolated, fragmented, and aggregated mitochondria [112, 114, 115]. Similarly, the expression of TDP-43 A315T in ALS mouse models and ALS patient fibroblasts has been associated with the loss of mitochondrial cristae [116, 117]. Over-expression of wild-type TDP-43, and in some cases ALS-mutant TDP-43 (Q331K and M337V), alters mitochondrial network dynamics by reducing mitochondrial length in primary motor neurons [114]. In transgenic mice and patient fibroblasts, expression of wild-type or ALS-mutant (M337V and A382T) TDP-43 exhibited altered expression levels of fusion and fission-associated genes, which correlated with anomalous mitochondrial morphology and aggregation [116, 118, 119].

The accumulation of wild-type and ALS-mutant (G298S, A315T and A382T) TDP-43 in mitochondria is mediated by internal mitochondrial targeting sequences in TDP-43 [121]. Wild-type TDP-43 and ALS-mutant TDP-43 (G298S, A315T and A382T) preferentially bind the mRNAs of mtDNA-encoded complex I subunits (ND3 and ND6), and cause the disassembly of complex I by impairing their transcription [121]. Cells expressing wild-type and ALS-mutant (Q331K and M337V) TDP-43 and primary motor neurons expressing TDP-43 M337V were shown to have a reduced mitochondrial membrane potential [124, 125, 120, 121]. TDP-43-associated mitochondrial depolarization is accompanied by a decrease in complex I activity [121]. While studies on patient fibroblasts agree that the expression of TDP-43 G298S and A382T decreases the mitochondrial membrane potential, there is contradictory evidence on whether there is also a decrease in complex I activity, oxygen consumption, and ATP levels [111, 116]. Overall, it is plausible that the gradual depletion of ATP from neurons, which are high energy demanding cells, leads to neuronal degeneration [122].

9.4   TDP-43 Low Complexity Domain and RNP Granules

The C-terminal domain of TDP-43 is a low complexity domain (LCD) sharing homology with prion-like domains, enriched with polar amino acids (asparagine, glutamine, tyrosine, and glycine) characteristic of yeast prions [60]. All but four (A90V, P112H, D169G, N259S) of the over 40 mutations in TDP-43 are clustered in the C-terminal domain (http://alsod.iop.kcl.ac.uk/) (Fig. 9.1b), indicating the importance of this domain in the pathogenesis of ALS/FTLD. A description of these mutations and their potential functional consequences from animal models and cell culture has been reviewed recently [123, 124]. The C-terminal domain is highly aggregation prone both in vitro and in cell culture, with many disease-associated mutations in this domain enhancing its aggregation propensity [125, 126]. LCDs are common to a number of RBPs, several of which are associated with ALS/FTLD, either through mutation and/or presence in disease pathology, including FUS, hnRNPA1, hnRNPA2, and TIA-1 [21, 127, 128]. RBPs associate with RNAs to form various types of RNP granules, membraneless compartments that have diverse roles in RNA processing, transport, storage, and degradation [129–131]. RNP granules include P-bodies, which store and degrade RNA [132]; stress granules, which triage stalled mRNA translation complexes under various stress conditions, promote translation of proteins necessary for cell survival [133, 134]; and neuronal transport granules, which deliver translationally silenced mRNAs to synapses [131, 135]. RNP granules exhibit properties of liquid droplets, being spherical in shape and undergoing fusion and dissolution, rapidly assembling and disassembling in response to environmental cues [136, 137]. The highly dynamic properties of these structures allow for free diffusion within the granules facilitating rapid exchange with the environment [134]. RBPs appear to form RNP granules by a process liquid-liquid phase separation, a condensed phase distinct from the aqueous phase, mediated by homo- or hetero-oligomerization of their LCDs ([129]. In vitro, the liquid droplets, or hydrogels, formed from the LCDs can undergo a process of molecular aging, or maturation, that appears to be concentration-dependent, transitioning from dynamic structures to amyloid-like fibers [138–143]. This process negatively affects RNA granule function [162]. In TDP-43, the underpinning of this change is linked to an amyloidogenic core region between residues 318–367, which undergoes a structural transformation from α-helix to β-sheet during aggregation [62, 66, 138]. This change in conformation may provide the nidus for pathological TDP-43 aggregation [62].

9.5 TDP-43 and Stress Granules

SGs are formed in response to a variety of stressors such as oxidative stress, osmotic stress, or mitochondrial stress [144–146]. Several hundred proteins are associated with SGs, compositions varying in a cell- and stress-type-specific manner [147, 148]. SGs are considered biphasic, comprising a condensed stable core formed by nucleation of non-translating mRNPs, and surrounded by a less concentrated and highly dynamic shell formed through liquid-liquid phase separation of the composite RNPs [130, 149, 150].TDP-43 is recruited to SGs under a range of environmental stressors in cell culture, including arsenite, paraquat, and sorbitol, and these SGs disassemble when the stressor is removed [151–155]. TDP-43 appears to influence SG assembly/disassembly by regulating expression of G3BP [156]. Moreover, TDP-43 mutations alter the frequency and size of SGs [152, 155, 157]. Interestingly, a study has shown that paraquat treatment induces TDP-43 SGs in HeLa cells [154]. After removal of paraquat the majority of SGs dissemble, however small amounts of TDP-43 persist, forming ubiquitinated aggregates [154]. This is reminiscent of molecular aging as seen from the in vitro studies of LCDs described above. It has been proposed that defects of SG dynamics are the precursor of TDP-43 pathology in ALS/FTLD. This is based on the observation that TDP-43 pathology in ALS spinal cord is co-labeled with SG markers, TIA-1 and/or eIF3 [158, 159]. However, other studies have failed to show colocalization of SG markers with TDP-43 inclusions [128, 151]. This issue remains unresolved. Stress granules are cleared by autophagy [160]; and it is interesting that a number of genes causing ALS/FTLD affect the autophagic machinery, including VCP, TBK1, SQSTM1, OPTN, and UBQLN2 [161]. This suggests that persistent SG pathology could result as a consequence of impaired autophagy.

9.6   Ataxin 2 Association with TDP-43 and ALS

Ataxin 2 is mainly a cytoplasmic protein with a diverse set of functions ranging from regulating RNA stability and translation to repressing fat and glycogen storage through mTORC1 signaling [162–164]. Polyglutamine tract expansions within the first exon of ataxin 2 cause spinocerebellar ataxia type 2 (SCA2) [165–169]. Normal expansion length is 22–23 glutamines (Q), with >34 Qs causing SCA2. Recently, intermediate length expansions of 27–33 Qs were identified as a risk factor for ALS [170, 171]. Ataxin 2 is a regulator of SG dynamics, with higher levels of ataxin 2 inducing SGs and lower levels reducing SGs [163, 164, 172], including TDP-43 SGs [172]. Downregulation or upregulation of ataxin 2 suppresses or enhances toxicity associated with TDP-43 expression in Drosophila and yeast [170]. Similarly, knockout or knockdown of ataxin 2 extended the life span and reduced pathology of a wild-type TDP-43 transgenic mouse model [172]. The normal life expectancy of the TDP-43 transgenic mouse model used in the study was P24 days with walking deficits apparent at P21 days. Complete knockout of ataxin 2 gave an 80% increase in survival with several animals surviving beyond a remarkable 300 days [172]. One hypothesis for the therapeutic effects of lowering levels of ataxin 2 is that it reduces the number of TDP-43 SGs, the proposed seeds of pathology. Indeed, loss of ataxin 2 reduced the number of TDP-43 aggregates (not defined as SGs) in the brain and spinal cord of the TDP-43 transgenic mouse model used for the study [172]. This supports the idea that abnormal SG dynamics is a key driver of TDP-43 pathology and associated neurodegenerative phenotypes, and that reducing ataxin 2 levels could have therapeutic utility. However, it is possible that ataxin 2 may also be acting through its role as a starvation response factor [162, 173].

9.7   TDP-43 C-Terminal Fragments

In the original studies identifying TDP-43 as a component of pathological inclusion bodies in ALS, a biochemical signature was also described in urea soluble fractions from FTLD-TDP tissues. This comprised the full-length protein at 43 kDa (TDP-43), a species of ~45 kDa corresponding to phosphorylated TDP-43 (P-TDP-43), a higher molecular weight smear corresponding to ubiquitinated TDP-43, and two lower molecular weight species of 24 and 26 kDa corresponding to the C-terminal domain of TDP-43, which separated into at least four species upon dephosphorylation [10, 11]. These lower molecular weight species have been collectively termed TDP-25. Subsequent studies have also identified an N-terminally truncated species of ~35 kDa in ALS or FTLD tissues [83, 174], called TDP-35. Studies have shown that species of 25 and 35 kDa can be generated by caspase 3 cleavage of TDP-43, both in vitro and in cell culture [83, 175]. There are three caspase-3 cleavage sites in TDP-43: DEND (residues 9–12), DETD (residues 86–89), and DVMD (residues 216–219) [83] (Fig. 9.1a). Cleavage at DEND generates a species of ~35 kDa, and cleavage at DETD generates a species of ~25 kDa [83]. Based on these findings, it has been hypothesized that TDP-25 and TDP-35 in ALS/FTLD are generated by caspase-3 cleavage, which has been activated during the neurodegenerative process. Although caspase-3 activation has not unequivocally been shown in disease tissues, this mechanism for the generation of low molecular weight species in ALS/FTLD has become dogma in the field. Using mass spectrometry analysis of disease tissues, additional proteolytic cleavage sites have also been identified, including R208 [176], N291 and N306 [177], and a series of peptides with differing lengths, corresponding to C-terminal domain regions [178, 179]. It is notable that the observed lower molecular weight species from ALS/FTLD tissue extracts correspond to C-terminal regions of TDP-43, and that N-terminal fragments are absent. It is possible that the N-terminal fragments are degraded and the C-terminal spared, perhaps because the C-terminal is pathologically phosphorylated, or that it contains the prion-like LCD, which may be aggregated/misfolded. Indeed, prior studies of human disease tissue have shown that residues 203–209 [180] and residues 341–346 or 341–360 of the Q/N-rich segment are protease resistant [178, 181]. TDP-43 species of 25 and 35 kDa are observed in caspase-3 knockout murine embryonic fibroblasts, indicating that species of these molecular weights can be generated by means other than proteolytic cleavage [182]. This led to the discovery of an alternative translation start site at Met 85, expression from which generates a protein product of 35 kDa [182]. Of note, Met 85 immediately precedes the DETD (residues 86–89) caspase 3 cleavage site (Fig. 9.1a) and as such there are two potential forms of TDP-35, caspase-3 cleaved TDP-35 (C3-TDP-35) and TDP-35 generated by expression from Met 85 (Met-TDP-35) [174]. TDP-35 (either caspase-3 cleaved or Met 85) has a disrupted NLS and mainly localizes to the cytoplasm in transfected cells [182, 183]. Interestingly, TDP-35 spontaneously forms cytoplasmic SGs that recruit nuclear TDP-43 [182]. The TDP-35 SGs are labeled with stress granule markers TIA-1 (Fig. 9.2a–c), G3BP, PABP, and HuR, but not with a P-body marker

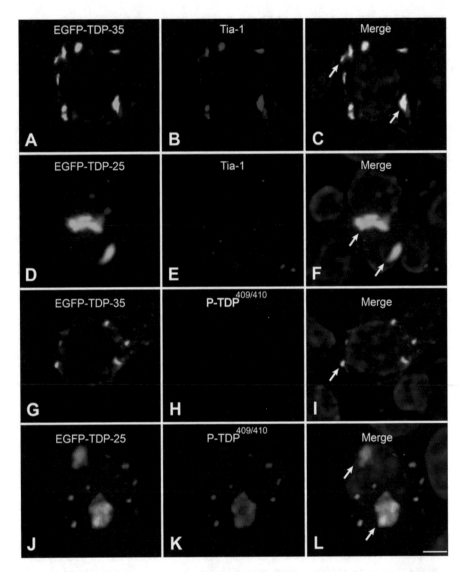

Fig. 9.2 TDP-35 forms stress granules and TDP-25 forms phosphorylated aggregates in transfected HEK 293 cells. HEK 293 cells expressing EGFP-TDP-35 (residues 85–414) (**a, g**) or EGFP-TDP-25 (residues 220–414) (**d** and **j**); double-labeling with antibody to TIA-1 (**b–e**), or P-TDP-43-409/410 antibody (**h** and **k**) (red); merge with DAPI stain (**c–l**). Note cytoplasmic EGFP-TDP-35 (**a–c**) but not EGFP-TDP-25 (**d, e**) colocalizes with Tia-1 (arrows in **c** and **f**), and EGFP-TDP-25 (**j–l**) but not EGFP-TDP-35 (**g–i**) colocalizes with P-TDP-43 (arrows in **i** and **l**). Scale bar = 5 μm

DCP1a. The TDP-35 SGs were not labeled with antibody to phosphorylated TDP-43 (phospho 409/410) (Fig. 9.2g–i), but interestingly TDP-35 SGs transformed to a phosphorylated and ubiquitinated aggregated state after treatment with the protease inhibitor MG132 [182]. This finding indicates that TDP-35 SG formation together with deficits in protein quality control can lead to cytoplasmic aggregates with features modeling TDP-43 inclusions in disease. This alludes to a maturation/molecular aging mechanism as described for LCDs above.

TDP-43 autoregulates its own expression and splicing, and a number of TDP-43 splice variants have been reported [70, 71, 184, 185]. A splice variant lacking 91 bp in exon was found upregulated in ALS spinal cord [174]. Expression of this splice variant in cell culture led to the use of the alternate translation start site at Met 85, generating cytoplasmic aggregates of Met 85-TDP-35 that were toxic in primary motor neurons [174]. To verify the genesis of TDP-35 in ALS, neoepitope antibodies were generated that could differentiate between C3-TDP-35 and Met 85-TDP-35 [174]. TDP-43 pathology in ALS spinal cord was labeled with the Met 85-TDP-35 antibody, but not by the caspase 3-TDP-35 antibody, supporting that TDP-35 in ALS is generated through the use of an alternate translation start site (Met 85) and not by caspase 3 cleavage at DETD. Since cytoplasmic TDP-35 can recruit full-length TDP-43, this suggests that TDP-35 could act as a seed for pathological TDP-43 aggregation [68, 174, 182, 186]. It is possible that abnormal splicing of TDP-43 at exon 2 (through means unknown) generates TDP-35 through use of Met 85, which over time forms SGs in a concentration-dependent manner, recruiting full-length TDP-43. As mentioned, TDP-43 functions in the nucleus through oligomerization of its N-terminal domains, thus preventing interaction between the C-terminal domains [73]. Loss of the N-terminal domain, as in TDP-35, would promote interaction between the C-terminal domains, causing aggregation and toxicity [73, 74, 76, 187, 188].

Various other regions and domains of TDP-43 have also been expressed in cell culture [62–64, 67, 69, 74, 75, 114, 151, 152, 189–200]. Expression of C-terminal domains representative of TDP-25 encompassing residues 220–414 [83] or 177–414, 187–414, 197–414 or 208–414 [176], generate cytoplasmic aggregates that are toxic. Unlike TDP-35, these aggregates do not appear to be SGs as they have irregular contours and only very minimally colocalize with SG markers [158, 201] (Fig. 9.2d–f). Instead, the TDP-25 aggregates are ubiquitinated and phosphorylated at 409/410 [83, 176] (Fig. 9.2j–l). Phosphorylation is not necessary for the aggregation of TDP-25 [83]. Instead, evidence suggests that phosphorylation is a defense mechanism against TDP-43 aggregation [202].

9.8 Concluding Remarks

Here, we have given an overview of some of the nuclear and cytoplasmic functions of TDP-43, and the potential consequences if either is perturbed. A major question is what causes TDP-43 to mislocalize from the nucleus to the cytoplasm? Is it a response to stress, causing nuclear clearing and stress granule formation, with some conformational change in TDP-43 causing aggregation?

Or is it cytoplasmic seeding by disease-associated variants of TDP-43, such as TDP-35? A current view is that TDP-43 pathology could be caused by defects in nucleocytoplasmic transport [203–206]. However, recent studies show that cytoplasmic aggregates of TDP-43 can initiate nucleocytoplasmic transport deficits [207, 208]. These issues remain to be resolved. Finally, there is great interest in prion-like propagation of misfolded/aggregated proteins as a means of spreading disease between cells and between different regions of the brain, as is the case for the microtubule associated protein tau [209, 210] and alpha synuclein [211, 212]. There is evidence that different misfolded conformers, or strains, of these proteins encode strain-specific information generating morphologically distinct types of pathologies [213, 214]. Recent studies suggest that TDP-43 may also spread in a similar fashion [215, 216]. It is tempting to speculate that the types A, B, C, and D TDP-43 neuropathologies observed in ALS/FTLD may be a consequence of different types of TDP-43 strains [215–218].

References

1. Kiernan MC, Vucic S, Cheah BC, Turner MR, Eisen A, Hardiman O, et al. Amyotrophic lateral sclerosis. Lancet. 2011;377(9769):942–55.
2. Chio A, Logroscino G, Traynor BJ, Collins J, Simeone JC, Goldstein LA, et al. Global epidemiology of amyotrophic lateral sclerosis: a systematic review of the published literature. Neuroepidemiology. 2013;41(2):118–30.
3. Chia R, Chio A, Traynor BJ. Novel genes associated with amyotrophic lateral sclerosis: diagnostic and clinical implications. Lancet Neurol. 2018;17(1):94–102.
4. Strong MJ. Revisiting the concept of amyotrophic lateral sclerosis as a multisystems disorder of limited phenotypic expression. Curr Opin Neurol. 2017;30(6):599–607.
5. Ng AS, Rademakers R, Miller BL. Frontotemporal dementia: a bridge between dementia and neuromuscular disease. Ann N Y Acad Sci. 2015;1338:71–93.
6. Bang J, Spina S, Miller BL. Frontotemporal dementia. Lancet. 2015;386(10004):1672–82.
7. Neary D, Snowden JS, Gustafson L, Passant U, Stuss D, Black S, et al. Frontotemporal lobar degeneration: a consensus on clinical diagnostic criteria. Neurology. 1998;51(6):1546–54.
8. Warren JD, Rohrer JD, Rossor MN. Clinical review. Frontotemporal dementia. BMJ. 2013;347:f4827.
9. Phukan J, Elamin M, Bede P, Jordan N, Gallagher L, Byrne S, et al. The syndrome of cognitive impairment in amyotrophic lateral sclerosis: a population-based study. J Neurol Neurosurg Psychiatry. 2012;83(1):102–8.
10. Arai T, Hasegawa M, Akiyama H, Ikeda K, Nonaka T, Mori H, et al. TDP-43 is a component of ubiquitin-positive tau-negative inclusions in frontotemporal lobar degeneration and amyotrophic lateral sclerosis. Biochem Biophys Res Commun. 2006;351(3):602–11.
11. Neumann M, Sampathu DM, Kwong LK, Truax AC, Micsenyi MC, Chou TT, et al. Ubiquitinated TDP-43 in frontotemporal lobar degeneration and amyotrophic lateral sclerosis. Science. 2006;314(5796):130–3.
12. Kabashi E, Valdmanis PN, Dion P, Spiegelman D, McConkey BJ, Vande Velde C, et al. TARDBP mutations in individuals with sporadic and familial amyotrophic lateral sclerosis. Nat Genet. 2008;40(5):572–4.

13. Sreedharan J, Blair IP, Tripathi VB, Hu X, Vance C, Rogelj B, et al. TDP-43 mutations in familial and sporadic amyotrophic lateral sclerosis. Science. 2008;319(5870):1668–72.
14. Buratti E, Baralle FE. The multiple roles of TDP-43 in pre-mRNA processing and gene expression regulation. RNA Biol. 2010;7(4):420–9.
15. Ratti A, Buratti E. Physiological functions and pathobiology of TDP-43 and FUS/TLS proteins. J Neurochem. 2016;138(Suppl 1):95–111.
16. Brandmeir NJ, Geser F, Kwong LK, Zimmerman E, Qian J, Lee VM, et al. Severe subcortical TDP-43 pathology in sporadic frontotemporal lobar degeneration with motor neuron disease. Acta Neuropathol. 2008;115(1):123–31.
17. Neumann M, Kwong LK, Truax AC, Vanmassenhove B, Kretzschmar HA, Van Deerlin VM, et al. TDP-43-positive white matter pathology in frontotemporal lobar degeneration with ubiquitin-positive inclusions. J Neuropathol Exp Neurol. 2007;66(3):177–83.
18. Sanelli T, Xiao S, Horne P, Bilbao J, Zinman L, Robertson J. Evidence that TDP-43 is not the major ubiquitinated target within the pathological inclusions of amyotrophic lateral sclerosis. J Neuropathol Exp Neurol. 2007;66(12):1147–53.
19. Hasegawa M, Arai T, Nonaka T, Kametani F, Yoshida M, Hashizume Y, et al. Phosphorylated TDP-43 in frontotemporal lobar degeneration and amyotrophic lateral sclerosis. Ann Neurol. 2008;64(1):60–70.
20. Ling SC, Polymenidou M, Cleveland DW. Converging mechanisms in ALS and FTD: disrupted RNA and protein homeostasis. Neuron. 2013;79(3):416–38.
21. Mackenzie IR, Neumann M. Reappraisal of TDP-43 pathology in FTLD-U subtypes. Acta Neuropathol. 2017b;134(1):79–96.
22. Mackenzie IR, Bigio EH, Ince PG, Geser F, Neumann M, Cairns NJ, et al. Pathological TDP-43 distinguishes sporadic amyotrophic lateral sclerosis from amyotrophic lateral sclerosis with SOD1 mutations. Ann Neurol. 2007;61(5):427–34.
23. Maekawa S, Leigh PN, King A, Jones E, Steele JC, Bodi I, et al. TDP-43 is consistently co-localized with ubiquitinated inclusions in sporadic and Guam amyotrophic lateral sclerosis but not in familial amyotrophic lateral sclerosis with and without SOD1 mutations. Neuropathology. 2009;29(6):672–83.
24. Robertson J, Sanelli T, Xiao S, Yang W, Horne P, Hammond R, et al. Lack of TDP-43 abnormalities in mutant SOD1 transgenic mice shows disparity with ALS. Neurosci Lett. 2007;420(2):128–32.
25. Braak H, Ludolph AC, Neumann M, Ravits J, Del Tredici K. Pathological TDP-43 changes in Betz cells differ from those in bulbar and spinal alpha-motoneurons in sporadic amyotrophic lateral sclerosis. Acta Neuropathol. 2017;133(1):79–90.
26. Mackenzie IR, Neumann M. Molecular neuropathology of frontotemporal dementia: insights into disease mechanisms from postmortem studies. J Neurochem. 2016;138(Suppl 1):54–70.
27. Josephs KA, Hodges JR, Snowden JS, Mackenzie IR, Neumann M, Mann DM, et al. Neuropathological background of phenotypical variability in frontotemporal dementia. Acta Neuropathol. 2011;122(2):137–53.
28. Mackenzie IR, Neumann M, Baborie A, Sampathu DM, Du Plessis D, Jaros E, et al. A harmonized classification system for FTLD-TDP pathology. Acta Neuropathol. 2011;122(1):111–3.
29. Mackenzie IRA, Neumann M. Fused in sarcoma neuropathology in neurodegenerative disease. Cold Spring Harb Perspect Med. 2017a;7(12). https://doi.org/10.1101/cshperspect. a024299.
30. Murray ME, DeJesus-Hernandez M, Rutherford NJ, Baker M, Duara R, Graff-Radford NR, et al. Clinical and neuropathologic heterogeneity of c9FTD/ALS associated with hexanucleotide repeat expansion in C9ORF72. Acta Neuropathol. 2011;122(6):673–90.
31. Ash PE, Bieniek KF, Gendron TF, Caulfield T, Lin WL, Dejesus-Hernandez M, et al. Unconventional translation of C9ORF72 GGGGCC expansion generates insoluble polypeptides specific to c9FTD/ALS. Neuron. 2013;77(4):639–46.

32. Mori K, Arzberger T, Grasser FA, Gijselinck I, May S, Rentzsch K, et al. Bidirectional transcripts of the expanded C9orf72 hexanucleotide repeat are translated into aggregating dipeptide repeat proteins. Acta Neuropathol. 2013;126(6):881–93.

33. Zu T, Liu Y, Banez-Coronel M, Reid T, Pletnikova O, Lewis J, et al. RAN proteins and RNA foci from antisense transcripts in C9ORF72 ALS and frontotemporal dementia. Proc Natl Acad Sci U S A. 2013;110(51):E4968–77.

34. Davidson YS, Barker H, Robinson AC, Thompson JC, Harris J, Troakes C, et al. Brain distribution of dipeptide repeat proteins in frontotemporal lobar degeneration and motor neurone disease associated with expansions in C9ORF72. Acta Neuropathol Commun. 2014;2:70.

35. Mackenzie IR, Frick P, Grasser FA, Gendron TF, Petrucelli L, Cashman NR, et al. Quantitative analysis and clinico-pathological correlations of different dipeptide repeat protein pathologies in C9ORF72 mutation carriers. Acta Neuropathol. 2015;130(6):845–61.

36. Davidson Y, Robinson AC, Liu X, Wu D, Troakes C, Rollinson S, et al. Neurodegeneration in frontotemporal lobar degeneration and motor neurone disease associated with expansions in C9orf72 is linked to TDP-43 pathology and not associated with aggregated forms of dipeptide repeat proteins. Neuropathol Appl Neurobiol. 2016;42(3):242–54.

37. Saberi S, Stauffer JE, Jiang J, Garcia SD, Taylor AE, Schulte D, et al. Sense-encoded poly-GR dipeptide repeat proteins correlate to neurodegeneration and uniquely co-localize with TDP-43 in dendrites of repeat-expanded C9orf72 amyotrophic lateral sclerosis. Acta Neuropathol. 2018;135(3):459–74.

38. Amador-Ortiz C, Lin WL, Ahmed Z, Personett D, Davies P, Duara R, et al. TDP-43 immunoreactivity in hippocampal sclerosis and Alzheimer's disease. Ann Neurol. 2007;61(5):435–45.

39. Higashi S, Iseki E, Yamamoto R, Minegishi M, Hino H, Fujisawa K, et al. Concurrence of TDP-43, tau and alpha-synuclein pathology in brains of Alzheimer's disease and dementia with Lewy bodies. Brain Res. 2007;1184:284–94.

40. Josephs KA, Murray ME, Whitwell JL, Parisi JE, Petrucelli L, Jack CR, et al. Staging TDP-43 pathology in Alzheimer's disease. Acta Neuropathol. 2014;127(3):441–50.

41. Josephs KA, Whitwell JL, Knopman DS, Hu WT, Stroh DA, Baker M, et al. Abnormal TDP-43 immunoreactivity in AD modifies clinicopathologic and radiologic phenotype. Neurology. 2008;70(19 Pt 2):1850–7.

42. Schwab C, Arai T, Hasegawa M, Yu S, McGeer PL. Colocalization of transactivation-responsive DNA-binding protein 43 and huntingtin in inclusions of Huntington disease. J Neuropathol Exp Neurol. 2008;67(12):1159–65.

43. Nakashima-Yasuda H, Uryu K, Robinson J, Xie SX, Hurtig H, Duda JE, et al. Co-morbidity of TDP-43 proteinopathy in Lewy body related diseases. Acta Neuropathol (Berl). 2007;114(3):221–9.

44. Lin WL, Dickson DW. Ultrastructural localization of TDP-43 in filamentous neuronal inclusions in various neurodegenerative diseases. Acta Neuropathol. 2008;116(2):205–13.

45. Arai T, Mackenzie IR, Hasegawa M, Nonoka T, Niizato K, Tsuchiya K, et al. Phosphorylated TDP-43 in Alzheimer's disease and dementia with Lewy bodies. Acta Neuropathol. 2009;117(2):125–36.

46. Fujishiro H, Uchikado H, Arai T, Hasegawa M, Akiyama H, Yokota O, et al. Accumulation of phosphorylated TDP-43 in brains of patients with argyrophilic grain disease. Acta Neuropathol. 2009;117(2):151–8.

47. McKee AC, Daneshvar DH, Alvarez VE, Stein TD. The neuropathology of sport. Acta Neuropathol. 2014;127(1):29–51.

48. Ou SH, Wu F, Harrich D, Garcia-Martinez LF, Gaynor RB. Cloning and characterization of a novel cellular protein, TDP-43, that binds to human immunodeficiency virus type 1 TAR DNA sequence motifs. J Virol. 1995;69(6):3584–96.

49. Buratti E, Baralle FE. Characterization and functional implications of the RNA binding properties of nuclear factor TDP-43, a novel splicing regulator of CFTR exon 9. J Biol Chem. 2001;276(39):36337–43.

50. Buratti E, Dork T, Zuccato E, Pagani F, Romano M, Baralle FE. Nuclear factor TDP-43 and SR proteins promote in vitro and in vivo CFTR exon 9 skipping. EMBO J. 2001;20(7):1774–84.

51. Ayala YM, Pantano S, D'Ambrogio A, Buratti E, Brindisi A, Marchetti C, et al. Human, Drosophila, and C. elegans TDP43: nucleic acid binding properties and splicing regulatory function. J Mol Biol. 2005;348(3):575–88.

52. Kraemer BC, Schuck T, Wheeler JM, Robinson LC, Trojanowski JQ, Lee VM, et al. Loss of murine TDP-43 disrupts motor function and plays an essential role in embryogenesis. Acta Neuropathol. 2010;119(4):409–19.

53. Sephton CF, Good SK, Atkin S, Dewey CM, Mayer P 3rd, Herz J, et al. TDP-43 is a developmentally regulated protein essential for early embryonic development. J Biol Chem. 2010;285(9):6826–34.

54. Wu LS, Cheng WC, Hou SC, Yan YT, Jiang ST, Shen CK. TDP-43, a neuro-pathosignature factor, is essential for early mouse embryogenesis. Genesis. 2010;48(1):56–62.

55. Lim L, Wei Y, Lu Y, Song J. ALS-causing mutations significantly perturb the self-assembly and interaction with nucleic acid of the intrinsically disordered prion-like domain of TDP-43. PLoS Biol. 2016;14(1):e1002338.

56. Qin H, Lim LZ, Wei Y, Song J. TDP-43 N terminus encodes a novel ubiquitin-like fold and its unfolded form in equilibrium that can be shifted by binding to ssDNA. Proc Natl Acad Sci U S A. 2014;111(52):18619–24.

57. Mompean M, Baralle M, Buratti E, Laurents DV. An amyloid-like pathological conformation of TDP-43 is stabilized by hypercooperative hydrogen bonds. Front Mol Neurosci. 2016b;9:125.

58. Budini M, Baralle FE, Buratti E. Targeting TDP-43 in neurodegenerative diseases. Expert Opin Ther Targets. 2014;18(6):617–32.

59. Buratti E, Brindisi A, Giombi M, Tisminetzky S, Ayala YM, Baralle FE. TDP-43 binds heterogeneous nuclear ribonucleoprotein A/B through its C-terminal tail: an important region for the inhibition of cystic fibrosis transmembrane conductance regulator exon 9 splicing. J Biol Chem. 2005;280(45):37572–84.

60. Cushman M, Johnson BS, King OD, Gitler AD, Shorter J. Prion-like disorders: blurring the divide between transmissibility and infectivity. J Cell Sci. 2010;123(Pt 8):1191–201.

61. D'Ambrogio A, Buratti E, Stuani C, Guarnaccia C, Romano M, Ayala YM, et al. Functional mapping of the interaction between TDP-43 and hnRNP A2 in vivo. Nucleic Acids Res. 2009;37(12):4116–26.

62. Jiang LL, Che MX, Zhao J, Zhou CJ, Xie MY, Li HY, et al. Structural transformation of the amyloidogenic core region of TDP-43 protein initiates its aggregation and cytoplasmic inclusion. J Biol Chem. 2013;288(27):19614–24.

63. Budini M, Buratti E, Stuani C, Guarnaccia C, Romano V, De Conti L, et al. Cellular model of TAR DNA-binding protein 43 (TDP-43) aggregation based on its C-terminal Gln/Asn-rich region. J Biol Chem. 2012;287(10):7512–25.

64. Fuentealba RA, Udan M, Bell S, Wegorzewska I, Shao J, Diamond MI, et al. Interaction with polyglutamine aggregates reveals a Q/N-rich domain in TDP-43. J Biol Chem. 2010;285(34):26304–14.

65. Mompean M, Buratti E, Guarnaccia C, Brito RM, Chakrabartty A, Baralle FE, et al. Structural characterization of the minimal segment of TDP-43 competent for aggregation. Arch Biochem Biophys. 2014;545:53–62.

66. Mompean M, Hervas R, Xu Y, Tran TH, Guarnaccia C, Buratti E, et al. Structural evidence of amyloid fibril formation in the putative aggregation domain of TDP-43. J Phys Chem Lett. 2015;6(13):2608–15.

67. Ayala YM, Zago P, Ambrogio A, Xu Y-F, Petrucelli L, Buratti E, et al. Structural determinants of the cellular localization and shuttling of TDP-43. J Cell Sci. 2008;121(22):3778.

68. Winton MJ, Igaz LM, Wong MM, Kwong LK, Trojanowski JQ, Lee VM. Disturbance of nuclear and cytoplasmic TAR DNA-binding protein (TDP-43) induces disease-like redistribution, sequestration, and aggregate formation. J Biol Chem. 2008a;283(19):13302–9.

69. Winton MJ, Van Deerlin VM, Kwong LK, Yuan W, Wood EM, Yu CE, et al. A90V TDP-43 vari-
    ant results in the aberrant localization of TDP-43 in vitro. FEBS Lett. 2008b;582(15):2252–6.
70. Ayala YM, De Conti L, Avendano-Vazquez SE, Dhir A, Romano M, D'Ambrogio A,
    et al. TDP-43 regulates its mRNA levels through a negative feedback loop. EMBO
    J. 2011;30(2):277–88.
71. Polymenidou M, Lagier-Tourenne C, Hutt KR, Huelga SC, Moran J, Liang TY, et al. Long
    pre-mRNA depletion and RNA missplicing contribute to neuronal vulnerability from loss of
    TDP-43. Nat Neurosci. 2011;14(4):459–68.
72. Lukavsky PJ, Daujotyte D, Tollervey JR, Ule J, Stuani C, Buratti E, et al. Molecular basis
    of UG-rich RNA recognition by the human splicing factor TDP-43. Nat Struct Mol Biol.
    2013;20(12):1443–9.
73. Afroz T, Hock EM, Ernst P, Foglieni C, Jambeau M, Gilhespy LAB, et al. Functional and
    dynamic polymerization of the ALS-linked protein TDP-43 antagonizes its pathologic aggre-
    gation. Nat Commun. 2017;8(1):45.
74. Jiang LL, Xue W, Hong JY, Zhang JT, Li MJ, Yu SN, et al. The N-terminal dimerization is
    required for TDP-43 splicing activity. Sci Rep. 2017;7(1):6196.
75. Shiina Y, Arima K, Tabunoki H, Satoh J. TDP-43 dimerizes in human cells in culture. Cell
    Mol Neurobiol. 2010;30(4):641–52.
76. Zhang YJ, Caulfield T, Xu YF, Gendron TF, Hubbard J, Stetler C, et al. The dual functions of
    the extreme N-terminus of TDP-43 in regulating its biological activity and inclusion forma-
    tion. Hum Mol Genet. 2013;22(15):3112–22.
77. Chang CK, Wu TH, Wu CY, Chiang MH, Toh EK, Hsu YC, et al. The N-terminus of TDP-
    43 promotes its oligomerization and enhances DNA binding affinity. Biochem Biophys Res
    Commun. 2012;425(2):219–24.
78. Kralovicova J, Patel A, Searle M, Vorechovsky I. The role of short RNA loops in recognition
    of a single-hairpin exon derived from a mammalian-wide interspersed repeat. RNA Biol.
    2015;12(1):54–69.
79. Moreno F, Rabinovici GD, Karydas A, Miller Z, Hsu SC, Legati A, et al. A novel muta-
    tion P112H in the TARDBP gene associated with frontotemporal lobar degeneration without
    motor neuron disease and abundant neuritic amyloid plaques. Acta Neuropathol Commun.
    2015;3:19.
80. Maurel C, Madji-Hounoum B, Thepault RA, Marouillat S, Brulard C, Danel-Brunaud V, et al.
    Mutation in the RRM2 domain of TDP-43 in Amyotrophic Lateral Sclerosis with rapid pro-
    gression associated with ubiquitin positive aggregates in cultured motor neurons. Amyotroph
    Lateral Scler Frontotemporal Degener. 2018;19(1–2):149–51.
81. Kuo PH, Chiang CH, Wang YT, Doudeva LG, Yuan HS. The crystal structure of TDP-43
    RRM1-DNA complex reveals the specific recognition for UG- and TG-rich nucleic acids.
    Nucleic Acids Res. 2014;42(7):4712–22.
82. Chiang CH, Grauffel C, Wu LS, Kuo PH, Doudeva LG, Lim C, et al. Structural analysis of
    disease-related TDP-43 D169G mutation: linking enhanced stability and caspase cleavage
    efficiency to protein accumulation. Sci Rep. 2016;6:21581.
83. Zhang YJ, Xu YF, Cook C, Gendron TF, Roettges P, Link CD, et al. Aberrant cleav-
    age of TDP-43 enhances aggregation and cellular toxicity. Proc Natl Acad Sci U S A.
    2009;106(18):7607–12.
84. Sephton CF, Cenik C, Kucukural A, Dammer EB, Cenik B, Han Y, et al. Identification of
    neuronal RNA targets of TDP-43-containing ribonucleoprotein complexes. J Biol Chem.
    2011;286(2):1204–15.
85. Tollervey JR, Wang Z, Hortobagyi T, Witten JT, Zarnack K, Kayikci M, et al. Analysis of
    alternative splicing associated with aging and neurodegeneration in the human brain. Genome
    Res. 2011;21(10):1572–82.
86. Xiao S, Sanelli T, Dib S, Sheps D, Findlater J, Bilbao J, et al. RNA targets of TDP-43 identi-
    fied by UV-CLIP are deregulated in ALS. Mol Cell Neurosci. 2011;47(3):167–80.

87. Prudencio M, Jansen-West KR, Lee WC, Gendron TF, Zhang YJ, Xu YF, et al. Misregulation of human sortilin splicing leads to the generation of a nonfunctional progranulin receptor. Proc Natl Acad Sci U S A. 2012;109(52):21510–5.

88. Pickering-Brown SM. Progranulin and frontotemporal lobar degeneration. Acta Neuropathol. 2007;114(1):39–47.

89. Rosenblum LT, Trotti D. EAAT2 and the molecular signature of amyotrophic lateral sclerosis. Adv Neurobiol. 2017;16:117–36.

90. Rothstein JD, Van Kammen M, Levey AI, Martin LJ, Kuncl RW. Selective loss of glial glutamate transporter GLT-1 in amyotrophic lateral sclerosis. Ann Neurol. 1995;38(1):73–84.

91. Colombrita C, Onesto E, Megiorni F, Pizzuti A, Baralle FE, Buratti E, et al. TDP-43 and FUS RNA-binding proteins bind distinct sets of cytoplasmic messenger RNAs and differently regulate their post-transcriptional fate in motoneuron-like cells. J Biol Chem. 2012;287(19):15635–47.

92. Honda D, Ishigaki S, Iguchi Y, Fujioka Y, Udagawa T, Masuda A, et al. The ALS/FTLD-related RNA-binding proteins TDP-43 and FUS have common downstream RNA targets in cortical neurons. FEBS Open Bio. 2013;4:1–10.

93. Wang IF, Wu LS, Chang HY, Shen CK. TDP-43, the signature protein of FTLD-U, is a neuronal activity-responsive factor. J Neurochem. 2008;105(3):797–806.

94. Fallini C, Bassell GJ, Rossoll W. The ALS disease protein TDP-43 is actively transported in motor neuron axons and regulates axon outgrowth. Hum Mol Genet. 2012;21(16):3703–18.

95. Coyne AN, Siddegowda BB, Estes PS, Johannesmeyer J, Kovalik T, Daniel SG, et al. Futsch/MAP1B mRNA is a translational target of TDP-43 and is neuroprotective in a Drosophila model of amyotrophic lateral sclerosis. J Neurosci. 2014;34(48):15962–74.

96. Ishiguro A, Kimura N, Watanabe Y, Watanabe S, Ishihama A. TDP-43 binds and transports G-quadruplex-containing mRNAs into neurites for local translation. Genes Cells. 2016;21(5):466–81.

97. Narayanan RK, Mangelsdorf M, Panwar A, Butler TJ, Noakes PG, Wallace RH. Identification of RNA bound to the TDP-43 ribonucleoprotein complex in the adult mouse brain. Amyotroph Lateral Scler Frontotemporal Degener. 2013;14(4):252–60.

98. Alami NH, Smith RB, Carrasco MA, Williams LA, Winborn CS, Han SSW, et al. Axonal transport of TDP-43 mRNA granules is impaired by ALS-causing mutations. Neuron. 2014;81(3):536–43.

99. Diaper DC, Adachi Y, Sutcliffe B, Humphrey DM, Elliott CJ, Stepto A, et al. Loss and gain of Drosophila TDP-43 impair synaptic efficacy and motor control leading to age-related neurodegeneration by loss-of-function phenotypes. Hum Mol Genet. 2013;22(8):1539–57.

100. Feiguin F, Godena VK, Romano G, D'Ambrogio A, Klima R, Baralle FE. Depletion of TDP-43 affects Drosophila motoneurons terminal synapsis and locomotive behavior. FEBS Lett. 2009;583(10):1586–92.

101. Romano G, Klima R, Buratti E, Verstreken P, Baralle FE, Feiguin F. Chronological requirements of TDP-43 function in synaptic organization and locomotive control. Neurobiol Dis. 2014;71:95–109.

102. Chand KK, Lee KM, Lee JD, Qiu H, Willis EF, Lavidis NA, et al. Defects in synaptic transmission at the neuromuscular junction precedes motor deficits in a TDP-43(Q331K) transgenic mouse model of amyotrophic lateral sclerosis. FASEB J. 2018. https://doi.org/10.1096/fj.201700835R.

103. Fogarty MJ, Klenowski PM, Lee JD, Drieberg-Thompson JR, Bartlett SE, Ngo ST, et al. Cortical synaptic and dendritic spine abnormalities in a presymptomatic TDP-43 model of amyotrophic lateral sclerosis. Sci Rep. 2016;6:37968.

104. Handley EE, Pitman KA, Dawkins E, Young KM, Clark RM, Jiang TC, et al. Synapse dysfunction of layer V pyramidal neurons precedes neurodegeneration in a mouse model of TDP-43 proteinopathies. Cereb Cortex. 2017;27(7):3630–47.

105. Medina DX, Orr ME, Oddo S. Accumulation of C-terminal fragments of transactive response DNA-binding protein 43 leads to synaptic loss and cognitive deficits in human TDP-43 transgenic mice. Neurobiol Aging. 2014;35(1):79–87.
106. Ling JP, Pletnikova O, Troncoso JC, Wong PC. TDP-43 repression of nonconserved cryptic exons is compromised in ALS-FTD. Science. 2015;349(6248):650–5.
107. Jeong YH, Ling JP, Lin SZ, Donde AN, Braunstein KE, Majounie E, et al. Tdp-43 cryptic exons are highly variable between cell types. Mol Neurodegener. 2017;12(1):13.
108. MacNair L, Xiao S, Miletic D, Ghani M, Julien JP, Keith J, et al. MTHFSD and DDX58 are novel RNA-binding proteins abnormally regulated in amyotrophic lateral sclerosis. Brain. 2016;139(Pt 1):86–100.
109. Swarup V, Phaneuf D, Bareil C, Robertson J, Rouleau GA, Kriz J, et al. Pathological hallmarks of amyotrophic lateral sclerosis/frontotemporal lobar degeneration in transgenic mice produced with TDP-43 genomic fragments. Brain. 2011;134(Pt 9):2610–26.
110. Bakkar N, Kovalik T, Lorenzini I, Spangler S, Lacoste A, Sponaugle K, et al. Artificial intelligence in neurodegenerative disease research: use of IBM Watson to identify additional RNA-binding proteins altered in amyotrophic lateral sclerosis. Acta Neuropathol. 2018;135(2):227–47.
111. Wang W, Wang L, Lu J, Siedlak SL, Fujioka H, Liang J, et al. The inhibition of TDP-43 mitochondrial localization blocks its neuronal toxicity. Nat Med. 2016;22(8):869–78.
112. Magrane J, Cortez C, Gan WB, Manfredi G. Abnormal mitochondrial transport and morphology are common pathological denominators in SOD1 and TDP43 ALS mouse models. Hum Mol Genet. 2014;23(6):1413–24.
113. Vande Velde C, McDonald KK, Boukhedimi Y, McAlonis-Downes M, Lobsiger CS, Bel Hadj S, et al. Misfolded SOD1 associated with motor neuron mitochondria alters mitochondrial shape and distribution prior to clinical onset. PLoS One. 2011;6(7):e22031.
114. Wang W, Li L, Lin WL, Dickson DW, Petrucelli L, Zhang T, et al. The ALS disease-associated mutant TDP-43 impairs mitochondrial dynamics and function in motor neurons. Hum Mol Genet. 2013;22(23):4706–19.
115. Hong K, Li Y, Duan W, Guo Y, Jiang H, Li W, et al. Full-length TDP-43 and its C-terminal fragments activate mitophagy in NSC34 cell line. Neurosci Lett. 2012;530(2):144–9.
116. Onesto E, Colombrita C, Gumina V, Borghi MO, Dusi S, Doretti A, et al. Gene-specific mitochondria dysfunctions in human TARDBP and C9ORF72 fibroblasts. Acta Neuropathol Commun. 2016;4(1):47.
117. Stribl C, Samara A, Trümbach D, Peis R, Neumann M, Fuchs H, et al. Mitochondrial dysfunction and decrease in body weight of a transgenic knock-in mouse model for TDP-43. J Biol Chem. 2014;289(15):10769–84.
118. Xu Y-F, Zhang Y-J, Lin W-L, Cao X, Stetler C, Dickson DW, et al. Expression of mutant TDP-43 induces neuronal dysfunction in transgenic mice. Mol Neurodegener. 2011;6:73.
119. Xu Y-F, Gendron TF, Zhang Y-J, Lin W-L, Alton S, Sheng H, et al. Wild-type human TDP-43 expression causes TDP-43 phosphorylation, mitochondrial aggregation, motor deficits, and early mortality in transgenic mice. J Neurosci. 2010;30(32):10851.
120. Duan W, Li X, Shi J, Guo Y, Li Z, Li C. Mutant TAR DNA-binding protein-43 induces oxidative injury in motor neuron-like cell. Neuroscience. 2010;169(4):1621–9.
121. Lu J, Duan W, Guo Y, Jiang H, Li Z, Huang J, et al. Mitochondrial dysfunction in human TDP-43 transfected NSC34 cell lines and the protective effect of dimethoxy curcumin. Brain Res Bull. 2012;89(5):185–90.
122. Smith EF, Shaw PJ, De Vos KJ. The role of mitochondria in amyotrophic lateral sclerosis. Neurosci Lett. 2017. https://doi.org/10.1016/j.neulet.2017.06.052.
123. Buratti E. Functional significance of TDP-43 mutations in disease. Adv Genet. 2015;91:1–53.
124. McGoldrick P, Joyce PI, Fisher EM, Greensmith L. Rodent models of amyotrophic lateral sclerosis. Biochim Biophys Acta. 2013;1832(9):1421–36.
125. Johnson BS, McCaffery JM, Lindquist S, Gitler AD. A yeast TDP-43 proteinopathy model: exploring the molecular determinants of TDP-43 aggregation and cellular toxicity. Proc Natl Acad Sci U S A. 2008;105(17):6439–44.

126. Johnson BS, Snead D, Lee JJ, McCaffery JM, Shorter J, Gitler AD. TDP-43 is intrinsically aggregation-prone, and amyotrophic lateral sclerosis-linked mutations accelerate aggregation and increase toxicity. J Biol Chem. 2009;284(30):20329–39.

127. King OD, Gitler AD, Shorter J. The tip of the iceberg: RNA-binding proteins with prion-like domains in neurodegenerative disease. Brain Res. 2012;1462:61–80.

128. Mackenzie IR, Nicholson AM, Sarkar M, Messing J, Purice MD, Pottier C, et al. TIA1 mutations in amyotrophic lateral sclerosis and frontotemporal dementia promote phase separation and alter stress granule dynamics. Neuron. 2017;95(4):808–16.e9.

129. Maziuk B, Ballance HI, Wolozin B. Dysregulation of RNA binding protein aggregation in neurodegenerative disorders. Front Mol Neurosci. 2017;10:89.

130. Protter DS, Parker R. Principles and properties of stress granules. Trends Cell Biol. 2016;26(9):668–79.

131. Sephton CF, Yu G. The function of RNA-binding proteins at the synapse: implications for neurodegeneration. Cell Mol Life Sci. 2015;72(19):3621–35.

132. Parker R, Sheth U. P bodies and the control of mRNA translation and degradation. Mol Cell. 2007;25(5):635–46.

133. Anderson P, Kedersha N. Stress granules. Curr Biol. 2009;19(10):R397–8.

134. Buchan JR, Parker R. Eukaryotic stress granules: the ins and outs of translation. Mol Cell. 2009;36(6):932–41.

135. Holt CE, Bullock SL. Subcellular mRNA localization in animal cells and why it matters. Science. 2009;326(5957):1212–6.

136. Brangwynne CP, Eckmann CR, Courson DS, Rybarska A, Hoege C, Gharakhani J, et al. Germline P granules are liquid droplets that localize by controlled dissolution/condensation. Science. 2009;324(5935):1729–32.

137. Brangwynne CP, Mitchison TJ, Hyman AA. Active liquid-like behavior of nucleoli determines their size and shape in Xenopus laevis oocytes. Proc Natl Acad Sci U S A. 2011;108(11):4334–9.

138. Conicella AE, Zerze GH, Mittal J, Fawzi NL. ALS mutations disrupt phase separation mediated by alpha-helical structure in the TDP-43 low-complexity C-terminal domain. Structure. 2016;24(9):1537–49.

139. Kato M, Han TW, Xie S, Shi K, Du X, Wu LC, et al. Cell-free formation of RNA granules: low complexity sequence domains form dynamic fibers within hydrogels. Cell. 2012;149(4):753–67.

140. Lin Y, Protter DS, Rosen MK, Parker R. Formation and maturation of phase-separated liquid droplets by RNA-binding proteins. Mol Cell. 2015;60(2):208–19.

141. Molliex A, Temirov J, Lee J, Coughlin M, Kanagaraj AP, Kim HJ, et al. Phase separation by low complexity domains promotes stress granule assembly and drives pathological fibrillization. Cell. 2015;163(1):123–33.

142. Murakami T, Qamar S, Lin JQ, Schierle GS, Rees E, Miyashita A, et al. ALS/FTD mutation-induced phase transition of FUS liquid droplets and reversible hydrogels into irreversible hydrogels impairs RNP granule function. Neuron. 2015;88(4):678–90.

143. Patel A, Lee HO, Jawerth L, Maharana S, Jahnel M, Hein MY, et al. A liquid-to-solid phase transition of the ALS protein FUS accelerated by disease mutation. Cell. 2015;162(5):1066–77.

144. McInerney GM, Kedersha NL, Kaufman RJ, Anderson P, Liljeström P. Importance of eIF2α phosphorylation and stress granule assembly in alphavirus translation regulation. Mol Biol Cell. 2005;16(8):3753–63.

145. White JP, Lloyd RE. Regulation of stress granules in virus systems. Trends Microbiol. 2012;20(4):175–83.

146. White JP, Cardenas AM, Marissen WE, Lloyd RE. Inhibition of cytoplasmic mRNA stress granule formation by a viral proteinase. Cell Host Microbe. 2007;2(5):295–305.

147. Aulas A, Vande Velde C. Alterations in stress granule dynamics driven by TDP-43 and FUS: a link to pathological inclusions in ALS? Front Cell Neurosci. 2015;9:423.

148. Markmiller S, Soltanieh S, Server KL, Mak R, Jin W, Fang MY, et al. Context-dependent and disease-specific diversity in protein interactions within stress granules. Cell. 2018;172(3):590–604.e13.
149. Jain S, Wheeler JR, Walters RW, Agrawal A, Barsic A, Parker R. ATPase-modulated stress granules contain a diverse proteome and substructure. Cell. 2016;164(3):487–98.
150. Wheeler JR, Matheny T, Jain S, Abrisch R, Parker R. Distinct stages in stress granule assembly and disassembly. Elife. 2016;5. https://doi.org/10.7554/eLife.18413.
151. Colombrita C, Zennaro E, Fallini C, Weber M, Sommacal A, Buratti E, et al. TDP-43 is recruited to stress granules in conditions of oxidative insult. J Neurochem. 2009;111(4):1051–61.
152. Dewey CM, Cenik B, Sephton CF, Dries DR, Mayer P 3rd, Good SK, et al. TDP-43 is directed to stress granules by sorbitol, a novel physiological osmotic and oxidative stressor. Mol Cell Biol. 2011;31(5):1098–108.
153. Meyerowitz J, Parker SJ, Vella LJ, Ng D, Price KA, Liddell JR, et al. C-Jun N-terminal kinase controls TDP-43 accumulation in stress granules induced by oxidative stress. Mol Neurodegener. 2011;6:57.
154. Parker SJ, Meyerowitz J, James JL, Liddell JR, Crouch PJ, Kanninen KM, et al. Endogenous TDP-43 localized to stress granules can subsequently form protein aggregates. Neurochem Int. 2012;60(4):415–24.
155. Liu-Yesucevitz L, Bilgutay A, Zhang YJ, Vanderweyde T, Citro A, Mehta T, et al. Tar DNA binding protein-43 (TDP-43) associates with stress granules: analysis of cultured cells and pathological brain tissue. PLoS One. 2010b;5(10):e13250.
156. McDonald KK, Aulas A, Destroismaisons L, Pickles S, Beleac E, Camu W, et al. TAR DNA-binding protein 43 (TDP-43) regulates stress granule dynamics via differential regulation of G3BP and TIA-1. Hum Mol Genet. 2011;20(7):1400–10.
157. Liu-Yesucevitz L, Lin AY, Ebata A, Boon JY, Reid W, Xu YF, et al. ALS-linked mutations enlarge TDP-43-enriched neuronal RNA granules in the dendritic arbor. J Neurosci. 2014;34(12):4167–74.
158. Liu-Yesucevitz L, Bilgutay A, Zhang YJ, Vanderwyde T, Citro A, Mehta T, et al. Tar DNA binding protein-43 (TDP-43) associates with stress granules: analysis of cultured cells and pathological brain tissue. PLoS One. 2010a;5(10):e13250.
159. Volkening K, Leystra-Lantz C, Yang W, Jaffee H, Strong MJ. Tar DNA binding protein of 43 kDa (TDP-43), 14-3-3 proteins and copper/zinc superoxide dismutase (SOD1) interact to modulate NFL mRNA stability. Implications for altered RNA processing in amyotrophic lateral sclerosis (ALS). Brain Res. 2009;1305:168–82.
160. Buchan JR, Kolaitis RM, Taylor JP, Parker R. Eukaryotic stress granules are cleared by autophagy and Cdc48/VCP function. Cell. 2013;153(7):1461–74.
161. Hardy J, Rogaeva E. Motor neuron disease and frontotemporal dementia: sometimes related, sometimes not. Exp Neurol. 2014;262(Pt B):75–83.
162. Auburger G, Sen N-E, Meierhofer D, Başak A-N, Gitler AD. Efficient prevention of neurodegenerative diseases by depletion of starvation response factor ataxin-2. Trends Neurosci. 2017;40(8):507–16.
163. Kaehler C, Isensee J, Nonhoff U, Terrey M, Hucho T, Lehrach H, et al. Ataxin-2-like is a regulator of stress granules and processing bodies. PLoS One. 2012;7(11):e50134.
164. Nonhoff U, Ralser M, Welzel F, Piccini I, Balzereit D, Yaspo M-L, et al. Ataxin-2 interacts with the DEAD/H-box RNA helicase DDX6 and interferes with P-bodies and stress granules. Mol Biol Cell. 2007;18(4):1385–96.
165. Lorenzetti D, Bohlega S, Zoghbi HY. The expansion of the CAG repeat in ataxin-2 is a frequent cause of autosomal dominant spinocerebellar ataxia. Neurology. 1997;49(4):1009.
166. Imbert G, Saudou F, Yvert G, Devys D, Trottier Y, Garnier J-M, et al. Cloning of the gene for spinocerebellar ataxia 2 reveals a locus with high sensitivity to expanded CAG/glutamine repeats. Nat Genet. 1996;14:285.
167. Sanpei K, Takano H, Igarashi S, Sato T, Oyake M, Sasaki H, et al. Identification of the spinocerebellar ataxia type 2 gene using a direct identification of repeat expansion and cloning technique, DIRECT. Nat Genet. 1996;14:277.

168. Pulst S-M, Nechiporuk A, Nechiporuk T, Gispert S, Chen X-N, Lopes-Cendes I, et al. Moderate expansion of a normally biallelic trinucleotide repeat in spinocerebellar ataxia type 2. Nat Genet. 1996;14:269.
169. Nanetti L, Fancellu R, Tomasello C, Gellera C, Pareyson D, Mariotti C. Rare association of motor neuron disease and spinocerebellar ataxia type 2 (SCA2): a new case and review of the literature. J Neurol. 2009;256(11):1926–8.
170. Elden AC, Kim H-J, Hart MP, Chen-Plotkin AS, Johnson BS, Fang X, et al. Ataxin-2 intermediate-length polyglutamine expansions are associated with increased risk for ALS. Nature. 2010;466:1069.
171. Bonini NM, Gitler AD. Model organisms reveal insight into human neurodegenerative disease: ataxin-2 intermediate-length polyglutamine expansions are a risk factor for ALS. J Mol Neurosci. 2011;45(3):676–83.
172. Becker LA, Huang B, Bieri G, Ma R, Knowles DA, Jafar-Nejad P, et al. Therapeutic reduction of ataxin-2 extends lifespan and reduces pathology in TDP-43 mice. Nature. 2017;544(7650):367–71.
173. Lastres-Becker I, Nonis D, Eich F, Klinkenberg M, Gorospe M, Kotter P, et al. Mammalian ataxin-2 modulates translation control at the pre-initiation complex via PI3K/mTOR and is induced by starvation. Biochim Biophys Acta. 2016;1862(9):1558–69.
174. Xiao S, Sanelli T, Chiang H, Sun Y, Chakrabartty A, Keith J, et al. Low molecular weight species of TDP-43 generated by abnormal splicing form inclusions in amyotrophic lateral sclerosis and result in motor neuron death. Acta Neuropathol. 2015;130(1):49–61.
175. Dormann D, Capell A, Carlson AM, Shankaran SS, Rodde R, Neumann M, et al. Proteolytic processing of TAR DNA binding protein-43 by caspases produces C-terminal fragments with disease defining properties independent of progranulin. J Neurochem. 2009;110(3):1082–94.
176. Igaz LM, Kwong LK, Chen-Plotkin A, Winton MJ, Unger TL, Xu Y, et al. Expression of TDP-43 C-terminal fragments in vitro recapitulates pathological features of TDP-43 proteinopathies. J Biol Chem. 2009;284(13):8516–24.
177. Herskowitz JH, Gozal YM, Duong DM, Dammer EB, Gearing M, Ye K, et al. Asparaginyl endopeptidase cleaves TDP-43 in brain. Proteomics. 2012;12(15–16):2455–63.
178. Kametani F, Obi T, Shishido T, Akatsu H, Murayama S, Saito Y, et al. Mass spectrometric analysis of accumulated TDP-43 in amyotrophic lateral sclerosis brains. Sci Rep. 2016;6:23281.
179. Tsuji H, Arai T, Kametani F, Nonaka T, Yamashita M, Suzukake M, et al. Molecular analysis and biochemical classification of TDP-43 proteinopathy. Brain. 2012a;135(Pt 11):3380–91.
180. Tsuji H, Nonaka T, Yamashita M, Masuda-Suzukake M, Kametani F, Akiyama H, et al. Epitope mapping of antibodies against TDP-43 and detection of protease-resistant fragments of pathological TDP-43 in amyotrophic lateral sclerosis and frontotemporal lobar degeneration. Biochem Biophys Res Commun. 2012b;417(1):116–21.
181. Mompean M, Romano V, Pantoja-Uceda D, Stuani C, Baralle FE, Buratti E, et al. The TDP-43 N-terminal domain structure at high resolution. FEBS J. 2016a;283(7):1242–60.
182. Nishimoto Y, Ito D, Yagi T, Nihei Y, Tsunoda Y, Suzuki N. Characterization of alternative isoforms and inclusion body of the TAR DNA-binding protein-43. J Biol Chem. 2010;285(1):608–19.
183. Che MX, Jiang YJ, Xie YY, Jiang LL, Hu HY. Aggregation of the 35-kDa fragment of TDP-43 causes formation of cytoplasmic inclusions and alteration of RNA processing. FASEB J. 2011;25(7):2344–53.
184. D'Alton S, Altshuler M, Lewis J. Studies of alternative isoforms provide insight into TDP-43 autoregulation and pathogenesis. RNA. 2015;21(8):1419–32.
185. Wang HY, Wang IF, Bose J, Shen CK. Structural diversity and functional implications of the eukaryotic TDP gene family. Genomics. 2004;83(1):130–9.
186. Che MX, Jiang LL, Li HY, Jiang YJ, Hu HY. TDP-35 sequesters TDP-43 into cytoplasmic inclusions through binding with RNA. FEBS Lett. 2015;589(15):1920–8.

187. Sasaguri H, Chew J, Xu YF, Gendron TF, Garrett A, Lee CW, et al. The extreme N-terminus of TDP-43 mediates the cytoplasmic aggregation of TDP-43 and associated toxicity in vivo. Brain Res. 2016;1647:57–64.

188. Wei Y, Lim L, Wang L, Song J. Inter-domain interactions of TDP-43 as decoded by NMR. Biochem Biophys Res Commun. 2016;473(2):614–9.

189. Yang C, Tan W, Whittle C, Qiu L, Cao L, Akbarian S, et al. The C-terminal TDP-43 fragments have a high aggregation propensity and harm neurons by a dominant-negative mechanism. PLoS One. 2010;5(12):e15878.

190. Budini M, Romano V, Quadri Z, Buratti E, Baralle FE. TDP-43 loss of cellular function through aggregation requires additional structural determinants beyond its C-terminal Q/N prion-like domain. Hum Mol Genet. 2015;24(1):9–20.

191. Brady OA, Meng P, Zheng Y, Mao Y, Hu F. Regulation of TDP-43 aggregation by phosphorylation andp62/SQSTM1. J Neurochem. 2011;116(2):248–59.

192. Kim SH, Shanware NP, Bowler MJ, Tibbetts RS. Amyotrophic lateral sclerosis-associated proteins TDP-43 and FUS/TLS function in a common biochemical complex to co-regulate HDAC6 mRNA. J Biol Chem. 2010;285(44):34097–105.

193. Nonaka T, Kametani F, Arai T, Akiyama H, Hasegawa M. Truncation and pathogenic mutations facilitate the formation of intracellular aggregates of TDP-43. Hum Mol Genet. 2009;18(18):3353–64.

194. Mashiko T, Sakashita E, Kasashima K, Tominaga K, Kuroiwa K, Nozaki Y, et al. Developmentally regulated RNA-binding protein 1 (Drb1)/RNA-binding motif protein 45 (RBM45), a nuclear-cytoplasmic trafficking protein, forms TAR DNA-binding protein 43 (TDP-43)-mediated cytoplasmic aggregates. J Biol Chem. 2016;291(29):14996–5007.

195. Chen H-J, Mitchell JC, Novoselov S, Miller J, Nishimura AL, Scotter EL, et al. The heat shock response plays an important role in TDP-43 clearance: evidence for dysfunction in amyotrophic lateral sclerosis. Brain. 2016;139(5):1417–32.

196. Yamashita M, Nonaka T, Arai T, Kametani F, Buchman VL, Ninkina N, et al. Methylene blue and dimebon inhibit aggregation of TDP-43 in cellular models. FEBS Lett. 2009;583(14):2419–24.

197. Voigt A, Herholz D, Fiesel FC, Kaur K, Müller D, Karsten P, et al. TDP-43-mediated neuron loss in vivo requires RNA-binding activity. PLoS One. 2010;5(8):e12247.

198. Cohen TJ, Hwang AW, Restrepo CR, Yuan C-X, Trojanowski JQ, Lee VMY. An acetylation switch controls TDP-43 function and aggregation propensity. Nat Commun. 2015;6:5845.

199. Scotter EL, Vance C, Nishimura AL, Lee Y-B, Chen H-J, Urwin H, et al. Differential roles of the ubiquitin proteasome system and autophagy in the clearance of soluble and aggregated TDP-43 species. J Cell Sci. 2014;127(6):1263–78.

200. Suzuki H, Shibagaki Y, Hattori S, Matsuoka M. Nuclear TDP-43 causes neuronal toxicity by escaping from the inhibitory regulation by hnRNPs. Hum Mol Genet. 2015;24(6):1513–27.

201. Bentmann E, Neumann M, Tahirovic S, Rodde R, Dormann D, Haass C. Requirements for stress granule recruitment of fused in sarcoma (FUS) and TAR DNA-binding protein of 43 kDa (TDP-43). J Biol Chem. 2012;287(27):23079–94.

202. Li HY, Yeh PA, Chiu HC, Tang CY, Tu BP. Hyperphosphorylation as a defense mechanism to reduce TDP-43 aggregation. PLoS One. 2011;6(8):e23075.

203. Boeynaems S, Bogaert E, Michiels E, Gijselinck I, Sieben A, Jovicic A, et al. Drosophila screen connects nuclear transport genes to DPR pathology in c9ALS/FTD. Sci Rep. 2016;6:20877.

204. Freibaum BD, Lu Y, Lopez-Gonzalez R, Kim NC, Almeida S, Lee KH, et al. GGGGCC repeat expansion in C9orf72 compromises nucleocytoplasmic transport. Nature. 2015;525(7567):129–33.

205. Jovicic A, Mertens J, Boeynaems S, Bogaert E, Chai N, Yamada SB, et al. Modifiers of C9orf72 dipeptide repeat toxicity connect nucleocytoplasmic transport defects to FTD/ALS. Nat Neurosci. 2015;18(9):1226–9.

206. Zhang K, Donnelly CJ, Haeusler AR, Grima JC, Machamer JB, Steinwald P, et al. The C9orf72 repeat expansion disrupts nucleocytoplasmic transport. Nature. 2015;525(7567):56–61.

207. Chou CC, Zhang Y, Umoh ME, Vaughan SW, Lorenzini I, Liu F, et al. TDP-43 pathology disrupts nuclear pore complexes and nucleocytoplasmic transport in ALS/FTD. Nat Neurosci. 2018;21(2):228–39.
208. Woerner AC, Frottin F, Hornburg D, Feng LR, Meissner F, Patra M, et al. Cytoplasmic protein aggregates interfere with nucleocytoplasmic transport of protein and RNA. Science. 2016;351(6269):173–6.
209. de Calignon A, Polydoro M, Suarez-Calvet M, William C, Adamowicz DH, Kopeikina KJ, et al. Propagation of tau pathology in a model of early Alzheimer's disease. Neuron. 2012;73(4):685–97.
210. Liu L, Drouet V, Wu JW, Witter MP, Small SA, Clelland C, et al. Trans-synaptic spread of tau pathology in vivo. PLoS One. 2012;7(2):e31302.
211. Luk KC, Kehm V, Carroll J, Zhang B, O'Brien P, Trojanowski JQ, et al. Pathological alpha-synuclein transmission initiates Parkinson-like neurodegeneration in nontransgenic mice. Science. 2012;338(6109):949–53.
212. Mougenot AL, Bencsik A, Nicot S, Vulin J, Morignat E, Verchere J, et al. Transmission of prion strains in a transgenic mouse model overexpressing human A53T mutated alpha-synuclein. J Neuropathol Exp Neurol. 2011;70(5):377–85.
213. Kaufman SK, Sanders DW, Thomas TL, Ruchinskas AJ, Vaquer-Alicea J, Sharma AM, et al. Tau prion strains dictate patterns of cell pathology, progression rate, and regional vulnerability in vivo. Neuron. 2016;92(4):796–812.
214. Sanders DW, Kaufman SK, DeVos SL, Sharma AM, Mirbaha H, Li A, et al. Distinct tau prion strains propagate in cells and mice and define different tauopathies. Neuron. 2014;82(6):1271–88.
215. Ishii T, Kawakami E, Endo K, Misawa H, Watabe K. Formation and spreading of TDP-43 aggregates in cultured neuronal and glial cells demonstrated by time-lapse imaging. PLoS One. 2017;12(6):e0179375.
216. Pokrishevsky E, Grad LI, Cashman NR. TDP-43 or FUS-induced misfolded human wild-type SOD1 can propagate intercellularly in a prion-like fashion. Sci Rep. 2016;6:22155.
217. Nonaka T, Masuda-Suzukake M, Arai T, Hasegawa Y, Akatsu H, Obi T, et al. Prion-like properties of pathological TDP-43 aggregates from diseased brains. Cell Rep. 2013;4(1):124–34.
218. Smethurst P, Sidle KC, Hardy J. Review: Prion-like mechanisms of transactive response DNA binding protein of 43 kDa (TDP-43) in amyotrophic lateral sclerosis (ALS). Neuropathol Appl Neurobiol. 2015;41(5):578–97.

Chapter 10
Senataxin, A Novel Helicase at the Interface of RNA Transcriptome Regulation and Neurobiology: From Normal Function to Pathological Roles in Motor Neuron Disease and Cerebellar Degeneration

Craig L. Bennett and Albert R. La Spada

Abstract Senataxin (SETX) is a DNA-RNA helicase whose C-terminal region shows homology to the helicase domain of the yeast protein Sen1p. Genetic discoveries have established the importance of SETX for neural function, as recessive mutations in the *SETX* gene cause Ataxia with Oculomotor Apraxia type 2 (AOA2) (OMIM: 606002), which is the third most common form of recessive ataxia, after Friedreich's ataxia and Ataxia-Telangiectasia. In addition, rare, dominant *SETX* mutations cause a juvenile-onset form of Amyotrophic Lateral Sclerosis (ALS), known as ALS4. SETX performs a number of RNA regulatory functions, including maintaining RNA transcriptome homeostasis. Over the last decade, altered RNA regulation and aberrant RNA-binding protein function have emerged as a central theme in motor neuron disease pathogenesis, with evidence suggesting that sporadic ALS disease pathology may overlap with the molecular pathology uncovered in familial ALS. Like other RNA processing proteins linked to ALS, the basis for SETX gain-of-function motor neuron toxicity remains ill-defined. Studies of yeast

C. L. Bennett
Department of Neurology, Duke University School of Medicine, Durham, NC, USA

Department of Neurobiology, Duke University School of Medicine, Durham, NC, USA

Department of Cell Biology, Duke University School of Medicine, Durham, NC, USA

A. R. La Spada (✉)
Duke Center for Neurodegeneration & Neurotherapeutics, Duke University School of Medicine, Durham, NC, USA

Department of Neurology, Duke University School of Medicine, Durham, NC, USA

Department of Neurobiology, Duke University School of Medicine, Durham, NC, USA

Department of Cell Biology, Duke University School of Medicine, Durham, NC, USA
e-mail: al.laspada@duke.edu

© Springer International Publishing AG, part of Springer Nature 2018
R. Sattler, C. J. Donnelly (eds.), *RNA Metabolism in Neurodegenerative Diseases*, Advances in Neurobiology 20,
https://doi.org/10.1007/978-3-319-89689-2_10

Sen1p and mammalian SETX protein have revealed a range of important RNA regulatory functions, including resolution of R-loops to permit transcription termination, and RNA splicing. Growing evidence suggests that *SETX* may represent an important genetic modifier locus for sporadic ALS. In cycling cells, SETX is found at nuclear foci during the S/G_2 cell-cycle transition phase, and may function at sites of collision between components of the replisome and transcription machinery. While we do not yet know which SETX activities are most critical to neurodegeneration, our evolving understanding of SETX function will undoubtedly be crucial for not only understanding the role of SETX in ALS and ataxia disease pathogenesis, but also for delineating the mechanistic biology of fundamentally important molecular processes in the cell.

Keywords Senataxin · Helicase · R-Loops · Nuclear exosome · RENT1 · IGHMBP2 · Sen1p · Exosc9 · Sumo · Nucleolus · tRNA

Senataxin (SETX) is now recognized as an important protein in the fields of molecular genetics and neurodegeneration. *SETX* gene mutations lead to two distinct neurological disorders, Ataxia with Oculomotor Apraxia type 2 (AOA2) and Amyotrophic Lateral Sclerosis type 4 (ALS4). AOA2 has uniformly early onset and leads to very severe disability, requiring life-long care [1]. ALS4 is moderate to severe with varying age of onset, but with its average onset at 17-years, ALS4 is considered a juvenile-onset form of familial ALS (FALS) [2]. The *SETX* gene has also attracted recent attention as a potential genetic modifier of sporadic ALS (SALS) [3].

The effort to define key functions of SETX continues as its roles in RNA processing and maintenance of genomic stability are now well established by the molecular genetics community. In early studies, the *SETX* gene was found to be ubiquitously expressed [4, 5], and many functional processes eventually attributed to SETX were originally described for its yeast orthologue, Sen1p. These SETX functions include: (1) RNA polymerase II (RNAP II) transcription termination; (2) the resolution of RNA/DNA hybrids, or R-Loops; (3) processing of noncoding RNAs and mRNA; (4) interaction with the nuclear exosome; and (5) the formation of replication stress-related foci during the S/G_2 transition phase. This last function suggests that SETX may be essential for cell cycling when long genes are being transcribed and RNAP II collides with the replisome. Other important roles for SETX that deserve attention include the regulation of the circadian rhythm genes, *Period* (PER) and *Cryptochrome* (CRY) [6]. Whether these functions are critical to neuron survival is unclear, but highlight the fact that the full spectrum of cellular processes for this helicase is extremely wide-ranging.

In this review, we attempt to clarify the many processes attributed to SETX, and evaluate if SETX gain-of-function toxicity impacts these functions and how this might contribute to motor neuron disease. Detailed proteomics studies have shown that human SETX, like Sen1p, has retained regulatory functions during gene transcription. Interactome analysis of purified TAP-tagged Sen1p identified the RNAP

II subunit Rpo21, along with subunits Rpb2 and Rpb4, as key interactors [7], confirming earlier yeast two-hybrid studies [8]. Furthermore, a number of RNAP I and RNAP III core subunits, elongation factors, and other key components were identified as protein interactors of Sen1p [7]. In contrast, when similar analyses were performed with SETX, no direct interaction with RNAP II was observed, but rather an enrichment within the chromatin fraction, and interaction with RNAP II-related core factors [7]. Importantly, these studies were undertaken with Flag/GFP-tagged SETX, used with HeLa cell stable integration, and at near endogenous levels of protein expression [7]. Hence, human SETX likely retains transcription-related functions, but the regulatory relationships may be quite different in comparison with budding yeast.

In regards to both SETX and Sen1p function, it is crucial to emphasize that these proteins are present at very low abundance; hence, overexpression can lead to aberrant cellular events. Gene duplication can uncover such sensitivity to protein levels as dosage-sensitivity [9, 10]. For example, increased levels of ataxin-2 can distort the cytoplasmic-to-nuclear ratio of TDP-43 and FUS proteins [11], both proteins that are critical in ALS disease. Sen1p is known to be maintained at very low levels in the cell, typically as low as 125 molecules per cell [12], which is much less than its known transcription termination partners, Nrd1p and Nab3p, present at ~19,000 and ~5800 molecules per cell respectively [13]. RNAP II is itself present in yeast at ~14,000 molecules/cell [14, 15]. In a detailed proteomics study using the human U2OS cell line, ~10,000 different proteins were quantified and found to span a concentration range of seven orders of magnitude up to 20,000,000 copies per cell [16]. From this study SETX was found to be in the lowest category of very low-abundant proteins at <500 molecules/cell [16]. Many SETX functional studies have been undertaken with tagged-constructs which will almost assuredly lead to cellular levels of recombinant SETX that are grossly elevated, which means that results of such studies must be interpreted with caution.

10.1   SETX Mutations Cause Both Ataxia with Oculomotor Apraxia and Motor Neuron Disease

Causal links between SETX protein defects and neurological disease were first reported in 2004. We discovered *SETX* gene mutations as the cause of ALS4, a rare, dominantly inherited, juvenile onset form of ALS (OMIM: 602433). Importantly, all 49 affected members of an extended American pedigree were found to carry a L389S mutation [4]. Other sizable European pedigrees were found to segregate R2136H and T3I *SETX* mutations. The L389S mutation was subsequently found in Italian and Dutch pedigrees, confirming the pathogenicity of this mutation in ALS4 [17, 18]. The phenotype of ALS4 is unique compared with classical ALS due to a number of factors, including normal patient life expectancy due to the sparing of the respiratory musculature; absence of bulbar involvement; and the presentation of symmetrical atrophy and weakness [2].

Recessive *SETX* mutations were also reported in 2004 as the cause of a severe Ataxia with Oculomotor Apraxia—type 2 (AOA2; OMIM: 606002) [5]. While dominant mutations are rare, *SETX* recessive mutations are not nearly so rare. AOA2 is considered the third most common autosomal recessive cerebellar ataxia [1, 5], with the most common being Friedreich's ataxia, closely followed by Ataxia-Telangiectasia (A-T). One unique feature shared between AOA2 and A-T is elevated serum levels of alpha-feto-protein (AFP), which are ~9-fold higher in AOA2 patients than normals [19]. A-T is associated with unique DNA repair defects and extra-neurologic features, including a greatly increased cancer risk and immune defects (OMIM: 208900). With AOA2 patients, some groups have reported sensitivity to DNA-damaging agents in patient cells [20], yet others report normal sensitivity [21, 22]. Nonetheless, AOA2 patients do not display an increased cancer risk nor immunological abnormalities, and thus the above features form a differential diagnosis for A-T versus AOA2. AOA2 patients show non-cancer-related extra-neurologic features, such as ovarian failure [23], suggesting tissues other than just the central nervous system (CNS) are susceptible to SETX loss. At the molecular level, Moreira et al. initially reported 15 different *SETX* mutations, ten of which predict premature protein termination. Thus, parental carriers of AOA2 null mutations were carefully examined and found to harbor no neurological phenotypes [1, 5]. Importantly, this suggests that ALS4 dominant mutations possess toxic gain-of-function properties. Now that greater than 150 different *SETX* mutations have been identified to date [24], it is known that missense mutations cluster within either the helicase domain or the amino-terminal domain, confirming the critical nature of these two protein regions [25].

It is of note that within the human genome, only two human proteins exist with significant homology within the helicase domain to SETX: RENT1 (46% similarity) and IGHMBP2 (45% similarity) (Fig. 10.1). RENT1 (the yeast Upf1 orthologue) is an essential component of the nonsense-mediated RNA decay (NMD) pathway for degrading incorrectly spliced or stop codon-containing mRNAs [26], and *Rent1* null mice show embryonic lethality [27], attesting to the importance of this pathway for transcriptome stability. Over-expression of RENT1 (hUpf1) can significantly rescue toxicity in ALS/FTD cell models (TDP-43/FUS) that are likely induced by uncharacterized RNA dysfunction [28]. Recessive mutations of the *IGHMBP2* gene cause a severe spinal muscle atrophy with respiratory distress (SMARD) [29]. This disease has overlap with AOA2 and ALS4, which will be discussed below. Thus, these three human helicase homologs cause either embryonic lethality (Rent1)—due to failure of NMD surveillance of mRNA splicing errors, or specific neuronal vulnerability and neurodegeneration as loss-of-function mutations (SETX and IGHMBP2) or gain-of-function mutations (SETX). Given the known low cellular levels of SETX protein, one could speculate that trace levels of SETX, produced by alternate splicing, might prevent lethality in the human, akin to the minimal levels of normal survival motor neuron protein produced by the SMN2 gene in spinal muscular atrophy [30]. In regards to IGHMBP2, human mutations are known to be homozygous hypomorphic loss-of-function alleles [29], in agreement with the naturally occurring mouse model of SMARD, where ~20% of correctly spliced *Ighmbp2* mRNA is produced [31].

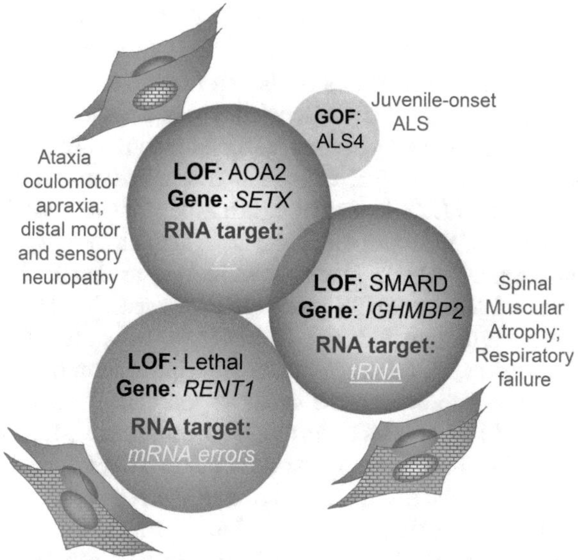

Fig. 10.1 SETX, RENT1, and IGHMBP2 form a helicase subfamily. While there are nearly 100 helicase proteins in the human genome, only two show significant homology to SETX. These three related homologues each perform critical roles in RNA processing, and are essential for survival or for normal central nervous system function. RENT1 is unique in that no human disease results from recessive mutation, likely due to embryonic lethality, and displays a cytosolic localization. The specific RNA targets linked to dysfunction or lethality are known for IGHMBP2 and RENT1, and are listed. In contrast, the specific RNA targets for SETX are unknown, and how altered RNA regulation results in AOA2 or ALS4 remains an open question. SETX resides in the nucleus, while IGHMBP2 localizes to both the nucleus and the cytosol, as shown

10.2 SETX Function: Similarities and Differences with Yeast Sen1p

Soon after *SETX* mutations were discovered, it was noted that SETX and Sen1p shared a highly conserved C-terminal helicase domain. Researchers hypothesized that key functions attributed to Sen1p would likely be retained by human SETX. The *Sen1* gene was named for its suspected function as a splicing endonuclease, but was thereafter found to function primarily in processing a diverse class of noncoding RNAs (ncRNAs) [32, 33]. In further studies, Sen1p was shown to process intron-containing tRNA precursors [34], rRNA precursors [33], 3′-extended forms of some small nucleolar RNAs (snoRNAs), and to a lesser extent, small nuclear RNAs (snRNAs) [33, 35, 36]. Sen1p directly interacts with the RNA-binding proteins Nrd1 and Nab3 [36, 37], and was confirmed as a functioning component of the NRD complex in transcription termination of RNAP II ncRNA transcripts [38, 39]. Taken together, years of dedicated yeast research portray a complex picture of RNA processing, which is essential to yeast survival and directed primarily to the regulation of ncRNAs, as well as a central role in RNAP II transcription termination that may be unique to yeast.

Early SETX work yielded divergent and contradictory views as to what might be its key disease-linked protein function [40]. Readers new to the field could easily be left confused by perusing summaries of the very broad array of proposed SETX functions. An example of one early discrepancy was whether cells lacking SETX were sensitive to DNA-damaging agents. A second function attributed to SETX is a generalized regulation of RNAP II transcription termination. While some reports strongly favor a role for SETX in mediating Xrn2-dependent transcription termination via the formation of R-loops (which are DNA-RNA hybrids formed during the process of transcription) [41], in much the same way as Sen1p, others have concluded that decreased Xrn2 or SETX levels yield only marginal effects on the regulation of transcription termination [42]. However, it is clear that for certain genes and biological processes, SETX is central to transcription termination regulation, as SETX has been shown to control the cyclic expression of the circadian rhythm genes PER and CRY [6]. This work has yielded a model in which recruitment of PER complexes to the elongating polymerase at *Per* and *Cry* termination sites inhibits SETX action, impeding RNAP II release and thereby repressing transcription re-initiation. A third SETX function, reported to be critical for neuron survival, is the resolution of RNAP II-mediated R-loops more generally (independent of termination), which will be discussed later. Fourth, a role for SETX in directing incomplete RNA transcripts to the nuclear exosome has been found in cycling cells when the DNA polymerase machinery collides with active RNAP II transcription.

10.3   R-Loop Resolution Is Not Defective in Setx Null Mice

Another avenue to address the question of which SETX RNA processing functions are most crucial to neuron survival is to create animal models. *Setx* knock-out mice, generated by gene targeted removal of exon four, resulted in near complete loss of SETX protein [43]. However, for a range of reasons possibly including differences in neuroanatomy and lifespan between mice and humans, *Setx* null mice show neither ataxia nor cerebellar degeneration, preventing the possibility to characterize mechanisms of neuron cell death. Failure to recapitulate human recessive ataxias in mice is not without precedent [44], yet a range of critical in vivo studies have been examined in these mice nonetheless. In post-mitotic neurons, R-loops were resolved normally in the cerebellum or brain of mice lacking SETX (similar to wild-type mice), as R-loops could not be detected in wild-type controls or SETX knock-outs. Furthermore, there was no evidence of cells undergoing apoptosis in these tissues in SETX null mice [45]. The authors postulated that the "*major clinical neurodegenerative phenotype seen in AOA2 patients is more likely to be due to a more general defect in RNA processing ...rather than a failure to resolve R-loops*". Recent studies from the Libri lab indicate that Sen1p has relatively low processivity on RNA [46]. This is relevant to R-loop removal, R-loops that form in mammals are believed to be very long (>1 kb in humans) [47]. Thus SETX, with its likely low processivity (based upon the Sen1p findings), would not be able to unwind such long structures [46].

10.4 SETX Is SUMO-Modified and Regulates RNAP II Transcription

The multiple RNA processing functions identified for Sen1p undoubtedly require the coordinated efforts of both the amino-terminal protein interaction domain for trafficking, as well as the carboxy-terminal helicase domain for RNA/DNA interaction and processing. In yeast, truncation of the amino-terminal region of Sen1p prevented its proper localization to the nucleolus, though only the helicase domain is required for survival [32]. To better define critical Sen1p protein-binding partners, a yeast two-hybrid (Y2H) screen was employed with follow-up co-immunoprecipitation validation. Only the first 565 residues of Sen1p were required for the identified interactions, again supporting the hypothesis of a crucial function for the amino-terminal domain. Specifically, Sen1p was shown to bind with Rpo21p, the large subunit of RNAP II; Rad2p, a deoxyribonuclease required in DNA repair; and Rnt1p, an endoribonuclease required for RNA maturation [8].

Interestingly, we noted that the Sen1p and SETX protein interaction domains are not conserved at the primary amino acid level [25]. The SETX orthologue in marine vertebrates, such as zebrafish, shows conservation within this domain, but not with fly [25]. However, based upon a hypothesis of functional conservation, we reasoned that a human-specific Y2H screen with the first 650 residues of SETX may identify overlap with Sen1p interactors. The results did bear out some overlap, but to a lesser degree than was expected [48], and was confirmed by a second independent SETX Y2H screen [49]. Our screen also included ALS4 mutants to potentially identify gain-of-function interactors [48], and revealed several important interactor groups that were confirmed with alternate techniques and in other independent proteomics screens [7, 50]. The key interactor groups were: (1) SETX self-interaction or dimerization; (2) critical Sumo/Ubiquitin posttranslational modification; and (3) DNA/RNA-binding proteins, including the exosome component 9 protein (Exosc9).

SETX amino-terminal domain self-interaction and dimerization were validated by purification, size exclusion chromatography, protein cross-linking, and Western blot analysis [48] (Fig. 10.2). Importantly, as ALS is one of a number of conditions in which protein aggregation may drive disease pathogenesis [51], we examined this further. Using techniques including targeted mammalian two-hybrid (M2H) analysis, we found that the SETX mutants L389S (ALS4) and W305C (AOA2) can still engage in self-interaction, and do not lead to excessive aggregation. SETX's ability to dimerize may thus set it apart from Sen1p. Additionally, we found five SETX interactors representing proteins in either the SUMO protein trafficking cascade or the ubiquitin protein degradation pathway. The key interaction with Exosc9 was shown to require SETX SUMOylation [49]. Dramatically, when the exosome components Exosc9 or Exosc10 were depleted by targeted siRNA, this yielded significant co-depletion of SETX [49].

As for the overlap of SETX and Sen1p interactors, several interesting distinctions were noted. The three major Sen1p interactors, Rpo21p, Rad2p, and Rnt1p, were not detected by our Y2H screen. Similarly, but this time in mammalian cells using targeted M2H analysis, we did not detect a direct interaction between SETX and the human

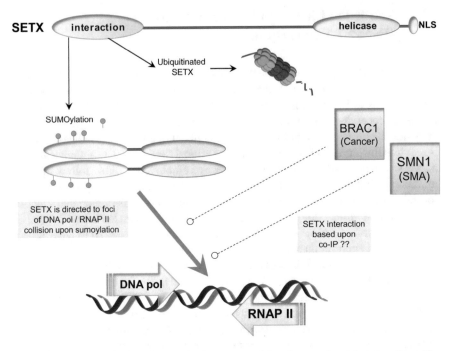

Fig. 10.2 SETX protein domain organization, proposed functions, and protein interactions. The SETX protein is ~303 kDa in molecular mass, and possesses just three known domains: the amino-terminal protein interaction domain, the carboxy-terminal helicase domain, and a nuclear localization signal (NLS) domain. SETX contains amino-terminal sequences that are targets of ubiquitination utilized to degrade SETX protein via the proteasome. The SETX amino-terminal domain possesses regions required for dimerization and for SUMOylation (gray ball and red stem). It is thought that SETX needs to be SUMOylated to direct it to sites of collision between the DNA polymerase-containing replisome and the RNA polymerase-containing transcription machinery. Other key proteins of interest that have been linked with SETX include BRCA1 and SMN1

orthologues to these three proteins (unpublished data). In another proteomics study of Sen1p, tandem affinity purification/mass spectrometry analysis defined this protein as a general transcription factor based upon interactions with RNAP I and RNAP III subunits, as well as with the classic mRNA polymerase complex, RNAP II [7]. This result is consistent with the types of ncRNAs that Sen1p has been previously shown to regulate. Alternatively, proteomics analysis of full-length, GFP/Flag-tagged SETX indicated a general association with the RNAP II complex. A direct interaction was not shown with the core RNAP II subunits (RPB1, RPB2, and RPB3). This suggests that while Sen1p may interact directly with RNAP I, II and III subunits, SETX likely interacts with RNAP II subunits via intermediary associations as periodically directed.

One can conclude that both SETX and Sen1p contain amino acid sequences within the relatively large 500–600 amino-terminal region that are targets for Sumo and Ubiquitin-mediated regulation (Fig. 10.2). For Sen1p, it has been clearly demonstrated that this region is required for signaling its degradation via the ubiquitin proteasome system to maintain low cellular protein levels [52]. Our Y2H screen identified ubiquitin pathway proteins, Ubc9 and UBC with the amino-terminal region as bait, suggesting similar regulation.

10.5   SETX and Sen1p: A Convincing Role in Connecting RNAP II and the Exosome

Characterizing a specific role for SETX in RNAP II termination has not been convincing, despite some initial reports [50]. A role for Sen1p in this process has been well characterized for ncRNAs, which employ a distinctive mechanism specific for these transcripts in yeast, and not likely used in higher eukaryotes (as noted above). For example, termination of the elongated snoRNA precursors relies upon different machinery than the cleavage and polyadenylation mechanism used for mRNA termination. Rather, ncRNA-specific processing relies on the NRD complex containing the Nrd1 and Nab3 RNA-binding proteins in association with Sen1p [36]. In this case, termination occurs downstream of tetranucleotide motifs, which form binding sites for Nrd1 and Nab3 on the nascent RNA [53]. The Nrd1–Nab3–Sen1p complex, which directly interacts with the RNAP II Carboxy-Terminal Domain (CTD), also directly interacts with the nuclear exosome [54]. Thus, Sen1p as part of the NRD complex forms a bridge between the RNAP II and the exosome to aid in the termination of ncRNA transcripts, such as snoRNAs. In these cases, transcription termination is coupled to $3'$–$5'$ exonuclease trimming by the TRAMP–exosome complex [54]. Extrapolation of the Sen1p termination process to mammals is not readily possible, as the RNA-binding protein Nab3, a critical protein bridging the interaction of Nrd1 and Sen1, has no human homologue [55], and Nab3 has no known role in poly-A-dependent termination [36]. In higher eukaryotes, RNAP II utilizes alternate means of transcription termination [56], and previous studies confirm that SETX is not required for snRNA termination [50].

With regard to transcription, despite likely divergence between the Sen1p and SETX regulation, there is significant evidence to suggest SETX has retained a role for linking RNAP II to the nuclear exosome. As noted above, some degree of SETX codepletion occurs when major components of the exosome, Exosc9 (Rrp6) and Exosc10 (Rrp45), are depleted. Thus, a model can be proposed that SETX needs to dimerize, and then be SUMOylated as a requirement for its interaction with Exosc9 (and the nuclear exosome) [49], and that transcription-related DNA damage directs the lowly abundant SETX to the exosome in response to such transcription pausing (Fig. 10.3).

10.6   SETX Localization and Function in Cycling Cells

SETX is a large 303 kDa protein that localizes to the nucleus in unsynchronized cell lines. But with different antibodies, several investigators have observed SETX clearly in the nucleolus. Initially, this was not unexpected, as yeast Sen1p was

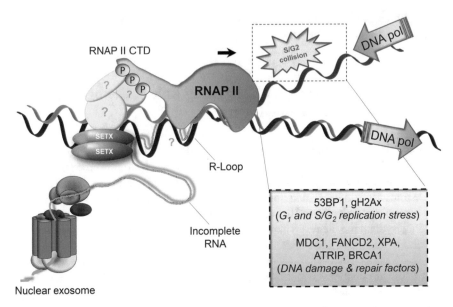

Fig. 10.3 SETX mediates RNA processing and degradation under specific circumstances. The SETX protein interacts with RNAP II via unknown intermediates, but only under certain circumstances, as SETX protein levels are exceedingly low, and SETX may require SUMOylation to direct it to specific foci. SETX foci form at times of replication stress and colocalize with markers such as 53BP1 and γH2Ax. Typical markers of DNA damage and repair factors are found in SETX foci at sites of DNA polymerase and RNAP II collision. According to this model, one likely SETX function is RNA processing via directing incompletely transcribed RNAs to the nuclear exosome for degradation

found to play a major role in the nucleolus, processing rRNA precursors and snoR-NAs [33]. Sen1p was also required to maintain the normal crescent shape of the yeast nucleolus, and the temperature sensitive mutant, *Sen1-1*, caused mislocalization of nucleolar proteins, fibrillarin, and Ssb1 [32]. Our initial studies with an affinity-purified SETX antibody revealed strong colocalization with fibrillarin in the nucleolus [57], but further studies are ongoing. A recent publication looking at the possibility of SETX mislocalization found near complete localization to the nucleolus in both control (*SETX*[+/+]) and patient (*SETX*[R332W/fs]) fibroblasts [58]. Despite these findings, immunocytochemistry analysis of tagged or endogenous SETX mostly shows a general nuclear localization.

A more detailed analysis of nuclear foci during the cell cycle was recently undertaken with GFP-tagged SETX. These investigators used double thymidine block to synchronize cells and automated wide-field microscopy to visualize SETX dynamic localization. They found that SETX foci were indeed present in the nucleolus at S-phase periodically, but as cells progressed into G_2-phase, SETX became distributed throughout the nucleoplasm [7]. In new studies, evidence was found to link the functions of SETX in potentially directing RNA to the exosome. How does this occur? It should be noted that transcription of large genes can take longer than the replication phase of the cell-cycle, such that the transcription and replication machinery may collide [59, 60], and SETX has been placed at the sites of these collisions (Fig. 10.3). Upon this backdrop, a range of studies were undertaken to define SETX nuclear foci, in response to phase transitions and drug treatments. In the nucleus, SETX distinct foci were found to be strongest during DNA replication or the S/G$_2$ phase. When cells were treated with aphidicolin to retard the replication fork, which is a form of replication stress, a two-fold increase in the number of SETX foci resulted [7]. These foci were perfectly colocalized with 53BP1 and γH2AX, markers of spontaneous DNA lesions and transcriptionally active nuclear bodies that form at fragile sites during replication [61] (Fig. 10.3). Then, after treatment of cycling cells with α-amanitin to inhibit RNAP II-mediated transcription, a significant reduction in SETX foci occurred, supporting the idea that coalescence of these distinct foci is dependent on RNAP II transcription. Such SETX-Exosc9 targeted interactions may represent one of the most pivotal roles of SETX, namely to bring functioning exosomes to sites of transcription—replication fork collisions, an interaction that is suggested to depend upon SUMO-2 and SUMO-3 SETX modification [49].

10.7 Lessons from the SETX Homologue IGHMBP2: Role of tRNA Regulation

What is missing from the study of SETX in neurological disease is a smoking gun pointing the way to the RNA pathways that are most affected. Interestingly, the study of SMARD was in a similar quandary, lacking knowledge of the affected RNA pathways; however, recent work has yielded a mechanistic understanding. In 2001, the *IGHMBP2* gene was identified as the molecular basis of SMARD, when key mutations in six families were reported [29]. SMARD is clinically distinct from SMA, but the IGHMBP2 protein, like SMN1, colocalizes with the RNA-processing machinery in both the cytosol and the nucleus [29]. The mouse model for SMARD is a spontaneous mutant discovered at The Jackson Laboratory known as *nmd* (for neuromuscular degeneration). An important clue to disease mechanism was provided with the discovery that the *nmd* phenotype is suppressed in a semi-dominant fashion by the presence of a modifier region on mouse Chromosome 13 from strain CAST/EiJ [31]. The critical region for rescue is limited to just 166 kb, defined by a BAC clone which contains several tRNA genes, including five tRNA^{Tyr} genes, one tRNA^{Ala} gene, and activator of basal transcription 1 (Abt1) [62]. The *nmd* mice are characterized by motor neuron degeneration with axonal loss leading to neurogenic muscle atrophy and death at 8–12 weeks of age. The phenotypic rescue of *nmd* mice by the Chromosome 13 modifier is dramatic with ventral nerve roots showing completely normal axonal morphology and density at 6–7 weeks of age [63]. Finally, IGHMBP2 has been shown to physically associate with tRNAs, and in particular with tRNA^{Tyr} and the tRNA transcription factor TFIIIC220 [62].

In 2013, further investigations revealed that aberrant tRNA processing can lead to neurodegeneration. The first mammalian RNA kinase to be identified was CLP1, and kinase-dead mice for this protein (*Clp1^{K/K}*) were generated [64]. On several genetic backgrounds, *Clp1^{K/K}* homozygous pups were nonviable, but on the CBA/J background, mice survived to ~23 weeks of age. *Clp1^{K/K}* mice display loss of spinal motor neurons associated with axonal degeneration in the peripheral nerves and denervation of neuromuscular junctions and respiratory failure [64]. Transgenic studies demonstrated that CLP1 functions in motor neurons, and that reduced CLP1 activity results in the accumulation of a novel set of small RNA fragments, derived from aberrant processing of pre-tRNA^{Tyr}. In 2014, a CLP1 R140H homozygous missense mutation was reported in five unrelated human families [65]. These patients suffered severe motor-sensory defects, cortical dysgenesis, and microcephaly. Biochemically, these presumed hypomorphic mutations lead to a loss of CLP1 interaction with the tRNA splicing endonuclease complex, greatly reduced pre-tRNA cleavage activity, and accumulation of linear tRNA introns [65].

Many other examples of tRNA biogenesis dysfunction leading to neurodegeneration exist. For brevity, we name just two: (1) an editing-defective tRNA synthetase causes protein misfolding and neurodegeneration in the *sticky* mouse [66]; and (2) a mutation of a CNS-specific tRNA causes neurodegeneration induced by ribosome stalling [67]. These examples serve to support the story of tRNA processing dysfunction in neurological disease and lend further credence to the mechanistic understanding of SMARD caused by IGHMBP2 recessive loss-of-function mutations. Similarly, methods and approaches that will reveal key insights into the most critical SETX RNA processing pathways for neuron health and survival are needed. While the lack of neurological phenotypes in *Setx* knock-out mice prevents the identification of similar modifier effects, other methodologies are likely to emerge to provide similar insight into RNA processing maintenance in neurodegeneration phenotypes caused by loss of SETX function.

10.8 SETX Gain-of-Function Motor Neuron Toxicity in ALS4 and Its Possible Role in Sporadic ALS

Here, we have considered two unique neurodegenerative disorders, AOA2 and ALS4, which represent the genotype/phenotype spectrum resulting from *SETX* mutation, and sought to underscore which RNA processing functions are most relevant to neurodegeneration. We began by recognizing that many functions attributed to SETX were extrapolated from its yeast orthologue Sen1p, which had been thoroughly studied long before *SETX* mutations were first discovered. Notably, Sen1p homology to SETX is restricted to an ~500 amino acid carboxy-terminal helicase domain, with no other regions of the large 303 kDa SETX protein conserved. The SETX amino-terminal protein interaction domain is divergent at the sequence level, but appears functionally conserved (with new protein interactions). Upon detailed examination, not all functions of yeast Sen1p were retained by mammalian SETX, which functions in a multicellular organism where cell cycling regulation has become much more elaborate. The well-characterized function of *general* transcription termination in Sen1p is not likely conserved in SETX; instead, SETX regulation of transcription termination is restricted to specific genes and cellular pathways, including interestingly circadian rhythm control. Whether disruption of this pathway is relevant to motor neuron health and ALS neurodegeneration remains to be studied. Insofar as future research is concerned, it is important to recognize that Sen1p and SETX proteins are present at exceedingly low levels in the cell. Thus, studies with Sen1p or SETX which utilize massive transient over-expression will likely generate results that are not physiologically relevant.

One key SETX function, conserved from Sen1p, is direct engagement with the nuclear exosome. Two studies demonstrate that SETX interacts with Exosc9 [48, 49], and is regulated by the SUMO and ubiquitin cascade pathways. One group has demonstrated that it is this SUMO-2/3 modification at the amino-terminus that is specifically required for interaction with the exosome and that co-depletion of SETX occurs with either Exosc9 or Exosc10 knock-down [49]. SETX was shown to be present in specific nuclear foci during S/G2-phase human synchronized cells coincident with collision of the DNA replication machinery and the RNA transcriptome [7]. These SETX foci were described as representing replication stress, and at these foci, the SETX interaction with the nuclear exosome was specifically present and enriched [49]. SETX thus appears to play a key role in directing incomplete RNA transcripts to the exosome for degradation (Fig. 10.3). The connection between SETX and exosome regulation deserves further consideration as a possible explanation for how SETX gain-of-function toxicity results in motor neuron disease. Interestingly, recessive loss-of-function mutations in Exosc3 yield infantile-onset motor neuron disease in human pontocerebellar hypoplasia with spinal muscular atrophy type 1B (PCH1B; OMIM 614678) [68], and Exosc3 interacts with matrin-3 [69], a known ALS gene. Furthermore, recessive loss-of-function mutations in Exosc8 yield infantile-onset motor neuron disease in human pontocerebellar hypoplasia with spinal muscular atrophy type 1C (PCH1C; OMIM 616081). These inherited motor neuron degeneration phenotypes highlight that alterations of exosome function are particularly poorly tolerated in cerebellar and motor neurons, two CNS regions where altered SETX function results in neuronal demise. However, another point to consider is that neurons are not cycling cells; hence, the role of SETX in resolving collisions between the replication machinery and the RNAP II transcription complex could actually play out in non-neuronal cells. As glia comprise the bulk of CNS cells, it seems reasonable to propose that neuron demise in ALS4 and AOA2 could be the result of a non-cell-autonomous process occurring in astrocytes or another non-neural CNS cell type.

A final important point to consider when seeking an explanation for SETX neurotoxicity is that SETX belongs to a group of just three homologous proteins, the other two being IGHMBP2 and RENT1. Of this trio, RENT1 is specifically implicated in NMD [70], and its role in NMD appears critical, as loss of function of RENT1 leads to embryonic lethality in mice, with no known human disease correlate. The importance of this helicase protein for RNA toxicity in neurons is suggested by its ability to rescue TDP-43 and FUS ALS-linked cellular pathology [28].

The other member of the trio is IGHMBP2, recessive mutations of which cause SMARD. In this disorder, key processing events for tRNA appear to be the responsible RNA pathway affected. As SMARD is related to autosomal spinal muscular atrophy (SMA), the theme of altered RNA function is reinforced, as SMN protein is essential to spliceosomal snRNP biogenesis and thus the integrity of RNA splicing, and therefore has become a model for understanding RNA dysfunction in neurodegeneration [71]. RNA-binding proteins, such as TDP-43, have also been centrally implicated in ALS, but understanding the role of TDP-43 in motor neuron neurodegeneration is proving to be challenging. There may be multiple dominant, recessive, and toxic mechanisms at play throughout the disease process. However, an intriguing theory based upon loss of function, which necessarily occurs with nuclear clearance of TDP-43, is that of impaired repression of non-conserved cryptic exons [72]. SETX gain-of-function mutations cause ALS4, which is a rare disease, and while several dominant mutations have been linked to ALS4, by far the most penetrant mutation to study is the L389S substitution. Clues to disease mechanism based upon SETX L389S toxic gain-of-function await the description of new mouse models that have been produced and are being characterized. Furthermore, based upon independent SALS exome sequencing reports [3, 73–75], SETX is emerging as a common target for mutation, especially in SALS patients who carry mutations in established pathogenic genes, including C9orf72 repeat expansion carriers [75]. These observations support the hypothesis that these recently discovered disease-linked polymorphisms in SETX could be modifiers of SALS. Hence, future research into SETX normal function and altered action upon gain-of-function mutation holds great potential for advancing our understanding of not just ALS4 motor neuron disease but also for much more common sporadic ALS as well.

Acknowledgments Our SETX research is supported by a grant from the Robert Packard Center for ALS Research at the Johns Hopkins School of Medicine.

References

1. Le Ber I, et al. Frequency and phenotypic spectrum of ataxia with oculomotor apraxia 2: a clinical and genetic study in 18 patients. Brain. 2004;127:759–67.
2. Rabin BA, et al. Autosomal dominant juvenile amyotrophic lateral sclerosis. Brain. 1999;122:1539–50.
3. Cady J, et al. Amyotrophic lateral sclerosis onset is influenced by the burden of rare variants in known amyotrophic lateral sclerosis genes. Ann Neurol. 2015;77:100–13.
4. Chen YZ, et al. DNA/RNA helicase gene mutations in a form of juvenile amyotrophic lateral sclerosis (ALS4). Am J Hum Genet. 2004;74:1128–35.
5. Moreira MC, et al. Senataxin, the ortholog of a yeast RNA helicase, is mutant in ataxia-ocular apraxia 2. Nat Genet. 2004;36:225–7.
6. Padmanabhan K, Robles MS, Westerling T, Weitz CJ. Feedback regulation of transcriptional termination by the mammalian circadian clock PERIOD complex. Science. 2012;337:599–602.

7. Yuce O, West SC. Senataxin, defective in the neurodegenerative disorder ataxia with oculomotor apraxia 2, lies at the interface of transcription and the DNA damage response. Mol Cell Biol. 2013;33:406–17.
8. Ursic D, Chinchilla K, Finkel JS, Culbertson MR. Multiple protein/protein and protein/RNA interactions suggest roles for yeast DNA/RNA helicase Sen1p in transcription, transcription-coupled DNA repair and RNA processing. Nucleic Acids Res. 2004;32:2441–52.
9. Lupski JR. Charcot-Marie-tooth polyneuropathy: duplication, gene dosage, and genetic heterogeneity. Pediatr Res. 1999;45:159–65.
10. Warner LE, Roa BB, Lupski JR. Absence of PMP22 coding region mutations in CMT1A duplication patients: further evidence supporting gene dosage as a mechanism for Charcot-Marie-Tooth disease type 1A. Hum Mutat. 1996;8:362.
11. Nihei Y, Ito D, Suzuki N. Roles of ataxin-2 in pathological cascades mediated by TAR DNA-binding protein 43 (TDP-43) and Fused in Sarcoma (FUS). J Biol Chem. 2012;287:41310–23.
12. Kim HD, Choe J, Seo YS. The sen1(+) gene of Schizosaccharomyces pombe, a homologue of budding yeast SEN1, encodes an RNA and DNA helicase. Biochemistry. 1999;38:14697–710.
13. Ghaemmaghami S, et al. Global analysis of protein expression in yeast. Nature. 2003;425:737–41.
14. Borggrefe T, Davis R, Bareket-Samish A, Kornberg RD. Quantitation of the RNA polymerase II transcription machinery in yeast. J Biol Chem. 2001;276:47150–3.
15. Svejstrup JQ, et al. Evidence for a mediator cycle at the initiation of transcription. Proc Natl Acad Sci U S A. 1997;94:6075–8.
16. Beck M, et al. The quantitative proteome of a human cell line. Mol Syst Biol. 2011;7:549.
17. Avemaria F, et al. Mutation in the senataxin gene found in a patient affected by familial ALS with juvenile onset and slow progression. Amyotroph Lateral Scler. 2011;12:228–30.
18. Rudnik-Schoneborn S, Arning L, Epplen JT, Zerres K. SETX gene mutation in a family diagnosed autosomal dominant proximal spinal muscular atrophy. Neuromuscul Disord. 2012;22:258–62.
19. Anheim M, et al. Ataxia with oculomotor apraxia type 2: clinical, biological and genotype/phenotype correlation study of a cohort of 90 patients. Brain. 2009;132:2688–98.
20. Suraweera A, et al. Senataxin, defective in ataxia oculomotor apraxia type 2, is involved in the defense against oxidative DNA damage. J Cell Biol. 2007;177:969–79.
21. Vantaggiato C, et al. Novel SETX variants in a patient with ataxia, neuropathy, and oculomotor apraxia are associated with normal sensitivity to oxidative DNA damaging agents. Brain Dev. 2014;36:682–9.
22. De Amicis A, et al. Role of senataxin in DNA damage and telomeric stability. DNA Repair. 2011;10:199–209.
23. Lynch DR, Braastad CD, Nagan N. Ovarian failure in ataxia with oculomotor apraxia type 2. Am J Med Genet A. 2007;143:1775–7.
24. Vance C, et al. Mutations in FUS, an RNA processing protein, cause familial amyotrophic lateral sclerosis type 6. Science. 2009;323:1208–11.
25. Bennett CL, La Spada AR. Unwinding the role of senataxin in neurodegeneration. Discov Med. 2015;19:127–36.
26. Weng Y, Czaplinski K, Peltz SW. Genetic and biochemical characterization of mutations in the ATPase and helicase regions of the Upf1 protein. Mol Cell Biol. 1996;16:5477–90.
27. Medghalchi SM, et al. Rent1, a trans-effector of nonsense-mediated mRNA decay, is essential for mammalian embryonic viability. Hum Mol Genet. 2001;10:99–105.
28. Barmada SJ, et al. Amelioration of toxicity in neuronal models of amyotrophic lateral sclerosis by hUPF1. Proc Natl Acad Sci U S A. 2015;112:7821–6.
29. Grohmann K, et al. Mutations in the gene encoding immunoglobulin mu-binding protein 2 cause spinal muscular atrophy with respiratory distress type 1. Nat Genet. 2001;29:75–7.
30. Lefebvre S, Burglen L, Frezal J, Munnich A, Melki J. The role of the SMN gene in proximal spinal muscular atrophy. Hum Mol Genet. 1998;7:1531–6.

31. Cox GA, Mahaffey CL, Frankel WN. Identification of the mouse neuromuscular degeneration gene and mapping of a second site suppressor allele. Neuron. 1998;21:1327–37.
32. Ursic D, DeMarini DJ, Culbertson MR. Inactivation of the yeast Sen1 protein affects the localization of nucleolar proteins. Mol Gen Genet. 1995;249:571–84.
33. Ursic D, Himmel KL, Gurley KA, Webb F, Culbertson MR. The yeast SEN1 gene is required for the processing of diverse RNA classes. Nucleic Acids Res. 1997;25:4778–85.
34. Winey M, Culbertson MR. Mutations affecting the tRNA-splicing endonuclease activity of *Saccharomyces cerevisiae*. Genetics. 1988;118:609–17.
35. Rasmussen TP, Culbertson MR. The putative nucleic acid helicase Sen1p is required for formation and stability of termini and for maximal rates of synthesis and levels of accumulation of small nucleolar RNAs in *Saccharomyces cerevisiae*. Mol Cell Biol. 1998;18:6885–96.
36. Steinmetz EJ, Conrad NK, Brow DA, Corden JL. RNA-binding protein Nrd1 directs poly(A)-independent 3′-end formation of RNA polymerase II transcripts. Nature. 2001;413:327–31.
37. Steinmetz EJ, Brow DA. Repression of gene expression by an exogenous sequence element acting in concert with a heterogeneous nuclear ribonucleoprotein-like protein, Nrd1, and the putative helicase Sen1. Mol Cell Biol. 1996;16:6993–7003.
38. Rondon AG, Mischo HE, Kawauchi J, Proudfoot NJ. Fail-safe transcriptional termination for protein-coding genes in *S. cerevisiae*. Mol Cell. 2009;36:88–98.
39. Steinmetz EJ, et al. Genome-wide distribution of yeast RNA polymerase II and its control by Sen1 helicase. Mol Cell. 2006;24:735–46.
40. Reynolds JJ, Stewart GS. A single strand that links multiple neuropathologies in human disease. Brain. 2013;136:14–27.
41. Skourti-Stathaki K, Proudfoot NJ, Gromak N. Human senataxin resolves RNA/DNA hybrids formed at transcriptional pause sites to promote Xrn2-dependent termination. Mol Cell. 2011;42:794–805.
42. Banerjee A, Sammarco MC, Ditch S, Wang J, Grabczyk E. A novel tandem reporter quantifies RNA polymerase II termination in mammalian cells. PLoS One. 2009;4:e6193.
43. Becherel OJ, et al. Senataxin plays an essential role with DNA damage response proteins in meiotic recombination and gene silencing. PLoS Genet. 2013;9:e1003435.
44. Elson A, et al. Pleiotropic defects in ataxia-telangiectasia protein-deficient mice. Proc Natl Acad Sci U S A. 1996;93:13084–9.
45. Yeo AJ, et al. R-loops in proliferating cells but not in the brain: implications for AOA2 and other autosomal recessive ataxias. PLoS One. 2014;9:e90219.
46. Han Z, Libri D, Porrua O. Biochemical characterization of the helicase Sen1 provides new insights into the mechanisms of non-coding transcription termination. Nucleic Acids Res. 2017;45:1355–70.
47. Yu K, Chedin F, Hsieh CL, Wilson TE, Lieber MR. R-loops at immunoglobulin class switch regions in the chromosomes of stimulated B cells. Nat Immunol. 2003;4:442–51.
48. Bennett CL, et al. Protein interaction analysis of senataxin and the ALS4 L389S mutant yields insights into senataxin post-translational modification and uncovers mutant-specific binding with a brain cytoplasmic RNA-encoded peptide. PLoS One. 2013;8:e78837.
49. Richard P, Feng S, Manley JL. A SUMO-dependent interaction between Senataxin and the exosome, disrupted in the neurodegenerative disease AOA2, targets the exosome to sites of transcription-induced DNA damage. Genes Dev. 2013;27:2227–32.
50. Suraweera A, et al. Functional role for senataxin, defective in ataxia oculomotor apraxia type 2, in transcriptional regulation. Hum Mol Genet. 2009;18:3384–96.
51. Deng HX, et al. Amyotrophic lateral sclerosis and structural defects in Cu,Zn superoxide dismutase. Science. 1993;261:1047–51.
52. DeMarini DJ, et al. The yeast SEN3 gene encodes a regulatory subunit of the 26S proteasome complex required for ubiquitin-dependent protein degradation in vivo. Mol Cell Biol. 1995;15:6311–21.

53. Carroll KL, Pradhan DA, Granek JA, Clarke ND, Corden JL. Identification of cis elements directing termination of yeast nonpolyadenylated snoRNA transcripts. Mol Cell Biol. 2004;24:6241–52.
54. Vasiljeva L, Buratowski S. Nrd1 interacts with the nuclear exosome for 3′ processing of RNA polymerase II transcripts. Mol Cell. 2006;21:239–48.
55. Kuehner JN, Pearson EL, Moore C. Unravelling the means to an end: RNA polymerase II transcription termination. Nat Rev Mol Cell Biol. 2011;12:283–94.
56. Richard P, Manley JL. Transcription termination by nuclear RNA polymerases. Genes Dev. 2009;23:1247–69.
57. Chen YZ, et al. Senataxin, the yeast Sen1p orthologue: characterization of a unique protein in which recessive mutations cause ataxia and dominant mutations cause motor neuron disease. Neurobiol Dis. 2006;23:97–108.
58. Roda RH, Rinaldi C, Singh R, Schindler AB, Blackstone C. Ataxia with oculomotor apraxia type 2 fibroblasts exhibit increased susceptibility to oxidative DNA damage. J Clin Neurosci. 2014;21:1627–31.
59. Helmrich A, Ballarino M, Nudler E, Tora L. Transcription-replication encounters, consequences and genomic instability. Nat Struct Mol Biol. 2013;20:412–8.
60. Helmrich A, Ballarino M, Tora L. Collisions between replication and transcription complexes cause common fragile site instability at the longest human genes. Mol Cell. 2011;44:966–77.
61. Lukas C, et al. 53BP1 nuclear bodies form around DNA lesions generated by mitotic transmission of chromosomes under replication stress. Nat Cell Biol. 2011;13:243–53.
62. de Planell-Saguer M, Schroeder DG, Rodicio MC, Cox GA, Mourelatos Z. Biochemical and genetic evidence for a role of IGHMBP2 in the translational machinery. Hum Mol Genet. 2009;18:2115–26.
63. Maddatu TP, Garvey SM, Schroeder DG, Hampton TG, Cox GA. Transgenic rescue of neurogenic atrophy in the nmd mouse reveals a role for Ighmbp2 in dilated cardiomyopathy. Hum Mol Genet. 2004;13:1105–15.
64. Hanada T, et al. CLP1 links tRNA metabolism to progressive motor-neuron loss. Nature. 2013;495:474–80.
65. Karaca E, et al. Human CLP1 mutations alter tRNA biogenesis, affecting both peripheral and central nervous system function. Cell. 2014;157:636–50.
66. Lee JW, et al. Editing-defective tRNA synthetase causes protein misfolding and neurodegeneration. Nature. 2006;443:50–5.
67. Ishimura R, et al. RNA function. Ribosome stalling induced by mutation of a CNS-specific tRNA causes neurodegeneration. Science. 2014;345:455–9.
68. Wan J, et al. Mutations in the RNA exosome component gene EXOSC3 cause pontocerebellar hypoplasia and spinal motor neuron degeneration. Nat Genet. 2012;44:704–8.
69. Erazo A, Goff SP. Nuclear matrix protein Matrin 3 is a regulator of ZAP-mediated retroviral restriction. Retrovirology. 2015;12:57.
70. Frischmeyer PA, Dietz HC. Nonsense-mediated mRNA decay in health and disease. Hum Mol Genet. 1999;8:1893–900.
71. Li DK, Tisdale S, Lotti F, Pellizzoni L. SMN control of RNP assembly: from post-transcriptional gene regulation to motor neuron disease. Semin Cell Dev Biol. 2014;32:22–9.
72. Ling JP, Pletnikova O, Troncoso JC, Wong PC. TDP-43 repression of nonconserved cryptic exons is compromised in ALS-FTD. Science. 2015;349:650–5.
73. Cirulli ET, et al. Exome sequencing in amyotrophic lateral sclerosis identifies risk genes and pathways. Science. 2015;347:1436–41.
74. Couthouis J, Raphael AR, Daneshjou R, Gitler AD. Targeted exon capture and sequencing in sporadic amyotrophic lateral sclerosis. PLoS Genet. 2014;10:e1004704.
75. Kenna KP, et al. Delineating the genetic heterogeneity of ALS using targeted high-throughput sequencing. J Med Genet. 2013;50:776–83.

Chapter 11
Lost in Translation: Evidence for Protein Synthesis Deficits in ALS/FTD and Related Neurodegenerative Diseases

Erik M. Lehmkuhl and Daniela C. Zarnescu

Abstract Cells utilize a complex network of proteins to regulate translation, involving post-transcriptional processing of RNA and assembly of the ribosomal unit. Although the complexity provides robust regulation of proteostasis, it also offers several opportunities for translational dysregulation, as has been observed in many neurodegenerative disorders. Defective mRNA localization, mRNA sequatration, inhibited ribogenesis, mutant tRNA synthetases, and translation of hexanucleotide expansions have all been associated with neurodegenerative disease. Here, we review dysregulation of translation in the context of age-related neurodegeneration and discuss novel methods to interrogate translation. This review primarily focuses on amyotrophic lateral sclerosis (ALS) and frontotemporal dementia (FTD), a spectrum disorder heavily associated with RNA metabolism, while also analyzing translational inhibition in the context of related neurodegenerative disorders such as Alzheimer's disease and Huntington's disease and the translation-related pathomechanisms common in neurodegenerative disease.

Keywords mRNA · Translation · Ribosome · ALS · FTD · TDP-43 · c9orf72 · RNA-binding proteins

E. M. Lehmkuhl
Department of Molecular and Cellular Biology, University of Arizona, Tucson, AZ, USA
e-mail: eriklehmkuhl@email.arizona.edu

D. C. Zarnescu (✉)
Department of Molecular and Cellular Biology, University of Arizona, Tucson, AZ, USA

Department of Neuroscience, University of Arizona, Tucson, AZ, USA

Department of Neurology, University of Arizona, Tucson, AZ, USA
e-mail: zarnescu@email.arizona.edu

© Springer International Publishing AG, part of Springer Nature 2018
R. Sattler, C. J. Donnelly (eds.), *RNA Metabolism in Neurodegenerative Diseases*, Advances in Neurobiology 20,
https://doi.org/10.1007/978-3-319-89689-2_11

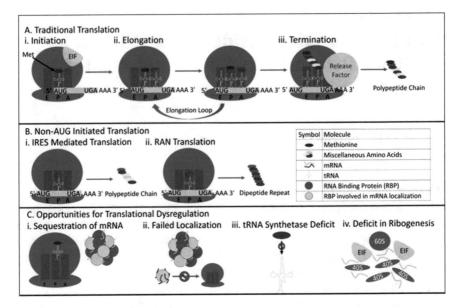

Fig. 11.1 Translation in normal conditions and disease. (**a**) (i) The process of polypeptide synthesis begins with the attachment of the AUG start codon within the mRNA template to the small ribosomal subunit. A transfer RNA (tRNA) specific to the AUG codon positions a methionine (Met) residue on the P (Peptidyl) site of the small ribosomal subunit, forming a translation initiation complex with the aid of translation initiation factors. (ii) Elongation of the peptide chain occurs through the A (Acceptor) site where a tRNA specific to the next codon within the mRNA template recruits the proper amino acid, which is bound to the initial Met at the P site with a peptide bond. (iii) Using this stepwise strategy, the peptide growth continues in an elongation loop until a stop codon (UAA, UAG, or UGA) is reached, causing specialized proteins called release factors to free the mRNA template as well as the newly formed polypeptide. (**b**) Non-AUG Initiated Translation. (i) Internal Ribosome Entry Site (IRES) translation generates normal polypeptide chains. (ii) Repeat Associated Non-AUG (RAN) translation forms toxic dipeptides. (**c**) Errors in translation. (i) Sequestration of mRNA in protein-mRNA granules prevents integration into a ribosome complex and therefore translation. (ii) Trafficking proteins are necessary for proper mRNA trafficking; their absence or dysfunction leads to a lack of mRNA in specific cellular locations. (iii) Deficient synthesis of tRNA prevents polypeptide addition, even with proper translation initiation. (iv) Deficits in ribogenesis reduce the number of actively translating ribosomal complexes

11.1   Introduction

In eukaryotes, normal translation relies on the assembly of a small (40S) and a large (60S) ribosomal subunit into fully assembled (80S) ribosomes. Each subunit comprises several ribosomal proteins and RNAs (rRNAs) that work together to catalyze protein synthesis using messenger RNA (mRNA) as a template (Fig. 11.1a). Translation is a major contributor to protein homeostasis (proteostasis) and its dysfunction has the potential to affect all cellular functions [1].

In addition to canonical AUG-dependent translation, a non-AUG version, known as Internal Ribosome Entry Site (IRES)-mediated translation, can take place via ribosomal attachment and elongation independent of a start codon (Fig. 11.1bi). Although no consensus sequence is known, the majority of IRESs are located near the 5′ end of the mRNA [2]. Interestingly, tau, one of the major proteins implicated in Alzheimer's disease, has been shown to undergo IRES translation, although the contribution, if any, of this mechanism to pathogenesis remains unknown [3]. A version of IRES translation is Repeat Associated Non-AUG (RAN) translation (Fig. 11.1bii), which has been associated with mutant microsatellite expansions in traditionally noncoding portions of the genome such as untranslated regions (UTRs) or introns, and more recently with coding regions (recently reviewed in [4, 5]). In the absence of an AUG codon, translation is possible because the hairpin structure formed by microsatellite mRNA can mimic the methionine tRNA that initiates translation [6]. Although dipeptide repeats generated via RAN translation and their distinct contribution to disease will be briefly discussed in this chapter, a more in-depth review has recently been published [7].

While the majority of translation occurs within the cytosol or ER bound ribosomes, 13 vital components of the oxidative phosphorylation complex are translated within the mitochondria, which host their own translational machinery including different ribosomal subunits, initiation factors, and tRNAs [8].

Regardless of its type, the complexity of translation as a highly regulated stepwise process provides numerous opportunities for errors caused by inhibition or deficits at any of these stages (see Fig. 11.1). Translation dysregulation has been implicated in several hereditary neurological disorders (reviewed in [9]). For example, loss-of-function mutations in one of five genes encoding subunits for the eukaryotic translation initiation factor eIF2B result in childhood ataxia, characterized by infant encephalopathy and later onset cognitive and motor impairment [10]. In Charcot-Marie-Tooth disease, an inherited neurodegenerative disorder, mutations in glycyl-tRNA synthetase (*GARS*), one of five tRNA synthetases linked to disease, disrupts translation [11]. FMRP, a protein implicated in Fragile X syndrome, regulates translation, in part by associating with ribosomes via direct binding to L5 protein [12]. Furthermore, mutations in *RPS19*, a small ribosomal subunit, have been associated with Blackfan Diamond Anemia, a deficiency in red blood cells that also leads to cognitive dysfunction [13]. Deficient mitochondrial translation has also been implicated in several neurological disorders; non-functional mitochondrial aspartyl or glutamyl-tRNA synthetases lead to leukoencephalopathy associated with ataxia, spasticity, and cognitive decline [9, 14].

Additionally, inhibition of translation can contribute to age-related neurodegenerative disorders. For example, in amyotrophic lateral sclerosis (ALS), TDP-43 sequesters mRNA away from translating ribosomes [15]. In Alzheimer's disease, microRNA-29, which regulates the expression of memory associated mRNAs (e.g., *BACE1*, a secretase implicated in the formation of pathogenic amyloid plaques) is downregulated causing uptranslation of its targets [16]. Translation is also expected to be affected by nucleolar stress and reduced rRNA biogenesis that have recently been associated with hexanucleotide repeat expansion (G4C2 HRE) within the first intron of *c9orf72*, the most common cause of ALS/FTD [17]. The role of translation in progressive neurodegeneration disorders is an exciting, developing field that is poised to uncover new therapeutic strategies. This chapter will focus on the mechanisms by which errors in translation contribute to age-related neurodegenerative disorders with references to modern methodologies for probing translation deficits in vivo.

11.2   Translational Alterations in Different Types of ALS and ALS/FTD

ALS is a progressive neurodegenerative disorder causing death of motor neurons [18]. Although 90% of ALS cases are sporadic, several genetic loci have been linked to both familial and sporadic cases and have been shown to be involved in a plethora of biological process ranging from ribostasis (e.g., *TARDP*, *FUS*, *senataxin*, *Gle1*) to proteostasis (e.g., *ubiquilin*, *SOD1*, *VCP*) [19]. Although these different gene products contribute directly to various specific aspects of cellular function, translation has been found to be directly or indirectly altered in the context of disease causing mutations, or in the context of wild-type TDP-43 pathology, which represents 97% of ALS and 45% of FTD cases. Recently, several genes linked to ALS were also shown to cause FTD leading to reframing of ALS and FTD as a spectrum disorder [20]. Here, we summarize the current state of the field in regards to translation dysregulation in different types of ALS/FTD.

SOD1—*SOD1* (Superoxide Dismutase) was the first gene associated with ALS in 1993 and remains solely associated with motor neuron disease across the ALS/FTD continuum. In a hallmark paper, Bruijn et al. demonstrated that a gain of toxic function rather than loss of enzymatic activity is responsible for neurodegeneration [21]. Although to date, the pathogenic mechanism of SOD1 has remained elusive, a great deal of evidence provides support for oxidative stress caused by mitochondrial dysfunction. Several hypotheses exist to explain the mechanism by which mutant SOD1 affects mitochondrion dysfunction (reviewed in [22]). Among these, Tan et al. showed that in $SOD1^{G93A}$ ALS mice, mutant SOD1 induces a conformational change in Bcl-2, which leads to reduced permeability of the mitochondrial membrane through altered interactions with Voltage Dependent Anion Channel 1 [23]. The loss of mitochondrial membrane polarity and subsequent oxidative stress lead to an increase in protein misfolding, causing a cascade of secondary effects including an unfolded protein response, which in turn inhibits global translation (reviewed in [24]).

Recently, Gal et al. discovered a novel role for mutant SOD1 pathogenesis [25]. In both $SOD1^{G93A}$ mice and patient-derived fibroblasts, mutated SOD1 was identified in inclusions containing TIA1 and G3BP1, two core components of stress granules [26]. Additional disease-associated variants $SOD1^{A4V}$ and $SOD1^{G85R}$ co-precipitated with G3BP1, indicating that stress granule interaction affects the pathogenesis of multiple SOD1 mutants. Co-precipitation occurs even following RNAse treatment suggesting that the G3BP1-mutant SOD1 interaction is not RNA dependent. Indeed, co-precipitation of truncated G3BP1 mutants and $SOD1^{A4V}$ indi-

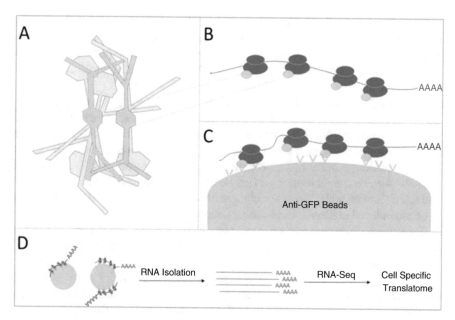

Fig. 11.2 Tagged ribosome affinity purification. (**a, b**) Model system-specific expression systems allow expression of tagged ribosomal subunit (RPL10-GFP) to be specifically expressed in cell types of interest (motor neurons or glia, green). (**c**) Using anti-GFP antibodies, the tagged ribosomal subunits are immunoprecipitated out of the whole body lysate. (**d**) Immunoprecipitated mRNA is isolated and subjected to RNA sequencing and bioinformatics to identify cell-specific translatomes normalized to input mRNA levels

cated that the RNA Recognition Motif (RRM) of G3BP1 is necessary and sufficient for the G3BP1-*SOD1^{A4V}* binding. Most importantly, the presence of *SOD^{A4V}* negatively correlated with stress granule formation, indicating a causal relationship between mutant SOD1-G3BP1 binding and stress granule dynamics [25]. Although the mechanism remains unclear, this work establishes a novel relationship between SOD1 mutants and stress granule dynamics, and suggests alterations in translation.

This possibility was addressed in a 2015 study, which examined translational changes that occur in mouse motor neurons, astrocytes, and oligodendrocytes during SOD1 driven ALS pathogenesis [27]. Previous cell-type-specific studies have relied on physical separation of the cell type of interest; this has numerous limitations including potential contamination and exclusion of axons and dendrites. This innovative study employed tagged ribosome affinity purification (TRAP), which allows the identification of cell-specific translatomes from intact, whole organisms (Fig. 11.2).

To define cell-type-specific translatomes, Sun et al. expressed GFP-RPL10 in motor neurons, astrocytes, and oligodendrocytes of *SOD1^{G37R}* mice using cell-specific promoters *Chat*, *Aldh1l1*, and *Cnp1*, respectively [27]. Importantly, the *SOD1^{G37R}* mutant line recapitulated the expression levels of endogenous SOD1, which is expressed in astrocytes and oligodendrocytes at 30% and 40% of motor neuron levels, respectively. The spinal cords of the mice were isolated at 8 months of age, corresponding to disease onset when muscle denervation has begun but phenotypes are not overtly present. Immunoprecipitation of GFP-RPL10 and subsequent RNA-seq followed by bioinformatics defined the translatome of each cell type.

At 8 months, Sun et al. observed that motor neurons exhibit upregulated translation of the components of the PERK (PRKR-like ER kinase)-mediated unfolded protein response (UPR) [27]. Protein misfolding induces PERK to phosphorylate eukaryotic initiation factor 2 alpha (eIF2α), which in turn upregulates translation of ATF4, a transcription factor that enhances the UPR response [28]. Translation of both ATF4 and its target transcripts (e.g., heat shock proteins *HSF1* and *HSF2*) was increased. Interestingly, the other components of UPR, namely ATF6 and IRE1, were not induced in disease onset motor neurons.

Laser microdissection was used to isolate motor neurons at the early symptomatic age of 10.5 months. Quantiative PCR experiments showed elevated ATF4 expression indicating that the UPR response continues through disease progression and is not limited to onset. A parallel experiment using $SOD1^{G85R}$ mice concluded that PERK-mediated UPR was also upregulated in these mutants, consistent with the idea that ER stress is common to SOD1 mediated pathogenesis.

Using the TRAP approach, 8 months old mouse astrocytes revealed an upregulation of inflammation related proteins (e.g., transcription factors Cedpb and Cedpd) while mRNAs encoding transcriptional co-activators for metabolic genes (e.g., PRRX1 and SERTAD2) experienced decreased translation. Upregulation of inflammation is characteristic of an astrocyte response to neuron damage. However, the increased translation of transcription factor PGC1α, related to metabolism and nuclear receptors, was also observed. Since upregulation of PGC1α is not characteristic of astrogliosis, it suggests that at least in part, SOD^{G37R} pathogenesis in astrocytes is independent of neuronal damage and may reflect a cell autonomous response to mutations in *SOD1* by astrocytes.

In contrast, oligodendrocytes exhibited minimal translational changes at 8 months, when mice were presymptomatic. However, profiling their translation again at the early symptomatic age of 10.5 months revealed that oligodendrocytes exhibit increased translation of transcripts involved in phagocytosis (e.g., *Rac2* and *Phosophoinositide Phospholipase C*) accompanied by a predicted decrease in proteins involved in myelination (e.g., CAMK2β).

From their findings, Sun et al. propose a model of SOD1-mediated pathogenesis where mutant SOD1 first induces motor neuron damage through the induction of ER stress [27]. Motor neurons may be selectively vulnerable because of high SOD1 expression and low ER chaperone presence. The effects of motor neuron damage are then amplified by subsequent damage to astrocytes and oligodendrocytes. The translational profiling conducted by Sun et al. provides elegant insights into translatome alterations in vivo, in a cell type and temporal-specific fashion, that highlight the central role of motor neurons in disease.

TDP-43—Encoded by the TAR DNA-Binding (*TARDP*) gene, TDP-43 is an RNA-binding protein comprising two RRM domains [29]. Remarkably, >97% of patients, regardless of etiology (with a couple of exceptions, including *SOD1* and *FUS* mutations) exhibit proteinaceous aggregates containing the RNA-binding protein TDP-43 [30]. TDP-43 has been implicated in several aspects of RNA processing including mRNA transport and localization to the distal ends of neurites including synapses [31–33]. While TDP-43 is normally required for RNA processing (e.g., splicing) [29], RNA binding is also required for toxicity [34], highlighting the involvement of RNA-based mechanisms in TDP-43 pathogenesis.

Recent studies have shed light into the mechanism by which TDP-43 contributes to ALS pathogenesis. In 2014, Coyne et al. [32] used a previously described *Drosophila* model of ALS [35, 36] based on TDP-43 overexpression to identify *futsch* as physiologically relevant target of TDP-43 regulation. *Futsch* mRNA was shown to be increased in motor neuron cell bodies, but decreased at neuromuscular synapses, consistent with failed mRNA localization. Polysome fractionations of ALS larvae indicated a shift of *futsch* mRNA from actively translating ribosomes to untranslated ribonucleoprotein particle (RNP) fractions consistent with translation inhibition [32]. This combination of defects in RNA localization and translation leads to increased levels of Futsch protein in motor neuron cell bodies, which was confirmed to also occur for its mammalian homolog, MAP1B, in spinal cords from ALS patients. This pathological alteration in Futsch/MAP1B, a microtubule stabilizing protein is consistent with neuromuscular junction (NMJ) instability, which was observed in the fly model. Notably, restoration of *futsch* levels by genetic overexpression mitigates ALS phenotypes including locomotor defects, TDP-43 aggregation, and reduced lifespan suggesting that *futsch* is an important mediator of TDP-43 toxicity in vivo.

TDP-43 knock-down studies in mouse hippocampal neurons have indicated that *Rac1* levels increase at the translational level and this affects spine morphogenesis in dendrites [37]. Together with observations that AMPAR clustering is increased following synaptic stimulation, these findings support a role for TDP-43 in plasticity. The human disease relevance of these observations remains to be established in future studies.

Recently, an interesting mechanistic connection has been identified between ribostasis and proteostasis [15]. Using the same *Drosophila* model of ALS [35, 36] the authors identified *hsc70-4* mRNA as a candidate target of mutant but not wild-type TDP-43 [15]. Hsc70-4 is a conserved member of the Hsc70 family of constitutive chaperones with several roles in protein folding, degradation and various cellular processes including stress response, and chaperone-mediated autophagy [38]. Specifically, Hsc70-4 regulates synaptic vesicle cycling, and just like its cognate mRNA was found to associate preferentially with mutant TDP-43. The consequence of this preferential association with mutant TDP-43 is the sequestration of *hsc70-4* mRNA accompanied by translation inhibition, which in turn leads to defects in the synaptic vesicle endocytosis. A similar post-transcriptional reduction was observed in C9 ALS fly and patient-derived motor neurons, although it remains to be determined whether this is caused by translation inhibition as was the case with TDP-43 models. Notably, restoration of Hsc70-4 through genetic overexpression mitigated ALS phenotypes in a variant-dependent manner suggesting that although both wild-type and mutant TDP-43 contribute to ALS pathogenesis, they do so through distinct mechanisms [15].

Additional links between TDP-43 and protein synthesis have been uncovered by biochemical studies showing its association with several RNA-binding proteins involved in translation including eukaryotic initiation factors and Fragile X Mental Retardation Protein (FMRP) [39–43]. FMRP overexpression was found to attenuate locomotor dysfunction and increase lifespan in a fly model of ALS based on TDP-43 [42]. Genetic interactions and fractionation experiments collectively led to a model whereby FMRP remodels TDP-43/RNA complexes and releases sequestered mRNA, which can subsequently be translated and mitigate TDP-43 toxicity. Interestingly, FMRP and TDP-43 appear to share translation targets including *Rac1* and *futsch* mRNAs, highlighting previously unknown common mechanisms between neurodevelopmental conditions such as Fragile X syndrome and neurodegenerative diseases like ALS/FTD.

TDP-43 has also been found to regulate translation globally [44]. Using an Affymetrix exon array, Fiesel et al. [44] evaluated splicing variants in HEK293E human embryonic kidney cells following knockdown of TDP-43 with small interfering RNA (siRNA). This study showed that loss of TDP-43 induced alternative splicing of *S6 kinase 1 Aly/REF-like target* (*SKAR*). In addition to being previously associated with spliced mRNA, SKAR also recruits S6 Kinase 1 to protein-mRNA granules to promote translation downstream of mTOR signaling [45]. Knock-down of TDP-43 causes exon 3 exclusion and generation of SKARβ, which results in increased phosphorylation of S6 K1 and its targets, leading to increased global translation [44]. It remains to be determined whether global translation is also altered in patients. The most compelling evidence so far of global translation dysregulation comes from findings that genetic and pharmacological inhibition of eIF2α phosphorylation mitigates ALS phenotypes in fly and cultured cells models [46]. However, given the intimate connections between UPR and global translation that eIF2α mediates, more studies are needed to determine the extent of translation dysregulation in disease pathogenesis.

Given its known interactions with protein partners, TDP-43 appears to be involved in the regulation of translation at multiple steps. Studies have identified both a normal role for TDP-43 in the regulation of global translation in cutured cells, and specific mRNAs targets, including mutant-specific targets in motor neurons, in the context of disease. A distinction needs to be made between TDP-43's normal role in various cell types and how that role changes in disease and more studies are needed to address this important question. The variety of ways in which TDP-43 interacts with the translational machinery leads to additional questions regarding the role of TDP-43 and RNA in ALS pathogenesis.

FUS—Mutations in *Fused in Sarcoma* (FUS), which encodes a nuclear RNA-binding protein, have also been associated with 4% of familial ALS with autopsy showing cytoplasmic inclusions of FUS [47]. Although no specific alterations have been identified in translation in the context of FUS ALS, given the aggregation of mutant FUS[P525L] in complexes containing nuclear-cytoplasmic shuttling proteins (e.g., hnRNP A1 and A2), spliceosome assembling proteins (e.g., SMN1), and mRNA [48], it is reasonable to predict indirect changes in the translatome caused by aberrant protein and protein-RNA interactions.

c9orf72—Hexanucleotide repeat expansions (HRE) in *c9orf72* have been recently identified as the most common cause of familial ALS [49, 50]. These G4C2 repeats lie within the first intron of *c9orf72* and range from 2 to 10 in the normal population, and 90 to several hundreds in disease. Repeat expansions as low as 20 have been identified in ALS cases, but a causal relationship has not been established between the size of expansions and disease, and evidence for multiple gene mutation contributions has been found in carrier families [51]. Much research has focused on discerning the normal function of c9orf72, a putative DENN protein [52], and the contribution of hexanucleotide repeat expansions to disease. Although evidence exists to support several disease mechanisms including haploinsufficiency, RNA foci, and dipeptide repeat (DPR)-mediated toxicity, the specific pathomechanism of *c9orf72* remains unclear and subject to controversy [53]. A most remarkable discovery made in regards to *c9orf72* pathomechanism is the finding of nuclear pore alterations and defects in nucleo-cytoplasmic shuttling [17, 54, 55]. Accompanying these phenotypes are defects in RNA SG assembly, translation, and ribosome biogenesis, discussed later.

RNA foci and toxicity—Overexpression of G4C2 HREs of various lengths led to reduced transcription of stress granule proteins TIA-1 and HuR indicating a role for RNA foci in stress granule assembly [56]. However, because these HREs also generated DPRs, the contribution of the latter cannot be excluded. Using elegant live imaging approaches, Schweizer Burguete et al. observed that *c9orf72* HREs foci colocalize with FMRP and translocate bi-directionally within neurites. Interestingly, the presence of HRE increased protein levels for both FMRP and PSD-95, a protein whose translation is facilitated by FMRP, indicating that the *c9orf72* HRE may alter local protein translation [57]. These studies propose that the HREs induce neurodegenerative phenotypes by altering mRNA localization to synapses, a phenotype previously associated with other types of ALS [15, 32, 33].

RAN translation—As mentioned above, RAN (Repeat Associated Non-AUG translation) is a version of IRES-based translation mechanism that causes the repeat expansions to be translated into dipeptide repeats (DPRs). In *c9orf72*-mediated ALS, RAN translation of the G4C2 repeat and its anti-sense mRNA result in poly(GA), poly(GP), poly(GR), poly(PA), and poly(PR) dipeptides, all of which have been detected in patient tissues. One study implicated poly(GA) dipeptides as the primary aggregate inducing DPR [58], however another study concluded that expression of poly(GR) and poly(PR) induced neurodegenerative phenotypes while the three dipeptide products lacking arginine did not [59]. Although which DPRs are toxic remains an actively investigated question in the field, their effect on translation is undisputed.

In the context of ALS, DPRs have been associated with the disruption of translation and nuclear-cytoplasmic transport. Using Surface Sensing of Translation (SUnSET) [60] (Fig. 11.3), Kanekura et al. observed that poly(PR) and poly(GR) DPRs inhibit global translation in NSC34 motor neuron like cells [61]. The arginine-containing DPRs were found to form aggregates containing RNA-binding proteins

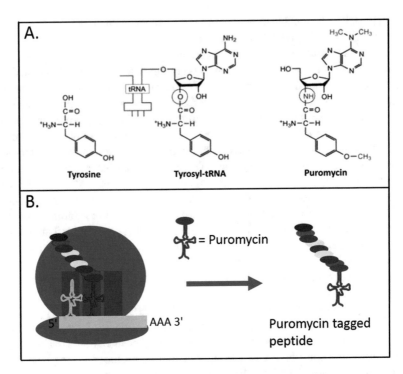

Fig. 11.3 Surface Sensing of Translation (SUnSET). (**a**) Molecular structure of tyrosine, tyrosyl-tRNA, and puromycin. (**b**) Puromycin is a structural analog of tyrosyl-tRNA from the bacterium *Streptomyces alboniger* and can be incorporated into elongating peptide chains. Puromycin attachment releases the peptide chain from the ribosome due to puromycin's non-hydrolyzable amid bond, yielding a puromycin tagged peptide chain. Fluorescent puromycin antibodies can then be used to track translation rates in real time. Traditional sulfur isotope assays were used to verify that puromycin expression does not significantly alter translation rates [60]

(e.g., FUS and TDP-43) along with RNA. The model emerging from this study is that the hydrophobic DPR aggregates block translation by preventing initiation factors from interacting with mRNA.

Poly(PR) and poly(GR) DPRs have also been implicated in post-transcriptional processing, nuclear cytoplasmic transport, and rRNA biogenesis, which can ultimately affect translation. Jovicic et al. also identified several genes related to rRNA biogenesis (e.g., *efg1* and *nsr1*) as significant modifiers of poly(PR) toxicity [17]. Additionally, poly(PR) and poly(GR) peptides colocalize with nucleoli, the site of rRNA synthesis [62]. Although the dysregulation of ribogenesis has been associated with *c9orf72* HRE in multiple studies, the mechanism remains poorly understood. Overall, although the role of *c9orf72* in both healthy and neurodegenerative individuals has been heavily studied in recent years, significant work remains to be done regarding the molecular mechanism and to precisely determine the contributions of G4C2 expanded RNA or the translated dipeptide products to disease. The mixed spectrum of results to date may reflect heterogenous responses to HREs and DPRs among different cell types in the nervous system (Fig. 11.4).

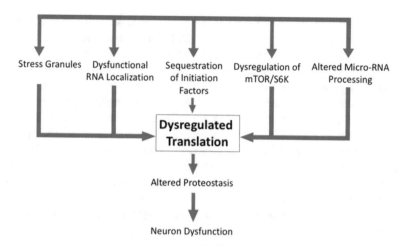

Fig. 11.4 Causes of Translational Inhibition. Cells regulate translation through a robust, complex integration of multiple pathways. The dysregulation of such pathways can alter proteostasis within a cell and lead to neuronal dysfunction

11.3 RAN Translation Beyond ALS/FTD

The first reports of RAN translation were made in association with CAG expanded transcripts in *SCA8* and Muscular Dystrophy type I (DM1) [6]. DM1 is caused by CTG repeat expansions within the myotonic dystrophy protein kinase (*DMPK*). Once transcribed, the expanded CUG mRNA forms a double-stranded structure that sequesters muscleblind (MBNL1), an RNA-binding protein involved in splicing [63]. As a result, CUGBP1 is upregulated, which together with MBNL1 sequestration leads to defects in the fetal to adult splicing transition and disease pathogenesis. Recent studies, however, have proposed an additional pathomechanism whereby the repeat expansions undergo RAN translation [6], albeit the mechanism by which RAN products contribute to disease is unknown. A recent review on this topic provides an excellent overview of the increasingly complex mechanisms behind myotonic dystrophy [64].

Since the initial discovery of RAN translation, additional microsatellite expansion disorders including Fragile X Tremors Ataxia Syndrome (FXTAS) and Huntingtin's disease have been added to the list of conditions in which bidirectional expanded transcripts produce RAN proteins [4, 5]. These novel and unexpected peptides contribute to toxicity challenging existing paradigms about disease mechanisms wherever they are found.

Huntington's disease (HD)—HD is an autosomal dominant neurodegenerative disorder [65]. The expansion of a CAG repeat region within the coding region of the huntingtin gene (*HTT*) leads to disease onset between the ages of 30 and 50 and causes progressive loss of neuron function [65]. The protein product of *HTT*, Huntingtin, is associated with microtubule-based trafficking of vesicles and mRNAs within neurons [66]. Initial suggestions that protein synthesis may be altered in HD came from fibroblasts showing that in cells cultured from Huntington's patients, RNA accumulates in the nucleus and is not properly translated [67]. A more recent report shows that *HTT* repeat expansions also undergo RAN translation that can drive neurodegeneration through the dysregulation of nuclear-cytoplasmic transport [68]. Similar phenotypes including nuclear envelope morphology, pore architecture, and RNA export defects were found in a parallel study, although no RAN translation products were reported [69]. The discovery of RAN translation by Grima et al. [68] led to proposing a mechanism whereby its products cause these newly discovered phenotypes by specifically altering nuclear pores and inhibiting RANGAP1, a GTPase-activating protein necessary for nuclear cytoplasmic shuttling. Further substantiating this model is the fact that RANGAP1 overexpression or pharmacological restoration of nuclear transport rescued HD phenotypes across multiple model system [68].

Fragile X tremor ataxia syndrome (FXTAS)—FXTAS is caused by the expansion of CGG repeats in the 5′ UTR of the Fragile X Mental Retardation gene 1 (FMR1) [70–72]. The length of the CGG expansions determines the phenotypic outcome, with repeats >200 causing complete loss of transcription and absence of FMRP, while intermediate length repeats (50–200) lead to increased transcript but reduced protein production [73]. The intermediate expansion was associated with intention tremors, ataxia, dementia, and parkinsonism occurring in aging individuals [74]. Mechanistically, CGG repeats were shown to sequester RNA-binding proteins thereby dysregulating their activity within cells [75, 76]. An example relevant to translation regulation is DGCR8, a miRNA processing factor, which binds to expanded *FMR1* mRNA; this leads to decreased levels of mature microRNAs, which in turn can impact the translatome by imparing the translation of their mRNA targets [77] . Several recent studies identify RAN translation products, specifically polyG peptides produced from CGG expanded repeats using an ACG codon as start [78–80]. Elegant experiments in mouse models and patient-derived cells demonstrate that RAN translation-derived polyG peptides but not CGG RNA alone are responsible for FXTAS phenotypes [80].

11.4   Translation Dysregulation in Alzheimer's Disease

Recent studies have associated dysregulation of protein-mRNA complexes with Alzheimer's disease pathology [81]. Tau is a microtubule-associated protein whose aggregation and hyper-phosphorylation is a hallmark of Alzheimer's with pathology predicted to be driven in part by failed axonal transport [82]. Recent studies identified pathological tau in complex with TIA1 [83], a core component of stress granules. Interestingly, tau-TIA1 binding was found to have a positive correlation with stress granule formation suggesting possible consequences on translation that will have to be elucidated in future studies. Additionally, TIA1 mediates translational inhibition of many stress response genes including P53, a major regulator of DNA damage repair [84]. It remains to be determined if susceptibility to, or DNA damage itself, may be involved in AD pathogenesis. Additionally, a dichotomous relationship was recently observed between TDP-43 and tau levels during Alzheimer's pathogenesis, with TDP-43 being shown to regulate tau protein expression by destabilizing its cognate mRNA [85].

Translation efficiency was also shown to be deficient in Alzheimer's brains and was noted to be an early event in disease. Ding et al. showed that in patient brain extracts, although the same quantity of polyribosome material was produced in control and disease cases, the translational efficiency of the polyribosomes was reduced by >60% in the inferior parietal lobe (IP) and superior middle temporal gyri (SMTG), albeit no significant reduction was observed in the cerebellum [86]. Regarding the mechanism of translation deficiency, $tRNA^{Asn}$ and 5S rRNA were significantly reduced in the IP of Alzheimer's patients together with increased oxidation of 28S rRNA. In contrast, the cerebellum of Alzheimer's patients exhibited increased phosphorylation of $eIF2\alpha$ and p70S6 [86]. While the former is associated with increased unfolded protein response and stress granule formation, the latter is associated with activation of the mTOR pathway and increased translation. Collectively, these findings provide intriguing links between Alzheimer's disease pathogenesis and translation through stress granules, initiation factors, and rRNA; however, the precise involvement of translation in Alzheimer's remains unknown.

11.5 Micro RNAs (miRNAs) and Translation Regulation

miRNAs are noncoding RNAs that can control gene expression by inhibiting mRNA translation or by selective degradation of transcripts (reviewed in [87]). It has been shown that TDP-43 aggregates sequester Dicer and DROSHA, two key RNA-binding proteins required for generating functional miRNA, thus implicating miRNA maturation in ALS pathogenesis [88]. DROSHA was also shown to form aggregates with RAN translation derived DPRs in patient tissues [89]. Additionally, XP05, which is required for precursor miRNA export from the nucleus, was identified as modifier of *c9orf72* HRE and TDP-43-based pathogenesis among other nuclear-cytoplasmic transport proteins [17, 54, 90]. Consequent reduced Dicer and DROSHA activity, and inhibited nuclear cytoplasmic transport potentially explain decreased global miRNA levels observed in multiple forms of ALS (recently reviewed in [91]). Several miRNAs required for synaptic development and maintenance including *miR-9* and *miR-124* were found to be altered in ALS/FTD patient derived cells and tissues suggesting the possibility that specific miRNAs may mediate aspects of toxicity in disease [92, 93]. Although the precise role of miRNAs in disease remains unclear, existing evidence supports the possibility of both a global and target-specific inhibition of miRNA synthesis as a contributor to ALS/FTD pathogenesis (recently reviewed in [91, 94]).

11.6   Concluding Remarks

Recent studies have provided compelling evidence that ALS/FTD and even Alzheimer's disease exhibit defects in multiple steps of RNA processing including protein synthesis. Several neurodegeneration-associated proteins are involved in mRNA export, trafficking, localization, and translation. These processes offer a plausible explanation for the unique pathogeneses observed in neurons, which have dendritic and axonal extremities requiring local translation and mRNA transport across distances vastly larger than the cell body (recently reviewed in [95–97]). The recent discovery that certain ribosomal subunits preferentially translate subsets of mRNA [98] provides an additional layer to the complex mosaic that is translation within neurons.

Evidence exists to support both global and target-specific dysregulation of translation. Globally, activation of the PERK pathway including phosphorylation of eIF2α inhibits global translation at the initiation step by inhibiting the incorporation of eIF2α into ribosomal complexes [99]. The inhibition of PERK has provided encouraging results in attenuating neurodegenerative phenotypes as it mitigated TDP-43-dependent phenotypes in flies and cultured mouse cells [46], and it restored synaptic protein levels and motor function in models of prion-mediated neurodegeneration [100]. Given findings that PERK is also upregulated in SOD1 mice [27], this approach could be extended to additional ALS types. However, pharmacological inhibition of translation did not rescue memory defects in mouse models of Alzheimer's disease [101, 102].

The evidence for specific mRNA targets further substantiates existing hypotheses that axonal transport, neuronal cytoskeleton, and synaptic vesicle cycling are underlying the synaptopathy associated with neurodegeneration. While these provide more specificity to any future interventions, it remains difficult to prioritize which mRNA target carries more physiological significance, and targeting multiple targets simultaneously poses significant challenges. Clearly, Aristotle's famous quote "The more you know, the more you know you don't know" remains relevant today; a lot is yet to be learned about the intricacies of translation dysregulation in neurodegeneration, specific RNA targets and processes, from disease onset to motor neuron failure and death.

References

1. Aitken CE, Lorsch JR. A mechanistic overview of translation initiation in eukaryotes. Nat Struct Mol Biol. 2012;19(6):568–76.
2. Komar AA, Hatzoglou M. Cellular IRES-mediated translation: the war of ITAFs in pathophysiological states. Cell Cycle. 2011;10(2):229–40.
3. Veo BL, Krushel LA. Translation initiation of the human tau mRNA through an internal ribosomal entry site. J Alzheimers Dis. 2009;16(2):271–5.

4. Cleary JD, Ranum LP. New developments in RAN translation: insights from multiple diseases. Curr Opin Genet Dev. 2017;44:125–34.
5. Gao FB, Richter JD, Cleveland DW. Rethinking unconventional translation in neurodegeneration. Cell. 2017;171(5):994–1000.
6. Zu T, Gibbens B, Doty NS, Gomes-Pereira M, Huguet A, Stone MD, et al. Non-ATG-initiated translation directed by microsatellite expansions. Proc Natl Acad Sci U S A. 2011;108(1):260–5.
7. Freibaum BD, Taylor JP. The role of dipeptide repeats in C9ORF72-related ALS-FTD. Front Mol Neurosci. 2017;10:35.
8. Boczonadi V, Horvath R. Mitochondria: impaired mitochondrial translation in human disease. Int J Biochem Cell Biol. 2014;48:77–84.
9. Scheper GC, van der Knaap MS, Proud CG. Translation matters: protein synthesis defects in inherited disease. Nat Rev Genet. 2007;8(9):711–23.
10. van der Knaap MS, Leegwater PA, Konst AA, Visser A, Naidu S, Oudejans CB, et al. Mutations in each of the five subunits of translation initiation factor eIF2B can cause leukoencephalopathy with vanishing white matter. Ann Neurol. 2002;51(2):264–70.
11. Antonellis A, Ellsworth RE, Sambuughin N, Puls I, Abel A, Lee-Lin SQ, et al. Glycyl tRNA synthetase mutations in Charcot-Marie-Tooth disease type 2D and distal spinal muscular atrophy type V. Am J Hum Genet. 2003;72(5):1293–9.
12. Chen E, Sharma MR, Shi X, Agrawal RK, Joseph S. Fragile x mental retardation protein regulates translation by binding directly to the ribosome. Mol Cell. 2014;54(3):407–17.
13. Kubik-Zahorodna A, Schuster B, Kanchev I, Sedlacek R. Neurological deficits of an Rps19(Arg67del) model of Diamond-Blackfan anaemia. Folia Biol (Praha). 2016;62(4):139–47.
14. Steenweg ME, Ghezzi D, Haack T, Abbink TE, Martinelli D, van Berkel CG, et al. Leukoencephalopathy with thalamus and brainstem involvement and high lactate 'LTBL' caused by EARS2 mutations. Brain. 2012;135(Pt 5):1387–94.
15. Coyne AN, Lorenzini I, Chou CC, Torvund M, Rogers RS, Starr A, et al. Post-transcriptional inhibition of Hsc70-4/HSPA8 expression leads to synaptic vesicle cycling defects in multiple models of ALS. Cell Rep. 2017;21(1):110–25.
16. Hebert SS, Horre K, Nicolai L, Papadopoulou AS, Mandemakers W, Silahtaroglu AN, et al. Loss of microRNA cluster miR-29a/b-1 in sporadic Alzheimer's disease correlates with increased BACE1/beta-secretase expression. Proc Natl Acad Sci U S A. 2008;105(17):6415–20.
17. Jovicic A, Mertens J, Boeynaems S, Bogaert E, Chai N, Yamada SB, et al. Modifiers of C9orf72 dipeptide repeat toxicity connect nucleocytoplasmic transport defects to FTD/ALS. Nat Neurosci. 2015;18(9):1226–9.
18. Cleveland DW, Rothstein JD. From Charcot to Lou Gehrig: deciphering selective motor neuron death in ALS. Nat Rev Neurosci. 2001;2(11):806–19.
19. Schymick JC, Traynor BJ. Expanding the genetics of amyotrophic lateral sclerosis and frontotemporal dementia. Alzheimers Res Ther. 2012;4(4):30.
20. Ling SC, Polymenidou M, Cleveland DW. Converging mechanisms in ALS and FTD: disrupted RNA and protein homeostasis. Neuron. 2013;79(3):416–38.
21. Bruijn LI, Houseweart MK, Kato S, Anderson KL, Anderson SD, Ohama E, et al. Aggregation and motor neuron toxicity of an ALS-linked SOD1 mutant independent from wild-type SOD1. Science. 1998;281(5384):1851–4.
22. Tafuri F, Ronchi D, Magri F, Comi GP, Corti S. SOD1 misplacing and mitochondrial dysfunction in amyotrophic lateral sclerosis pathogenesis. Front Cell Neurosci. 2015;9:336.
23. Tan W, Naniche N, Bogush A, Pedrini S, Trotti D, Pasinelli P. Small peptides against the mutant SOD1/Bcl-2 toxic mitochondrial complex restore mitochondrial function and cell viability in mutant SOD1-mediated ALS. J Neurosci. 2013;33(28):11588–98.
24. Hetz C, Mollereau B. Disturbance of endoplasmic reticulum proteostasis in neurodegenerative diseases. Nat Rev Neurosci. 2014;15(4):233–49.

25. Gal J, Kuang L, Barnett KR, Zhu BZ, Shissler SC, Korotkov KV, et al. ALS mutant SOD1 interacts with G3BP1 and affects stress granule dynamics. Acta Neuropathol. 2016;132(4):563–76.
26. Kedersha N, Anderson P. Regulation of translation by stress granules and processing bodies. Prog Mol Biol Transl Sci. 2009;90:155–85.
27. Sun S, Sun Y, Ling SC, Ferraiuolo L, McAlonis-Downes M, Zou Y, et al. Translational profiling identifies a cascade of damage initiated in motor neurons and spreading to glia in mutant SOD1-mediated ALS. Proc Natl Acad Sci U S A. 2015;112(50):E6993–7002.
28. Harding HP, Calfon M, Urano F, Novoa I, Ron D. Transcriptional and translational control in the Mammalian unfolded protein response. Annu Rev Cell Dev Biol. 2002;18:575–99.
29. Buratti E, Dork T, Zuccato E, Pagani F, Romano M, Baralle FE. Nuclear factor TDP-43 and SR proteins promote in vitro and in vivo CFTR exon 9 skipping. EMBO J. 2001;20(7):1774–84.
30. Neumann M, Sampathu DM, Kwong LK, Truax AC, Micsenyi MC, Chou TT, et al. Ubiquitinated TDP-43 in frontotemporal lobar degeneration and amyotrophic lateral sclerosis. Science. 2006;314(5796):130–3.
31. Ishiguro A, Kimura N, Watanabe Y, Watanabe S, Ishihama A. TDP-43 binds and transports G-quadruplex-containing mRNAs into neurites for local translation. Genes Cells. 2016;21(5):466–81.
32. Coyne AN, Siddegowda BB, Estes PS, Johannesmeyer J, Kovalik T, Daniel SG, et al. Futsch/MAP1B mRNA is a translational target of TDP-43 and is neuroprotective in a Drosophila model of amyotrophic lateral sclerosis. J Neurosci. 2014;34(48):15962–74.
33. Alami NH, Smith RB, Carrasco MA, Williams LA, Winborn CS, Han SS, et al. Axonal transport of TDP-43 mRNA granules is impaired by ALS-causing mutations. Neuron. 2014;81(3):536–43.
34. Voigt A, Herholz D, Fiesel FC, Kaur K, Muller D, Karsten P, et al. TDP-43-mediated neuron loss in vivo requires RNA-binding activity. PLoS One. 2010;5(8):e12247.
35. Estes PS, Boehringer A, Zwick R, Tang JE, Grigsby B, Zarnescu DC. Wild-type and A315T mutant TDP-43 exert differential neurotoxicity in a Drosophila model of ALS. Hum Mol Genet. 2011;20(12):2308–21.
36. Estes PS, Daniel SG, McCallum AP, Boehringer AV, Sukhina AS, Zwick RA, et al. Motor neurons and glia exhibit specific individualized responses to TDP-43 expression in a Drosophila model of amyotrophic lateral sclerosis. Dis Model Mech. 2013;6(3):721–33.
37. Majumder P, Chen YT, Bose JK, Wu CC, Cheng WC, Cheng SJ, et al. TDP-43 regulates the mammalian spinogenesis through translational repression of Rac1. Acta Neuropathol. 2012;124(2):231–45.
38. Liu T, Daniels CK, Cao S. Comprehensive review on the HSC70 functions, interactions with related molecules and involvement in clinical diseases and therapeutic potential. Pharmacol Ther. 2012;136(3):354–74.
39. Sephton CF, Cenik C, Kucukural A, Dammer EB, Cenik B, Han Y, et al. Identification of neuronal RNA targets of TDP-43-containing ribonucleoprotein complexes. J Biol Chem. 2011;286(2):1204–15.
40. Freibaum BD, Chitta RK, High AA, Taylor JP. Global analysis of TDP-43 interacting proteins reveals strong association with RNA splicing and translation machinery. J Proteome Res. 2010;9(2):1104–20.
41. Majumder P, Chu JF, Chatterjee B, Swamy KB, Shen CJ. Co-regulation of mRNA translation by TDP-43 and Fragile X Syndrome protein FMRP. Acta Neuropathol. 2016;132(5):721–38.
42. Coyne AN, Yamada SB, Siddegowda BB, Estes PS, Zaepfel BL, Johannesmeyer JS, et al. Fragile X protein mitigates TDP-43 toxicity by remodeling RNA granules and restoring translation. Hum Mol Genet. 2015;24(24):6886–98.
43. Wang JW, Brent JR, Tomlinson A, Shneider NA, McCabe BD. The ALS-associated proteins FUS and TDP-43 function together to affect Drosophila locomotion and life span. J Clin Invest. 2011;121(10):4118–26.

44. Fiesel FC, Weber SS, Supper J, Zell A, Kahle PJ. TDP-43 regulates global translational yield by splicing of exon junction complex component SKAR. Nucleic Acids Res. 2012;40(6):2668–82.
45. Ma XM, Yoon SO, Richardson CJ, Julich K, Blenis J. SKAR links pre-mRNA splicing to mTOR/S6K1-mediated enhanced translation efficiency of spliced mRNAs. Cell. 2008;133(2):303–13.
46. Kim HJ, Raphael AR, LaDow ES, McGurk L, Weber RA, Trojanowski JQ, et al. Therapeutic modulation of eIF2alpha phosphorylation rescues TDP-43 toxicity in amyotrophic lateral sclerosis disease models. Nat Genet. 2014;46(2):152–60.
47. Vance C, Rogelj B, Hortobagyi T, De Vos KJ, Nishimura AL, Sreedharan J, et al. Mutations in FUS, an RNA processing protein, cause familial amyotrophic lateral sclerosis type 6. Science. 2009;323(5918):1208–11.
48. Yamaguchi A, Takanashi K. FUS interacts with nuclear matrix-associated protein SAFB1 as well as Matrin3 to regulate splicing and ligand-mediated transcription. Sci Rep. 2016;6:35195.
49. DeJesus-Hernandez M, Mackenzie IR, Boeve BF, Boxer AL, Baker M, Rutherford NJ, et al. Expanded GGGGCC hexanucleotide repeat in noncoding region of C9ORF72 causes chromosome 9p-linked FTD and ALS. Neuron. 2011;72(2):245–56.
50. Renton AE, Majounie E, Waite A, Simon-Sanchez J, Rollinson S, Gibbs JR, et al. A hexanucleotide repeat expansion in C9ORF72 is the cause of chromosome 9p21-linked ALS-FTD. Neuron. 2011;72(2):257–68.
51. van Blitterswijk M, van Es MA, Hennekam EA, Dooijes D, van Rheenen W, Medic J, et al. Evidence for an oligogenic basis of amyotrophic lateral sclerosis. Hum Mol Genet. 2012;21(17):3776–84.
52. Levine TP, Daniels RD, Gatta AT, Wong LH, Hayes MJ. The product of C9orf72, a gene strongly implicated in neurodegeneration, is structurally related to DENN Rab-GEFs. Bioinformatics. 2013;29(4):499–503.
53. Gitler AD, Tsuiji H. There has been an awakening: emerging mechanisms of C9orf72 mutations in FTD/ALS. Brain Res. 2016;1647:19–29.
54. Freibaum BD, Lu Y, Lopez-Gonzalez R, Kim NC, Almeida S, Lee KH, et al. GGGGCC repeat expansion in C9orf72 compromises nucleocytoplasmic transport. Nature. 2015;525(7567):129–33.
55. Zhang K, Donnelly CJ, Haeusler AR, Grima JC, Machamer JB, Steinwald P, et al. The C9orf72 repeat expansion disrupts nucleocytoplasmic transport. Nature. 2015;525(7567):56–61.
56. Maharjan N, Kunzli C, Buthey K, Saxena S. C9ORF72 regulates stress granule formation and its deficiency impairs stress granule assembly, hypersensitizing cells to stress. Mol Neurobiol. 2017;54(4):3062–77.
57. Schweizer Burguete A, Almeida S, Gao FB, Kalb R, Akins MR, Bonini NM. GGGGCC microsatellite RNA is neuritically localized, induces branching defects, and perturbs transport granule function. Elife. 2015;4:e08881.
58. Mori K, Weng SM, Arzberger T, May S, Rentzsch K, Kremmer E, et al. The C9orf72 GGGGCC repeat is translated into aggregating dipeptide-repeat proteins in FTLD/ALS. Science. 2013;339(6125):1335–8.
59. Mizielinska S, Gronke S, Niccoli T, Ridler CE, Clayton EL, Devoy A, et al. C9orf72 repeat expansions cause neurodegeneration in Drosophila through arginine-rich proteins. Science. 2014;345(6201):1192–4.
60. Schmidt EK, Clavarino G, Ceppi M, Pierre P. SUnSET, a nonradioactive method to monitor protein synthesis. Nat Methods. 2009;6(4):275–7.
61. Kanekura K, Yagi T, Cammack AJ, Mahadevan J, Kuroda M, Harms MB, et al. Poly-dipeptides encoded by the C9ORF72 repeats block global protein translation. Hum Mol Genet. 2016;25(9):1803–13.
62. Kwon MS, Noh MY, Oh KW, Cho KA, Kang BY, Kim KS, et al. The immunomodulatory effects of human mesenchymal stem cells on peripheral blood mononuclear cells in ALS patients. J Neurochem. 2014;131(2):206–18.

63. Lee JE, Cooper TA. Pathogenic mechanisms of myotonic dystrophy. Biochem Soc Trans. 2009;37(Pt 6):1281–6.

64. Meola G, Cardani R. Myotonic dystrophy type 2 and modifier genes: an update on clinical and pathomolecular aspects. Neurol Sci. 2017;38(4):535–46.

65. Bates G. Huntingtin aggregation and toxicity in Huntington's disease. Lancet. 2003;361(9369):1642–4.

66. Savas JN, Ma B, Deinhardt K, Culver BP, Restituito S, Wu L, et al. A role for Huntington disease protein in dendritic RNA granules. J Biol Chem. 2010;285(17):13142–53.

67. de Mezer M, Wojciechowska M, Napierala M, Sobczak K, Krzyzosiak WJ. Mutant CAG repeats of Huntingtin transcript fold into hairpins, form nuclear foci and are targets for RNA interference. Nucleic Acids Res. 2011;39(9):3852–63.

68. Grima JC, Daigle JG, Arbez N, Cunningham KC, Zhang K, Ochaba J, et al. Mutant Huntingtin disrupts the nuclear pore complex. Neuron. 2017;94(1):93–107.e6.

69. Gasset-Rosa F, Chillon-Marinas C, Goginashvili A, Atwal RS, Artates JW, Tabet R, et al. Polyglutamine-expanded Huntingtin exacerbates age-related disruption of nuclear integrity and nucleocytoplasmic transport. Neuron. 2017;94(1):48–57.e4.

70. Fu YH, Kuhl DP, Pizzuti A, Pieretti M, Sutcliffe JS, Richards S, et al. Variation of the CGG repeat at the fragile X site results in genetic instability: resolution of the Sherman paradox. Cell. 1991;67(6):1047–58.

71. Kremer EJ, Pritchard M, Lynch M, Yu S, Holman K, Baker E, et al. Mapping of DNA instability at the fragile X to a trinucleotide repeat sequence p(CCG)n. Science. 1991;252(5013):1711–4.

72. Yu S, Pritchard M, Kremer E, Lynch M, Nancarrow J, Baker E, et al. Fragile X genotype characterized by an unstable region of DNA. Science. 1991;252(5010):1179–81.

73. Tassone F, Hagerman RJ, Chamberlain WD, Hagerman PJ. Transcription of the FMR1 gene in individuals with fragile X syndrome. Am J Med Genet. 2000;97(3):195–203.

74. Hagerman RJ, Leehey M, Heinrichs W, Tassone F, Wilson R, Hills J, et al. Intention tremor, parkinsonism, and generalized brain atrophy in male carriers of fragile X. Neurology. 2001;57(1):127–30.

75. Sofola OA, Jin P, Qin Y, Duan R, Liu H, de Haro M, et al. RNA-binding proteins hnRNP A2/B1 and CUGBP1 suppress fragile X CGG premutation repeat-induced neurodegeneration in a Drosophila model of FXTAS. Neuron. 2007;55(4):565–71.

76. Jin P, Duan R, Qurashi A, Qin Y, Tian D, Rosser TC, et al. Pur alpha binds to rCGG repeats and modulates repeat-mediated neurodegeneration in a Drosophila model of fragile X tremor/ataxia syndrome. Neuron. 2007;55(4):556–64.

77. Sellier C, Freyermuth F, Tabet R, Tran T, He F, Ruffenach F, et al. Sequestration of DROSHA and DGCR8 by expanded CGG RNA repeats alters microRNA processing in fragile X-associated tremor/ataxia syndrome. Cell Rep. 2013;3(3):869–80.

78. Todd PK, Oh SY, Krans A, He F, Sellier C, Frazer M, et al. CGG repeat-associated translation mediates neurodegeneration in fragile X tremor ataxia syndrome. Neuron. 2013;78(3):440–55.

79. Oh SY, He F, Krans A, Frazer M, Taylor JP, Paulson HL, et al. RAN translation at CGG repeats induces ubiquitin proteasome system impairment in models of fragile X-associated tremor ataxia syndrome. Hum Mol Genet. 2015;24(15):4317–26.

80. Sellier C, Buijsen RA, He F, Natla S, Jung L, Tropel P, et al. Translation of expanded CGG repeats into FMRpolyG is pathogenic and may contribute to fragile X tremor ataxia syndrome. Neuron. 2017;93(2):331–47.

81. Maziuk B, Ballance HI, Wolozin B. Dysregulation of RNA binding protein aggregation in neurodegenerative disorders. Front Mol Neurosci. 2017;10:89.

82. Iqbal K, Liu F, Gong CX, Grundke-Iqbal I. Tau in Alzheimer disease and related tauopathies. Curr Alzheimer Res. 2010;7(8):656–64.

83. Vanderweyde T, Apicco DJ, Youmans-Kidder K, Ash PEA, Cook C, Lummertz da Rocha E, et al. Interaction of tau with the RNA-binding protein TIA1 regulates tau pathophysiology and toxicity. Cell Rep. 2016;15(7):1455–66.

84. Daz-Muoz MDaKVYaNNLaCTaUJaTM. Tia1 dependent regulation of mRNA subcellular location and translation controls p53 expression in B cells. Nat Commun. 2017;8(1):530.

85. Gu J, Wu F, Xu W, Shi J, Hu W, Jin N, et al. TDP-43 suppresses tau expression via promoting its mRNA instability. Nucleic Acids Res. 2017;45(10):6177–93.

86. Ding Q, Markesbery WR, Chen Q, Li F, Keller JN. Ribosome dysfunction is an early event in Alzheimer's disease. J Neurosci. 2005;25(40):9171–5.

87. Carthew RW, Sontheimer EJ. Origins and mechanisms of miRNAs and siRNAs. Cell. 2009;136(4):642–55.

88. Kawahara Y, Mieda-Sato A. TDP-43 promotes microRNA biogenesis as a component of the Drosha and Dicer complexes. Proc Natl Acad Sci U S A. 2012;109(9):3347–52.

89. Porta S, Kwong LK, Trojanowski JQ, Lee VM. Drosha inclusions are new components of dipeptide-repeat protein aggregates in FTLD-TDP and ALS C9orf72 expansion cases. J Neuropathol Exp Neurol. 2015;74(4):380–7.

90. Chou CC, Zhang Y, Umoh ME, Vaughan SW, Lorenzini I, Liu F, et al. TDP-43 pathology disrupts nuclear pore complexes and nucleocytoplasmic transport in 1 ALS/FTD. Nat Neurosci. 2018;21(2):228–39.

91. Eitan C, Hornstein E. Vulnerability of microRNA biogenesis in FTD-ALS. Brain Res. 2016;1647:105–11.

92. Gascon E, Lynch K, Ruan H, Almeida S, Verheyden JM, Seeley WW, et al. Alterations in microRNA-124 and AMPA receptors contribute to social behavioral deficits in frontotemporal dementia. Nat Med. 2014;20(12):1444–51.

93. Zhang Z, Almeida S, Lu Y, Nishimura AL, Peng L, Sun D, et al. Downregulation of microRNA-9 in iPSC-derived neurons of FTD/ALS patients with TDP-43 mutations. PLoS One. 2013;8(10):e76055.

94. Rinchetti P, Rizzuti M, Faravelli I, Corti S. MicroRNA metabolism and dysregulation in amyotrophic lateral sclerosis. Mol Neurobiol. 2018;55(3):2617–30.

95. Costa CJ, Willis DE. To the end of the line: axonal mRNA transport and local translation in health and neurodegenerative disease. Dev Neurobiol. 2018;78(3):209–20.

96. Spaulding EL, Burgess RW. Accumulating evidence for axonal translation in neuronal homeostasis. Front Neurosci. 2017;11:312.

97. Cestra G, Rossi S, Di Salvio M, Cozzolino M. Control of mRNA translation in ALS proteinopathy. Front Mol Neurosci. 2017;10:85.

98. Shi Z, Fujii K, Kovary KM, Genuth NR, Rost HL, Teruel MN, et al. Heterogeneous ribosomes preferentially translate distinct subpools of mRNAs genome-wide. Mol Cell. 2017;67(1):71–83.e7.

99. Ernst V, Levin DH, London IM. In situ phosphorylation of the alpha subunit of eukaryotic initiation factor 2 in reticulocyte lysates inhibited by heme deficiency, double-stranded RNA, oxidized glutathione, or the heme-regulated protein kinase. Proc Natl Acad Sci U S A. 1979;76(5):2118–22.

100. Moreno JA, Halliday M, Molloy C, Radford H, Verity N, Axten JM, et al. Oral treatment targeting the unfolded protein response prevents neurodegeneration and clinical disease in prion-infected mice. Sci Transl Med. 2013;5(206):206ra138.

101. Briggs DI, Defensor E, Memar Ardestani P, Yi B, Halpain M, Seabrook G, et al. Role of endoplasmic reticulum stress in learning and memory impairment and Alzheimer's disease-like neuropathology in the PS19 and APPSwe mouse models of tauopathy and amyloidosis. eNeuro. 2017;4(4). https://doi.org/10.1523/ENEURO.0025-17.2017.

102. Johnson EC, Kang J. A small molecule targeting protein translation does not rescue spatial learning and memory deficits in the hAPP-J20 mouse model of Alzheimer's disease. PeerJ. 2016;4:e2565.

Index

A

Acetylation, 182

Acting on double-stranded RNA (ADARs)
ADAR1, ADAR2 and ADAR3, 64, 65
A/I editing, 67, 69
downregulation, 71
in GBMs, 76
and GluA2 Q/R editing, 75
GluA2 Q/R unedited mice, 73
long noncoding RNA, 72
in mammals, 64
micro RNAs, 72
siRNA knockdown, 71

Adenylate-uridylate-rich elements (AREs), 110

Affymetrix GeneChip Human Tiling 2.0R
Arrays, 10

ALS/FTD epigenetic studies, 4–5, 10–13
characterization, 2
C9orf72 transcript, 14–16
eRNAs, 8
genetic etiologies, 2
genetic mutations, 9
histone modifications, 8–9
lncRNAs, 8
mature miRNAs, 7
mature siRNAs, 8
mechanisms, 2–9
miRNAs, 13–14
mitochondrial DNA methylation, 5–6
nuclear DNA
demethylation, 5, 10–13
methylation, 4–5, 10–13
overview, 11
PARs, 8

pathogenic *C9orf72* G_4C_2 repeat
expansion, 10
patients, 2
piRNAs, 8
potential drivers, 16–17
RNA methylation, 6–7
schematic representation, 3
therapeutic potential, 18–19
transcriptome, 7

ALS/FTD types, 293, 294
c9orf72, 291
FUS, 290
RAN translation, 291, 292
CUGBP1, 293
FXTAS, 294
HD, 293
novel and unexpected peptides, 293
SOD1, 286–288
TDP-43, 288–290

Alternative splicing and polyadenylation, 220

Alzheimer's disease (AD), 69–71, 93, 95, 99,
242, 285, 294–296
translation dysregulation, 294–295

Amino-3-hydroxy-5-methyl-4-
isoxazolepropionic acid (AMPA)
ApoE gene, 70
Ca^{2+} permeability, 68
in CA1 pyramidal neurons, 74
CNS physiology, 66
functional properties, 67
GluA2 Q/R editing, 67

AMPA receptor, α-amino-3-hydroxy-5-
methyl-4-isoxazolepropionic acid
receptor (AMPAR), 14, 289

Printed in the United States
By Bookmasters